Number Theory with Applications

T0320623

SERIES ON UNIVERSITY MATHEMATICS

Published

Vol. 5: Algebra
T T Moh

Vol. 6: A Concise Introduction to Calculus
W Y Hsiang

Forthcoming

Vol. 1: Differential Geometry
S S Chern

Vol. 2: Lie Group
W Y Hsiang

Vol. 3: Classical Geometries
W Y Hsiang

Vol. 4: Foundation of Linear Algebra
W Y Hsiang & T T Moh

SERIES ON UNIVERSITY MATHEMATICS VOL. 7

NUMBER THEORY

WITH APPLICATIONS

W. C. Winnie Li
Department of Mathematics
Pennsylvania State University
USA

World Scientific
Singapore • New Jersey • London • Hong Kong

Published by

World Scientific Publishing Co. Pte. Ltd.

P O Box 128, Farrer Road, Singapore 912805

USA office: Suite 1B, 1060 Main Street, River Edge, NJ 07661

UK office: 57 Shelton Street, Covent Garden, London WC2H 9HE

Library of Congress Cataloging-in-Publication Data
Li, W. C. Winnie (Wen-Ching Winnie)
Number theory with applications / W. C. Winnie Li.
244 p. 23 cm -- (Series on university mathematics ; vol. 7)
Includes bibliographical references and index.
ISBN 9810222262
1. Number theory. I. Title. II. Series.
QA241.L59 1996
512'.7--dc20 95-49001
CIP

British Library Cataloguing-in-Publication Data
A catalogue record for this book is available from the British Library.

First published 1996
First reprint 1998

Printed in Singapore.

Preface

In the past decade, there have been important applications of number theory to network communications and computation complexity through explicit constructions of the so-called Ramanujan graphs. These are regular graphs whose nontrivial eigenvalues are small. (See Chapter 9 for more details.) All the known constructions are number-theoretic: one based on the estimates of Fourier coefficients of modular forms, namely, the former Ramanujan-Petersson conjecture established by Deligne, and one based on the estimates of certain character sums, which can be derived as consequences of the Riemann hypothesis for curves over finite fields proved by Weil. The common thread of both theoretic backgrounds is the celebrated Weil conjectures settled by Deligne in 1973. This is our starting point. The purpose of this book is to explain in detail the material in number theory involved in the applications discussed above, with the ultimate goal of giving the explicit constructions. In fact, this simple-minded goal serves as the guideline of the selection and the exposition of the material in this book. The reader will be amazed to find, as we take the journey along this line, how far and deep in mathematics we have gone through when we arrive at the destination.

The style of this book is semi-formal. It is written for advanced undergraduate students, graduate students and people interested in number theory and its applications. While it is desirable that the reader has some background in algebra and basic number theory, I try to make this book as self-contained as possible. More emphasis is given on basic concepts and results, while complicated proofs of hard theorems are only sketched in order to give the reader some flavor and idea of the approach. Occasionally statements without proofs are asserted for the sake of completeness and exposition. The reader can find the missing details and many untreated topics from the references listed at the end of each chapter. Exercises are scattered throughout the text, and sometimes used to prove theorems.

The material in this book is organized as follows. After reviewing the basic facts about finite fields in Chapter 1, we discuss in Chapter 2 the celebrated Weil conjectures on zeta functions attached to projective varieties over finite fields. We'll see how Weil arrived at his conjectures from computing the number of solutions of an equation over extensions of a finite field, and the ideas involved in the proof of these conjectures will also be sketched. Local and global fields are introduced and studied in Chapter 3. In this book adèlic language is employed for global fields. Chapters 4 and 5 concern function fields, where the Riemann-Roch theorem is proved and the analytic behaviour of the zeta and L-functions attached to idèle class characters is shown. By appealing to some results in class field theory (which we review briefly) and combining the results on L-functions and the Riemann hypothesis for curves over finite fields established in Chapters 5 and 2, we derive in Chapter 6 many character sum estimates, some of which will be used to construct Ramanujan graphs in Chapter 9. Chapter 7 deals with classical modular forms, which are a very rich subject, deeply intertwined with many branches of mathematics. We give a summary of the development of the theory, including Hecke operators, L-functions, converse theorems, and the theory of newforms. We also discuss main conjectures and consequences in this area, some of which are still outstanding and would have far-reaching consequences in number theory. One such example is the Taniyama-Shimura conjecture, which is recently established for semistable elliptic curves by Wiles and Taylor and Wiles. This result together with earlier works of Frey and Ribet implies the truth of Fermat's Last Theorem! Automorphic forms and representations are discussed in Chapter 8. There we give an adèlic interpretation of the classical modular forms, which naturally leads to the adèlic definition of automorphic forms and representations for $GL(2)$. Then we survey the Jacquet-Langlands theory of local and global representation theory for $GL(2)$ and quaternion groups. In particular, we show how the local representations of these groups are determined by the attached L- and ε- factors, from which correspondences of local representations of these two groups follow immediately. Finally in chapter 9 we see the interplay between number theory and combinatorics. On one hand, we apply what we have learned to give explicit constructions of Ramanujan graphs; on the other hand, we obtain some information on the distribution of eigenvalues of a Hecke operator from considering the limit of the measures attached to certain family of graphs arising from quaternion groups.

This book grew out of the one year graduate course in number theory I gave at the National Taiwan University during the year 1992-93. An outline of the material was given as a special one month summer course for graduate students in Sichuan University in the summer of 1992. I would like to thank both universities for their hospitality and support. The positive feedback from the audience has been a great source of encouragement to me. The main part of this book was written while I was spending my sabbatical year, 1992-93, visiting the National Taiwan University. Special thanks are due to the National Science Council in Taiwan and the National Security Agency in USA for their financial support, and to Ms. Shirley Wang for her superb typing job. The final part of the book was completed in the spring of

1995 while I was visiting the Mathematical Sciences Research Institute at Berkeley, California, to which I would like to express my sincere gratitude for its hospitality and support.

Wen-Ching Winnie Li
Berkeley, California
Spring, 1995

Contents

Number Theory with Applications

CHAPTER 1

Finite Fields

§1 The structure of a finite field

A finite field k is a finite commutative ring in which all nonzero elements have multiplicative inverse. Its characteristic, i.e., the smallest positive integer n such that $1 + 1 + \cdots + 1(n \text{ times}) = 0$, is a prime number p. Thus it contains $\mathbf{Z}/p\mathbf{Z}$ as a subfield, and it is a finite-dimensional vector space over $\mathbf{Z}/p\mathbf{Z}$. Its cardinality $|k| = q = p^d$ is a power of p, with exponent being the dimension of k over $\mathbf{Z}/p\mathbf{Z}$. This also indicates that the additive group of k is a direct sum of d copies of cyclic group of order p.

Next consider the multiplicative group $k^\times$, it has order $q-1$. So every nonzero element in k satisfies

$$x^{q-1} = 1,$$

and the order of an element in $k^\times$ divides $q-1$. For each positive divisor r of $q-1$, let

$$\Omega(r) = \{x \in k^\times : \text{ the order of } x \text{ is } r\}.$$

Then $k^\times$ is a disjoint union of $\Omega(r)$ as r runs through all positive divisors of $q-1$. We want to show that $\Omega(q-1)$ is nonempty, in other words,

Theorem 1. *$k^\times$ is cyclic of order $q-1$.*

To prove this theorem, observe first the general fact :

Lemma 1. *A polynomial $f(x)$ of degree n over a field F has at most n distinct roots in F.*

Proof. Let α be a root of $f(x)$ in F. Then $f(\alpha) = 0$, and

$$f(x) = f(x) - f(\alpha) = (x - \alpha)g(x)$$

for a polynomial $g(x)$ of degree $n-1$ over F. If β is a root of $f(x)$ in F different from α, then $0 = f(\beta) = (\beta - \alpha)f(\beta)$ and $\beta - \alpha \neq 0$ imply $g(\beta) = 0$. By induction,

$g(x)$ has at most $n-1$ distinct roots in F, so $f(x)$ has at most n distinct roots in F. □

It follows from Lemma 1 that if $\Omega(r)$ is nonempty, say, contains y, then y generates a cyclic subgroup of order r consisting of all solutions of $x^r = 1$ in k and $\Omega(r)$ is the set of generators in the cyclic group $< y >$. That is, $\Omega(r) = \{y^i : 1 \le i \le r,\ \gcd(i,r) = 1\}$. This shows that the cardinality of $\Omega(r)$ is either 0 or $\phi(r)$, where $\phi(n)$ is the Euler function which denotes the number of integers between 1 and n which are prime to n. We have

$$|k^\times| = q - 1 = \sum_{r|q-1} |\Omega(r)| \le \sum_{r|q-1} \phi(r).$$

To continue, we note another fact.

Lemma 2. *For every positive integer m, $\sum_{r|m} \phi(r) = m$.*

Granting Lemma 2, we conclude immediately from the above inequality that $|\Omega(r)| = \phi(r)$ for all $r \mid q-1$, and in particular, $|\Omega(r)| = \phi(r) \ge 1$, and the theorem follows.

To show Lemma 2, we partition the set $\{1, 2, \cdots, m\}$ as a disjoint union of

$$Y(r) = \{1 \le i \le m : \gcd(i,m) = \frac{m}{r}\}$$

as r runs through all positive divisors of m. For $i \in Y(r)$, write $i = j\frac{m}{r}$. Then $1 \le j \le r$ and $\gcd(i,m) = \gcd(j\frac{m}{r}, m) = \frac{m}{r}\gcd(j,r) = \frac{m}{r}$ implies $\gcd(j,r) = 1$. Hence $|Y(r)| = \phi(r)$. This proves

$$m = \sum_{r|m} |Y(r)| = \sum_{r|m} \phi(r).$$

□

Some immediate consequences of the above arguments are

Corollary 1. *The field k consists of the solutions to $x^q - x = 0$ in an algebraic closure of $\mathbf{Z}/p\mathbf{Z}$ containing k.*

Corollary 2. *There is an element $\xi \in k$ such that $k = (\mathbf{Z}/p\mathbf{Z})(\xi)$, that is, k is a simple extension of the prime field $\mathbf{Z}/p\mathbf{Z}$.*

Corollary 3. *For each positive divisor r of $q-1(= |k^\times|)$ there are exactly $\phi(r)$ elements in $k^\times$ of order r.*

Corollary 4. *Given a positive integer n, there is a unique field extension of $\mathbf{Z}/p\mathbf{Z}$ of degree n within an algebraic closure of $\mathbf{Z}/p\mathbf{Z}$.*

Proof. Corollary 1 shows that a degree n extension of $\mathbf{Z}/p\mathbf{Z}$, if exists, is unique, namely, it should consist of the roots of $x^{p^n} = x$ in the algebraic closure. On the other hand, one checks easily that if α, β are solutions to $x^{p^n} = x$, then so are $\alpha - \beta$ and $\alpha\beta^{-1}$(for $\beta \neq 0$), so the solutions to $x^{p^n} = x$ do form a field. □

Corollary 5. *Given any positive integer n, there exists an irreducible polynomial of degree n over $\mathbf{Z}/p\mathbf{Z}$.*

Proof. Let k be a finite field of degree n over $\mathbf{Z}/p\mathbf{Z}$. Then $k = (\mathbf{Z}/p\mathbf{Z})(\xi)$ by Corollary 2. Let $f(x)$ be the irreducible polynomial of ξ over $\mathbf{Z}/p\mathbf{Z}$. Then $k = (\mathbf{Z}/p\mathbf{Z})(\xi) = (\mathbf{Z}/p\mathbf{Z})[\xi] \cong (\mathbf{Z}/p\mathbf{Z})[x]/(f(x))$ shows $\deg f = [k : \mathbf{Z}/p\mathbf{Z}] = n$. □

---

§2 Extensions of a finite field

Let k be a finite field with q elements and let k_n be a degree n field extension of k. If k_m is an intermediate field of degree m over k, then k_n is a vector space over k_m, so m divides n. Conversely, any degree m extension of k within an algebraic closure of k_n with $m \mid n$ is a subfield of k_n by Corollary 1.

For an extension E of a field F, denote by $Gal(E/F)$ the group of automorphisms of E leaving F elementwise fixed. Consider the map σ on k_n which sends x to x^q. From

$$\sigma(x+y) = (x+y)^q = x^q + y^q = \sigma(x) + \sigma(y)$$

and

$$\sigma(xy) = (xy)^q = x^q y^q = \sigma(x)\sigma(y)$$

we see that σ is an endomorphism. Further, $\sigma(x) = x^q = 1$ together with $x^{q^n} = x$ implies $x = 1$. So σ is 1−1. As k_n is finite, we have shown that σ is an automorphism of k_n. Finally, $\sigma(x) = x^q = x$ for $x \in k$, this shows that $\sigma \in Gal(k_n/k)$, called the Frobenius automorphism. Let r be the order of σ. Then

$$\sigma^r(x) = x^{q^r} = x \textit{ for all } x \in k_n$$

implies $r = n$ since $k_n^\times$ is cyclic of order $q^n - 1$. Hence $Gal(k_n/k)$ contains the cyclic group $<\sigma>$ of order n. To determine $Gal(k_n/k)$ we notice following facts.

Each automorphism in $Gal(E/F)$ can be viewed as an F-linear transformation on E.

Lemma 3. *The automorphisms in $Gal(E/F)$ are E–linearly independent F–linear transformations.*

Proof. Suppose otherwise. Let $a_1\tau_1 + \cdots + a_r\tau_r = 0$ be a shortest nontrivial linear relation with $a_1, \cdots, a_r \in E^\times$ and $\tau_1, \cdots, \tau_r \in Gal(E/F)$. Then $r \geq 2$ and τ_i are distinct. Let $y \in E$ be such that $\tau_1(y) \neq \tau_2(y)$. From $\sum_{i=1}^r a_i\tau_i = 0$ we get $\sum_{i=1}^r a_i\tau_i(yx) = \sum_{i=1}^r a_i\tau_i(y)\tau_i(x) = 0$ for all $x \in k_n$, so $\sum_{i=1}^r a_i\tau_i(y)\tau_i = 0$. This yields another nontrivial relation

$$\sum_{i=1}^r a_i\tau_i(y)\tau_i - \tau_1(y)\sum_{i=1}^r a_i\tau_i = \sum_{i=2}^r a_i\big(\tau_i(y) - \tau_1(y)\big)\tau_i = 0,$$

which is shorter than the relation we started with, a contradiction. □

Lemma 4. *Let E be a degree n extension of a field F. Then there are at most n distinct automorphisms in $Gal(E/F)$.*

Proof. Suppose otherwise. Let $\tau_1, \cdots, \tau_m$, $m > n$, be distinct automorphisms in $Gal(E/F)$. Let $\{v_1, \cdots, v_n\}$ be a basis of E over F. Let $(a_1, \cdots, a_m)$ be a nontrivial solution in E to the $n \times m$ system of linear equations

$$\begin{pmatrix} \tau_1(v_1) & \tau_2(v_1) & \cdots & \tau_m(v_1) \\ \vdots & \vdots & \ddots & \vdots \\ \tau_1(v_n) & \cdots & \cdots & \tau_m(v_n) \end{pmatrix} \begin{pmatrix} x_1 \\ \vdots \\ x_m \end{pmatrix} = \begin{pmatrix} 0 \\ \vdots \\ 0 \end{pmatrix}.$$

Consider $\sum_{i=1}^m a_i\tau_i$. By construction, we have $\sum_{i=1}^m a_i\tau_i(v_j) = 0$ for $j = 1, \cdots, n$, hence $\sum_{i=1}^m a_i\tau_i(x) = 0$ for all $x \in E$. In other words, $\tau_1, \cdots, \tau_m$ are linearly dependent over E. This is impossible by Lemma 3. □

Therefore $|Gal(k_n/k)| = |<\sigma>| = n = [k_n : k]$, which is the maximal possible. In this case we say that the field k_n is Galois over k. We record this in

Theorem 2. *The field k_n is Galois over k with $Gal(k_n/k)$ cyclic of order n, generated by the Frobenius automorphism σ.*

Note that an element $x \in k_n$ lies in k if and only if it satisfies $x^q = x$, in other words, if and only if it is fixed by the Frobenius automorphism, or equivalently, by the group $Gal(k_n/k)$.

Using $G = Gal(k_n/k)$, we define two important maps, called trace and norm, denoted by $\mathrm{Tr}_{k_n/k}$ and $\mathrm{N}_{k_n/k}$, respectively, from k_n to k as follows:

$$\mathrm{Tr}_{k_n/k} : x \longmapsto \sum_{\tau \in G} \tau(x) = \sum_{i=1}^n \sigma^i(x),$$

$$\mathrm{N}_{k_n/k} : x \longmapsto \prod_{\tau \in G} \tau(x) = \prod_{i=1}^n \sigma^i(x).$$

One checks easily that the images of trace and norm maps are in k. It is clear that $\mathrm{Tr}_{k_n/k}$ is a homomorphism from the additive group k_n to the additive group k, and $\mathrm{N}_{k_n/k}$ is a homomorphism from $k_n^\times$ to $k^\times$. Next we study their images.

Theorem 3. *(Hilbert Theorem 90) The norm map $\mathrm{N}_{k_n/k}$ from $k_n^\times$ to $k^\times$ is surjective with the kernel consisting of $x/\sigma(x)$, $x \in k_n^\times$.*

Proof. Since $\mathrm{N}_{k_n/k}(\sigma(x)) = \sum_{i=1}^n \sigma^{i+1}(x) = \sum_{i=1}^n \sigma(x) = \mathrm{N}_{k_n/k}(x)$, so $x/\sigma(x)$ lies in the kernel of the norm map for all $x \in k_n^\times$. Further, $x/\sigma(x) = y/\sigma(y)$ if and only if $xy^{-1} \in k^\times$, hence the elements $x/\sigma(x)$ with $x \in k_n^\times$ form a subgroup of $k_n^\times$ of order $(q^n-1)/(q-1)$. Thus it is equal to the whole kernel if and only if the norm map is surjective. To see the surjectiveness of $\mathrm{N}_{k_n/k}$, observe that $\mathrm{N}_{k_n/k}(x) = \prod_{i=1}^n \sigma^i(x) = x \cdot x^q \cdot x^{q^2} \cdots x^{q^{n-1}} = x^{1+q+q^2+\cdots+q^{n-1}} = x^{(q^n-1)/(q-1)}$ for all $x \in k_n^\times$. Thus any generator x of $k_n^\times$ has $\mathrm{N}_{k_n/k}(x)$ of order $q-1$. □

Theorem 4. *(Hilbert Theorem 90) The trace map $\mathrm{Tr}_{k_n/k}$ from k_n to k is surjective with the kernel consisting of $x - \sigma(x)$, $x \in k_n$.*

Proof. As elements in $Gal(k_n/k)$ are k-linear maps, the image of $\mathrm{Tr}_{k_n/k}$ is a vector space over k, hence $\mathrm{Tr}_{k_n/k}(k_n) = 0$ or k. If $\mathrm{Tr}_{k_n/k}(k_n) = 0$, then $\sum_{i=1}^n \sigma^i = 0$, which is a nontrivial linear relation among elements of $Gal(k_n/k)$, hence impossible by Lemma 3. Therefore $\mathrm{Tr}_{k_n/k}$ is surjective. Then its kernel has order q^{n-1}. Clearly, $\mathrm{Tr}_{k_n/k}(\sigma(x)) = \mathrm{Tr}_{k_n/k}(x)$ so that kernel contains $x - \sigma(x)$ for all $x \in k_n$. Further, $x - \sigma(x) = y - \sigma(y)$ if and only if $x - y \in k$, so the group $\{x - \sigma(x) : x \in k_n\}$ has order q^n/q, thus is equal to the kernel. □

Remark. The Hilbert theorem 90 for norm and trace maps is usually proved using first cohomology group of the Galois group (à la Noether). When the base field is finite, we may use counting argument, as shown above.

Exercise 1. Let k be a finite field with finite extensions k_m and k_{mn} of degree m, mn, respectively. Show that

$$\mathrm{Tr}_{k_{mn}/k} = \mathrm{Tr}_{k_m/k} \circ \mathrm{Tr}_{k_{mn}/k_m} \quad \text{and} \quad \mathrm{N}_{k_{mn}/k} = \mathrm{N}_{k_m/k} \circ \mathrm{N}_{k_{mn}/k_m}.$$

Given $z \in k_n$, it defines a k-linear transformation L_z on k_n by $x \mapsto zx$, that is, multiplication by z. The trace and determinant of L_z are defined as the trace and determinant of any $n \times n$ matrix representing L_z. They are in fact given by $\mathrm{Tr}_{k_{mn}/k}$ and $\mathrm{N}_{k_{mn}/k}$ of z. More precisely, we have

Theorem 5. *Let $z \in k_n$ and define L_z as above. Then*

(1) $\operatorname{Tr} L_z = \operatorname{Tr}_{k_n/k}(z)$ *and* $\det L_z = \mathrm{N}_{k_n/k}(z)$

(2) *Suppose $k(z) = k_n$. Let $f(x) = x^n + a_1x^{n-1} + \cdots + a_n$ be the irreducible polynomial of z over k. Then*

$$a_1 = -\operatorname{Tr}_{k_n/k}(z) \quad \textit{and} \quad a_n = (-1)^n \mathrm{N}_{k_n/k}(z).$$

Proof. We shall prove (1) and (2) under the assumption (2) and leave (1) for the case $k(z)$ being a proper subfield of k_n as an exercise. For each τ in $Gal(k_n/k)$, $0 = \tau(f(z)) = f(\tau(z))$, hence $\tau(z)$ is also a root of $f(x)$. Further, if τ and τ' are two different elements in $Gal(k_n/k)$, then $\tau(z) \neq \tau'(z)$ (otherwise they would agree on $k(z) = k_n$). This shows that z has n distinct images under $Gal(k_n/k)$ and they are the roots of $f(x)$. Therefore,

$$-a_1 = \text{ the sum of roots of } f(x) = \operatorname{Tr}_{k_m/k}(z)$$

and

$$(-1)^n a_n = \text{ the product of roots of } f(x) = \mathrm{N}_{k_n/k}(z).$$

This proves (2). For (1), we know that L_z satisfies $f(x) = 0$. As $f(x)$ is irreducible over k and $[k_n : k] = n$, $f(x)$ is the characteristic polynomial of L_z. The companion matrix attached to L_z is

$$\begin{pmatrix} 0 & & & & -a_n \\ 1 & 0 & & & -a_{n-1} \\ & 1 & \cdot & & \cdot \\ & & \cdot & & \cdot \\ & & & 0 & \cdot \\ & & & 1 & -a_1 \end{pmatrix},$$

which has trace $= -a_1$ and determinant $= (-1)^n a_n$. This proves (1). □

Exercise 2. Let $z \in k_n$. Suppose $k(z) = k_m$ is a proper subfield of k_n. Prove that $\operatorname{Tr} L_z = \operatorname{Tr}_{k_n/k}(z) = \frac{n}{m} \operatorname{Tr}_{k_m/k}(z)$ and $\det L_z = \mathrm{N}_{k_m/k}(z)^{n/m}$.

Exercise 3. (1) (Normal Basis Theorem) There exists an element $z \in k_n$ such that $\{\tau(z) : \tau \in Gal(k_n/k)\}$ is basis of k_n over k. (Hint : Consider the minimal polynomial of the Frobenius automorphism σ.)

(2) For z in (1) we have $\operatorname{Tr}_{k_n/k}(z) \neq 0$. (Hint : Express an element in k_n as a k-linear combination of $\{\tau(z)\}$. Then show $\operatorname{Tr}_{k_n/k}(k_n) = k \operatorname{Tr}_{k_n/k}(z)$.)

§3 Characters

A character of a topological group G is a continuous homomorphism from G to the unit circle S^1 in the complex plane. If G is a finite group, then it is endowed with the discrete topology so that a character is simply a homomorphism from G to S^1. As S^1 is commutative, any character of G factors through the quotient of G by its commutator subgroup. The character sending G to 1 is called the trivial character of G.

All characters of G form an abelian group under pointwise multiplication, called the dual group of G and denoted by $\widehat{G}$.

Example 1. Compute $\widehat{G}$ for a finite cyclic group G.

Suppose G has order n. Let g be a generator of G and ζ be a primitive nth root of 1. The homomorphism η from G to S^1 sending g to ζ is a charactor of G of order n. Hence $\widehat{G}$ contains the cyclic group $<\eta>$. On the other hand, any character χ of G is determined by its value $\chi(g)$ at g, which is an nth root of 1. Thus $\chi(g) = \zeta^k$ for some integer k, and this shows that $\chi = \eta^k$. So $G =<\eta>$ is also a cyclic group of order n. We see that $\widehat{G}$ is isomorphic to G.

Proposition 1. *If G is a finite abelian group, then G is isomorphic to its dual group $\widehat{G}$.*

Proof. By the fundamental theorem of finite abelian groups, we may decompose G as a product of cyclic groups

$$G = G_1 \times \cdots \times G_r.$$

For a character χ of G, denote by χ_i its restriction to G_i. Thus χ is the product of $\chi_1, \cdots, \chi_r$ and $\widehat{G} = \widehat{G_1} \times \cdots \times \widehat{G_r}$. We have seen that each $\widehat{G_i}$ is isomorphic to G_i, hence $\widehat{G}$ is isomorphic to G. □

Remark. The above isomorphism $G \cong \widehat{G}$ is not canonical since it depends on the decomposition of G into a product of cyclic groups and for each cyclic group, the isomorphism depends on the choice of generators. However, the dual of $\widehat{G}$, namely, $\hat{\hat{G}}$, is naturally isomorphic to G. This follows from the non-degeneracy of the pairing

$$\xi : G \times \widehat{G} \longrightarrow S^1$$

given by

$$\xi(g, \chi) = \chi(g).$$

(When we fix one variable, ξ is a homomorphism with respect to the other variable.)

Exercise 4. (1) Show that ξ defined above is nondegenerate, that is,

(i) If g is not the identity element of G, then there is a character χ of G such that $\chi(g) \neq 1$.

(ii) If χ is a nontrivial character of G, then there exists an element g of G such that $\chi(g) \neq 1$.

(2) Show that the nondegeneracy of ξ implies that G and $\hat{\hat{G}}$ are naturally isomorphic.

For a closed subgroup H of G, denote by $H^{\perp} = \{\chi \in \widehat{G} : \chi(H) = 1\}$. When G is an abelian group, $H^{\perp}$ is canonically identified with $\widehat{G/H}$.

Theorem 6. *(Pontrjagin duality) Let G be an abelian topological group. The map* $H \mapsto H^{\perp}$ *establishes a bijection from the closed subgroups of G to those of* $\widehat{G}$. *Further,* $H^{\perp\perp}$ *is naturally isomorphic to H.*

We examine this theorem for the case where G is a finite abelian group, thus topology plays no role. Again, by the fundamental theorem of finite abelian groups, we may assume that G is cyclic of order n. The subgroups H of G are indexed by the positive divisors d of n such that $d =$ the order of H. Then $\widehat{G}$ is cyclic of order n and $H^{\perp}$ has order $\frac{n}{|H|}$. The bijection $H \mapsto H^{\perp}$ is obvious. Under the canonical isomorphism $\hat{\hat{G}} \cong G$, we may identity $H^{\perp\perp}$ with the group $\widetilde{H} = \{g \in G : \chi(g) = 1$ for all $\chi \in H^{\perp}\}$. Since every $\chi \in H^{\perp}$ is trivial on H, the group $\widetilde{H}$ contains H. On the other hand, $|\widetilde{H}| = |\widehat{G}|/|H^{\perp}|$ and $|H^{\perp}| = |G|/|H|$ imply $|\widehat{H}| = |H|$. Hence $\widehat{H} = H$, i.e., $H^{\perp\perp} \cong H$ naturally. □

Define an inner product $<,>$ on the space $\mathbf{C}[G]$ of complex-valued functions on a finite abelian group G by

$$< f, g >= \frac{1}{|G|} \sum_{x \in G} f(x)\overline{g(x)}.$$

Proposition 2. *Let G be a finite abelian group. Then the characters of G form an orthonormal basis of* $\mathbf{C}[G]$.

To prove this, we shall need

Lemma 5. *Let G be a finite abelian group,* $g \in G$ *and* $\chi \in$ *hatG. Then*

(1) $\sum_{x \in G} \chi(x) = \begin{cases} 0 & \text{if } \chi \text{ is nontrivial,} \\ |G| & \text{if } \chi \text{ is trivial.} \end{cases}$

(2) $\sum_{\eta \in \widehat{G}} \eta(g) = \begin{cases} 0 & \text{if } g \text{ is not the identity of } G, \\ |G| & \text{if } g \text{ is the identity of } G. \end{cases}$

Proof. (1) It is clear if χ is trivial. Suppose χ is nontrivial and let y be an element in G such that $\chi(y) \neq 1$. From

$$\sum_{x \in G} \chi(x) = \sum_{x \in G} \chi(yx) = \chi(y) \sum_{x \in G} \chi(x)$$

and $\chi(y) \neq 1$ we conclude immediately that $\sum_{x \in G} \chi(x) = 0$.

(2) follows from (1) by the isomorphism $G \cong \hat{\hat{G}}$. □

Now we prove Proposition 2. Let $\chi_1, \chi_2 \in \widehat{G}$. Then

$$< \chi_1, \chi_2 >= \frac{1}{|G|} \sum_{x \in G} \chi_1 \overline{\chi_2(x)} = \frac{1}{|G|} \sum_{x \in G} (\chi_1 \chi_2^{-1})(x) = \begin{cases} 1 & \text{if } \chi_1 = \chi_2, \\ 0 & \text{if } \chi_1 \neq \chi_2, \end{cases}$$

by Lemma 5. This shows that the characters of G form an orthonormal set, hence are linearly independent. On the other hand, $|\widehat{G}| = |G| = \dim_{\mathbf{C}} \mathbf{C}[G]$, so the characters of G form an orthonormal basis. □

---

§4 Characters of a finite field, Gauss sums

Let k be a finite field with q elements. We have seen in §1 that the additive group of k is a direct sum of d cyclic groups of order p, here $q = p^d$. Thus the group $\widehat{k}$ of additive characters of k has the same structure.

Example 2. $\widehat{\mathbf{Z}/p\mathbf{Z}} =< \psi >$, where $\psi : \mathbf{Z}/p\mathbf{Z} \to S^1$ is given by $\psi(x) = e^{2\pi i x/p}$.

We have seen that $\mathrm{Tr}_{k/(\mathbf{Z}/p\mathbf{Z})} : k \to \mathbf{Z}/p\mathbf{Z}$ is a homomorphism of additive groups, so for each $\psi \in \widehat{\mathbf{Z}/p\mathbf{Z}}$, $\psi \circ \mathrm{Tr}_{k/(\mathbf{Z}/p\mathbf{Z})}$ is an additive character of k. Since Tr is surjective, any nontrivial character of $\mathbf{Z}/p\mathbf{Z}$ yields a nontrivial character of k by composing with the trace map.

Proposition 3. *Let ψ be a nontrivial additive character of a finite field k. For $a \in k$, define $\psi^a : k \to k$ by $\psi^a(x) = \psi(ax)$. Then ψ^a is an additive character of k and $a \mapsto \psi^a$ gives rise to an isomorphism from k to $\widehat{k}$. In particular, $\widehat{k} = \{\psi^a : a \in k\}$ and $\overline{\psi(x)} = (\psi(x))^{-1} = \psi(-x) = \psi^{-1}(x)$ for all $x \in k$.*

Proof. Denote by L_a the linear transformation on k which is multiplication by a. Then $\psi^a = \psi \circ L_a$. If $a = 0$, then $L_a = 0$ and ψ^0 is the trivial character of k. If $a \neq 0$, L_a is an automorphism of the additive group k, so ψ^a is a nontrivial additive character. For $a, b \in k$, $\psi^{a+b} = \psi^a \psi^b$; hence $a \mapsto \psi^a$ is an injective homomorphism from k to $\widehat{k}$. As $|\widehat{k}| = |k|$, this homomorphism is surjective. □

Example 3. Denote by Φ the character $\psi \circ \mathrm{Tr}_{k/(\mathbf{Z}/p\mathbf{Z})}$, then $\widehat{k} = \{\Phi^a : a \in k\}$ by Proposition 3.

Exercise 5. Let V be a finite-dimensional vector space over k. Let $v_1, \cdots, v_n$ be a basis of V over k. Let $<,>$ be a bilinear form from $V \times V$ to k defined by $< x, y >= \sum_{i=1}^{n} x_i y_i$, where $x = x_1 v_1 + \cdots + x_n v_n$, $y = y_1 v_1 + \cdots + y_n v_n$, $x_i, y_i \in k$. Let ψ be a nontrivial additive character of k. For $v \in V$, define ψ^v to be the character $\psi^v(x) = \psi(< v, x >)$. Show that $\widehat{V} = \{\psi^v : v \in V\}$.

The group of multiplicative characters $\widehat{k^\times}$ of $k^\times$ is cyclic of order $q - 1$. A character χ of $k^\times$ will also be viewed as a function on k with value

$$\chi(0) = \begin{cases} 0 & \text{if } \chi \text{ is nontrivial,} \\ 1 & \text{if } \chi \text{ is trivial.} \end{cases}$$

Then $\chi \in \mathbf{C}[k]$, and hence is a linear combination of the additive characters of k by Proposition 2. The coefficients are called the normalized Gauss sums. More precisely, for a nontrivial character χ, we have

$$(1) \qquad \chi = \sum_{\psi \in \widehat{k}} < \chi, \overline{\psi} > \overline{\psi} = \sum_{\substack{\psi \in \widehat{k} \\ \psi \text{ nontrivial}}} < \chi, \overline{\psi} > \overline{\psi} = \frac{1}{|k|} \sum_{\substack{\psi \in \widehat{k} \\ \psi \text{ nontrivial}}} g(\chi, \psi) \overline{\psi},$$

where $g(\chi, \psi)$, called the Gauss sum of χ with respect to ψ, is

$$(2) \qquad g(\chi, \psi) = |k| < \chi, \overline{\psi} >= \sum_{x \in k} \chi(x)\psi(x) = \sum_{x \in k^\times} \chi(x)\psi(x).$$

(1) may be viewed as the Fourier expansion of χ with respect to the additive characters, and the Gauss sums are Fourier coefficients.

Proposition 4. *Let $\chi \in \widehat{k^\times}$ and $\psi \in \widehat{k}$ be nontrivial characters. Then*
(1) $g(\chi, \psi) g(\overline{\chi}, \psi) = |k| \chi(-1) = q\chi(-1)$.
(2) $\overline{g(\chi, \psi)} = \chi(-1) g(\overline{\chi}, \psi)$ so that $g(\chi, \psi)$ has absolute value $\sqrt{q}$.

Proof. (1) From definition we have

$$\begin{aligned} g(\chi, \psi) g(\overline{\chi}, \psi) &= \sum_{x \in k^\times} \chi(x)\psi(x) \sum_{y \in k^\times} \overline{\chi}(y)\psi(y) \\ &= \sum_{x \in k^\times} \chi(x)\psi(x) \sum_{z \in k^\times} \overline{\chi}(xz)\psi(xz) \quad \text{(write } y = xz) \\ &= \sum_{z \in k^\times} \overline{\chi}(z) \sum_{x \in k^\times} \psi((1+z)x) \\ &= \sum_{z \in k^\times} \overline{\chi}(z) (\sum_{x \in k} \psi((1+z)x) - \psi(0)) \\ &= \sum_{z \in k^\times} \overline{\chi}(z) (\sum_{x \in k} \psi^{1+z}(x) - 1) \\ &= \sum_{z \in k^\times} \overline{\chi}(z) \sum_{x \in k} \psi^{1+z}(x) \end{aligned}$$

since $\sum_{z\in k^\times}\overline{\chi}(z)=0$ for nontrivial character χ. Applying Lemma 5 to $G=k$ yields

$$\sum_{x\in k}\psi^{1+z}(x)=\begin{cases}0 & \text{if } \psi^{1+z} \text{ is nontrivial, i.e., if } 1+z\neq 0,\\ q & \text{if } \psi^{1+z} \text{ is trivial, i.e., if } 1+z=0.\end{cases}$$

Therefore, $g(\chi,\psi)g(\overline{\chi},\psi)=q\chi(-1)$, as desired.

(2) This is because

$$\overline{g(\chi,\psi)}=\sum_{x\in k^\times}\overline{\chi}(x)\overline{\psi(x)}=\sum_{x\in k^\times}\overline{\chi}(x)\psi(-x)=\chi(-1)g(\overline{\chi},\psi).$$

To compute $g(\chi,\psi)/\sqrt{q}$ is a hard problem in general.

Exercise 6. Let N be a positive integer. A character of $\mathbf{Z}$ mod N is a character of the group of units $(\mathbf{Z}/N\mathbf{Z})^\times$ in the ring $\mathbf{Z}/N\mathbf{Z}$. It is said to have conductor N if it does not factor through any quotient group $(\mathbf{Z}/m\mathbf{Z})^\times$ with m dividing N properly. Let ψ_N be the additive character of $\mathbf{Z}/N\mathbf{Z}$ given by $\psi_N(x)=e^{2\pi i x/N}$.

(1) Let χ be a character of $\mathbf{Z}$ mod N with conductor N. Show that the Gauss sum

$$g(\chi,\psi_N):=\sum_{\substack{x \bmod N\\ (x,N)=1}}\chi(x)\psi_N(x)$$

has absolute value $\sqrt{N}$.

(2) Let χ_1,χ_2 be characters of $\mathbf{Z}$ mod N_1,N_2, respectively, with conductor χ_i equal to N_i, $i=1,2$. Suppose that N_1 and N_2 are coprime. Then $(\mathbf{Z}/N_1N_2\mathbf{Z})^\times\cong(\mathbf{Z}/N_1\mathbf{Z})^\times\times(\mathbf{Z}/N_2\mathbf{Z})^\times$. Denote by $\chi_1\chi_2$ the character mod N_1N_2 whose restriction to $(\mathbf{Z}/N_i\mathbf{Z})^\times$ is equal to χ_i. Relate $g(\chi_1\chi_2,\psi_{N_1N_2})$ to $g(\chi_1,\psi_{N_1})$ and $g(\chi_2,\psi_{N_2})$.

The Gauss sum $g(\chi,\psi)$, when viewed as a function in χ and ψ, varies in a simple way with respect to ψ. Indeed, fix a nontrivial additive character ψ of k, then any other nontrivial additive character has the form $\psi^t, t\in k^\times$. We have

$$(3)\qquad g(\chi,\psi^t)=\sum_{t\in k^\times}\chi(x)\psi^t(x)=\sum_{t\in k^\times}\chi(x)\psi(tx)=\chi(t)^{-1}\sum_{t\in k^\times}\chi(tx)\psi(tx)$$
$$=\chi(t)^{-1}g(\chi,\psi).$$

Hence formula (1) can be written as

$$(4)\qquad \chi=\frac{1}{|k|}\sum_{t\in k^\times}g(\chi,\psi^t)\overline{\psi^t}=\frac{1}{|k|}g(\chi,\psi)\sum_{t\in k^\times}\chi(t)^{-1}\overline{\psi^t}.$$

As a function in χ, it is more complicated. One can show that

$$\frac{g(\chi_1,\psi)g(\chi_2,\psi)}{g(\chi_1\chi_2,\psi)} = \sum_{\substack{s+t=1\\ s,t\in k}} \chi_1(s)\chi_2(t) = \chi_1\chi_2(-1) \sum_{\substack{s,t\in k\\ s+t+1=0}} \chi_1(s)\chi_2(t)$$

if χ_1, χ_2 and $\chi_1\chi_2$ are nontrivial. Note that the right hand side is independent of ψ, as seen from the left hand side and (3). It is called the Jacobi sum attached to χ_1 and χ_2. We define Jacobi sum in general as follows. Let $\chi_1, \cdots, \chi_r$ be nontrivial characters of $k^\times$. The Jacobi sum attached to $\chi_1, \cdots, \chi_r$ is defined as

$$\begin{aligned} j(\chi_1, \cdots, \chi_r) &= \sum_{\substack{v_1,\cdots,v_r\in k\\ v_1+\cdots+v_r+1=0}} \chi_1(v_1)\cdots\chi_r(v_r) \\ &= (q-1)^{-1} \sum_{\substack{u_0,\cdots,u_r\in k^\times\\ u_0+\cdots+u_r=0}} \chi_0(u_1)\cdots\chi_r(u_r), \end{aligned}$$

where $\chi_0 = (\chi_1\cdots\chi_r)^{-1}$.

We shall need

Proposition 5. *Let $\chi_0, \chi_1, \cdots, \chi_r$ be nontrivial characters of $k^\times$ such that the product $\chi_0\chi_1\cdots\chi_r = 1$. Then*

$$j(\chi_1, \cdots, \chi_r) = \frac{1}{q} g(\chi_0, \psi)g(\chi_1, \psi)\cdots g(\chi_r, \psi)$$

for any nontrivial character ψ of k. In particular, $j(\chi_1, \cdots, \chi_r)$ has absolute value $q^{\frac{r-1}{2}}$.

Proof. Fix a nontrivial additive character ψ of k. Express χ_i as a linear combination of ψ^t like in (4):

$$\chi_i = \frac{1}{q} g(\chi_i, \psi) \sum_{t\in k^\times} \overline{\chi_i}(t)\overline{\psi}^t.$$

Substituting the above in the definition of $j(\chi_1, \cdots, \chi_r)$ yields

$$\begin{aligned} &(q-1)j(\chi_1, \cdots, \chi_r) \\ =&q^{-r-1} g(\chi_0, \psi)\cdots g(\chi_r, \psi) \sum_{\substack{u_i\in k\\ \Sigma u_i=0}} \sum_{t_i\in k^\times} \overline{\chi_0}(t_0)\cdots\overline{\chi_r}(t_r)\overline{\psi}(t_0u_0 + t_1u_1 + \cdots + t_ru_r). \end{aligned}$$

But for fixed $t_0, \cdots, t_r$,

$$\begin{aligned}
&\sum_{\substack{u_i \in k \\ \Sigma u_i = 0}} \overline{\psi}(t_0 u_0 + \cdots + t_r u_r) \\
&= \sum_{\Sigma u_i = 0} \overline{\psi}(t_0(u_0 + \cdots + u_r) + u_1(t_1 - t_0) + \cdots + u_r(t_r - t_0)) \\
&= \sum_{u_1, \cdots, u_r \in k} \overline{\psi}(u_1(t_1 - t_0) + \cdots + u_r(t_r - t_0)) \\
&= \begin{cases} q^r & \text{if } t_0 = t_1 = \cdots = t_r, \\ 0 & \text{otherwise.} \end{cases}
\end{aligned}$$

Thus

$$\begin{aligned}
(q-1)j(\chi_1, \cdots, \chi_r) &= \frac{1}{q} g(\chi_0, \psi) \cdots g(\chi_r, \psi) \sum_{t_0 \in k^\times} \overline{\chi_0} \cdots \overline{\chi_r}(t_0) \\
&= \frac{q-1}{q} g(\chi_0, \psi) \cdots g(\chi_r, \psi)
\end{aligned}$$

since $\chi_0 \cdots \chi_r = 1$. This proves the proposition. □

---

§5 Davenport-Hasse identity

As in §4, let k denote a finite field, let χ be a nontrivial multiplicative character of $k^\times$ and let ψ be nontrivial additive character of k. For a positive integer ν, denote by k_ν a degree ν extension of k. Recall the trace map $\mathrm{Tr}_{k_\nu/k}$ and the norm map $\mathrm{N}_{k_\nu/k}$ defined in §2, they are surjective homomorphisms with respect to the additive and multiplicative groups of the fields, respectively. Hence by composition we obtain

$$X = \chi \circ \mathrm{N}_{k_\nu/k} \quad \text{and} \quad \varPhi = \psi \circ \mathrm{Tr}_{k_\nu/k},$$

which are nontrivial characters of $k_\nu^\times$, k_ν, respectively. The two Gauss sums

$$\begin{aligned}
g(\chi, \psi) &= \sum_{y \in k^\times} \chi(y)\psi(y) \quad \text{and} \\
g(X, \varPhi) &= \sum_{y \in k_\nu^\times} X(y)\varPhi(y)
\end{aligned}$$

are related as follows.

Theorem 6. *(Davenport-Hasse)* $-g(X,\Phi) = (-g(\chi,\psi))^{\nu}$.

The proof below is due to *A.* Weil [5]. To each monic polynomial

$$f(x) = x^n + a_{n-1}x^{n-1} + \cdots + a_0, \quad a_i \in k,$$

with $a_0 = f(0) \neq 0$, define

$$\lambda(f) = \chi(a_0)\psi(a_{n-1}).$$

It is easily seen that if f_1, f_2 are two such polynomials, then $\lambda(f_1 f_2) = \lambda(f_1)\lambda(f_2)$. Since every monic polynomial with nonzero constant is a product of monic irreducible polynomials with nonzero constant, we have, formally,

$$\sum_{\substack{f(x)\in k[x] \\ f \text{ monic, } f(0)\neq 0}} \lambda(f)u^{\deg f} = \prod_{\substack{p(x)\in k[x] \\ \text{irred, monic,} \\ p(0)\neq 0}} (1-\lambda(p)u^{\deg p})^{-1}.$$

First we compute the sum on the left hand side. Fix a degree $d \geq 2$. Consider monic polynomials in $k[x]$ of degree d and with fixed $a_{d-2}, \cdots, a_0 \neq 0$. The coefficient a_{d-1} of x^{d-1} can be any element in k. Thus the summation of the corresponding $\lambda(f)$ over such f is zero. Hence the coefficient of u^d for $d \geq 2$ is zero. The coefficient of u^0 is 1, and the monic degree 1 polynomials occurring on the left hand side are $x + a$, $a \in k^{\times}$ so that the coefficient of u is

$$\sum_{a\in k^{\times}} \lambda(x+a) = \sum_{a\in k^{\times}} \chi(a)\psi(a) = g(\chi,\psi).$$

We have shown

$$1 + g(\chi,\psi)u = \prod_{\substack{p(x)\in k[x] \\ \text{monic, irred, } p(0)\neq 0}} (1-\lambda(p)u^{\deg p})^{-1}. \tag{5}$$

Similarly, for a polynomial $F(x) = x^n + a_{n-1}x^{n-1} + \cdots + a_0 \in k_{\nu}[x]$ with $a_0 \neq 0$, define

$$\Lambda(F) = X(a_{n-1})\Phi(a_0),$$

and we have

$$1 + g(X,\Phi)U = \prod_{\substack{P(x)\in k_{\nu}[x] \\ \text{monic, irred, } P(0)\neq 0}} (1-\lambda(P)U^{\deg P})^{-1}. \tag{6}$$

We investigate the infinite product in (5) and (6). Let $p(x)$ be a monic irreducible polynomial in $k[x]$. Let $P(x)$ be a monic irreducible polynomial in $k_{\nu}[x]$ dividing $p(x)$. Then for any $\tau \in Gal(k_{\nu}/k)$, $\tau(P(x))$ divides $\tau(p(x)) = p(x)$, and $\tau(P(x))$ is also monic and irreducible in $k_\nu[x]$. If $\tau_1(P(x)), \cdots, \tau_r(P(x))$ are the distinct images of $P(x)$ under $Gal(k_\nu/k)$, then they are distinct irreducible factors of $p(x)$ in $k_\nu[x]$, and the product $q(x) = \tau_1(P(x)) \cdots \tau_r(P(x))$ divides $p(x)$ and is invariant under the action of $Gal(k_\nu/k)$, hence $q(x)$ lies in $k[x]$. As $p(x)$ is irreducible and $p(x)$ and $q(x)$ are both monic, we conclude that $q(x) = p(x)$, that is,

$$p(x) = \tau_1(P(x))\tau_2(P(x)) \cdots \tau_r(P(x))$$

is a product of conjugates of $P(x)$ under $Gal(k_\nu/k)$. This shows that each monic irreducible polynomial of $k[x]$ factors as a product of distinct monic irreducible polynomials in $k_\nu[x]$, which are conjugate under $Gal(k_\nu/k)$, and each monic irreducible polynomial in $k_\nu[x]$ divides a unique monic irreducible polynomial in $k[x]$. So we may write

$$1 + g(X, \Phi)U = \prod_{\substack{p(x)\in k[x] \\ \text{monic, irred,} \\ p(0)\neq 0}} \prod_{\substack{P(x)\in k_\nu[x] \\ P(x)|p(x) \\ P \text{ monic, irred}}} (1 - \Lambda(P)U^{\deg P})^{-1}.$$

Next we fix a monic irreducible polynomial $p(x) \in k[x]$ of degree n and with $p(0) \neq 0$. Let $P(x)$ be a monic irreducible factor of $p(x)$ in $k_\nu[x]$. We study the relation between $\lambda(p)$ and $\Lambda(P)$. Let $-\xi$ be a root of $P(x)$ in an algebraic closure $\overline{k_\nu}$ of k_ν. Then $p(x)$ is the irreducible polynomial of $-\xi$ over k and $P(x)$ is that of $-\xi$ over k_ν. Thus the degree of $k(\xi)$ over k is n and the degree of $k_\nu(\xi)$ over k_ν is $\deg P(x)$. Suppose the intersection $k(\xi) \cap k_\nu$ has degree d over k. Since in $\overline{k_\nu}$, given any finite degree there is only one field of that degree over k, we conclude that $d = \gcd(n, \nu)$. (Thus $\frac{n}{d}$ and $\frac{\nu}{d}$ are coprime.) Therefore $[k_\nu(\xi) : k_\nu] = [k(\xi) : k(\xi) \cap k_\nu] = \frac{n}{d} = \deg P$, and $p(x)$ has d distinct monic irreducible factors in $k_\nu[x]$.

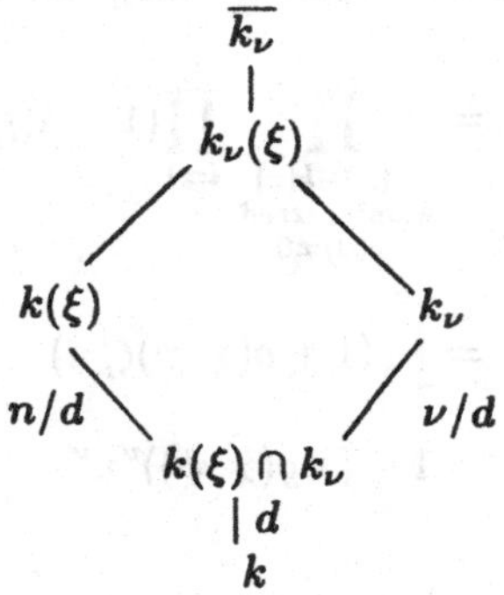

Write $p(x) = x^n + bx^{n-1} + \cdots + a$ and $P(x) = x^{n/d} + Bx^{n/d-1} + \cdots + A$. By Theorem 5, we have

$$b = -\operatorname{Tr}_{k(\xi)/k}(-\xi) = \operatorname{Tr}_{k(\xi)/k}(\xi), \qquad a = (-1)^n \operatorname{N}_{k(\xi)/k}(-\xi) = \operatorname{N}_{k(\xi)/k}(\xi),$$
$$B = -\operatorname{Tr}_{k_\nu(\xi)/k_\nu}(-\xi) = \operatorname{Tr}_{k_\nu(\xi)/k_\nu}(\xi), \quad A = (-1)^{n/d} \operatorname{N}_{k_\nu(\xi)/k_\nu}(-\xi) = \operatorname{N}_{k_\nu(\xi)/k_\nu}(\xi).$$

Therefore

$$\lambda(p) = \chi(a)\psi(b) = \chi(\mathrm{N}_{k(\xi)/k}(\xi))\psi(\mathrm{Tr}_{k(\xi)/k}(\xi))$$

and

$$\begin{aligned}\Lambda(P) &= X(A)\Phi(B)\\ &= \chi(\mathrm{N}_{k_\nu/k}(A))\psi(\mathrm{Tr}_{k_\nu/k}(B))\\ &= \chi(\mathrm{N}_{k_\nu(\xi)/k}(\xi))\psi(\mathrm{Tr}_{k_\nu(\xi)/k}(\xi))\\ &= \chi(\mathrm{N}_{k(\xi)/k}(\xi))^{\nu/d}\psi(\mathrm{Tr}_{k(\xi)/k}(\xi))^{\nu/d}\\ &= \lambda(p)^{\nu/d}.\end{aligned}$$

Consequently,

$$\prod_{\substack{P(x)\in k_\nu[x]\\ P(x)|p(x)\\ \text{monic, irred}}} (1-\Lambda(P)U^{\deg P})^{-1} = (1-\lambda(p)^{\nu/d}U^{n/d})^{-d}.$$

Replacing U by u^ν, we can write

$$\begin{aligned}(1-\lambda(p)^{\nu/d}U^{n/d})^{-d} &= (1-\lambda(p)^{\nu/d}u^{n\nu/d})^{-d}\\ &= \prod_{i=1}^{\nu/d}(1-\lambda(p)\zeta_{\nu/d}^{i}u^{n})^{-d}\\ &= \prod_{i=1}^{\nu}(1-\lambda(p)(\zeta_\nu^i u)^n)^{-1},\end{aligned}$$

where ζ_m denotes a primitive mth root of unity. We have shown

$$\begin{aligned}1+g(X,\Phi)u^\nu &= \prod_{\substack{p(x)\in k[x]\\ \text{monic, irred,}\\ p(0)\neq 0}}\prod_{i=1}^{\nu}(1-\lambda(p)(\zeta_\nu^i u)^{\deg p})^{-1}.\\ &= \prod_{i=1}^{\nu}(1+g(\chi,\psi)\zeta_\nu^i u)\\ &= 1-(-g(\chi,\psi))^\nu u^\nu.\end{aligned}$$

This proves the theorem. □

References

[1] H. Davenport and H. Hasse : Die Nullstellen der Kongruenzetafünctionen in gewissen zyklischen Fällen, J. für reine und ang. Math. 172 (1935), 151-182.

[2] N. Jacobson: Basic Algebra I, Freeman, San Francisco (1980).
[3] S. Lang: Algebra, Addison-Wesley, Reading, Mass (1967).
[4] R. Lidl and H. Niederreiter : Finite Fields, Addison–Wesley, Reading, Mass (1983).
[5] A. Weil: Numbers of solutions of equations in finite fields, Bulletin of Amer. Math. Soc. 55 (1949), 497-508.

CHAPTER 2

Weil Conjectures

§1 Numbers of solutions of equations in finite fields

In this section we discuss the work by Hua-Vandiver [12] and independently by A. Weil [14] on estimating the number of rational points on varieties defined over finite fields. We shall follow the approach in [14], which leads to the statement of the famous Weil conjectures.

As before, k denotes a finite field with q elements. Consider an equation of type

$$a_0x_0^{n_0} + a_1x_1^{n_1} + \cdots + a_rx_r^{n_r} = b,$$

where $a_0, a_1, \cdots, a_r$ are nonzero elements in k, $b \in k$, and $n_0, n_1, \cdots, n_r$ are positive integers. We want to estimate the number of solutions in k.

First we deal with the case $b = 0$. Given $u \in k$, denote by $N_i(u)$ the number of solutions to $x^{n_i} = u$ in k. Thus $N_i(0) = 1, N_i(u) = d_i$, the greatest common divisor of n_i and $q-1$, if u is a d_ith power in $k^\times$, and $N_i(u) = 0$ otherwise. This is because $k^\times$ is cyclic of order $q-1$. Denote by N the total number of solutions in k to

$$a_0x_0^{n_0} + a_1x_1^{n_1} + \cdots + a_rx_r^{n_r} = 0.$$

Then

$$N = \sum_{\substack{u_0,\cdots,u_r \in k \\ \sum a_iu_i=0}} N_0(u_0)N_1(u_1)\cdots N_r(u_r).$$

Note that the set of $(r+1)$ tuples $(u_0, \cdots, u_r)$ satisfying $\sum a_iu_i = 0$ form a vector space over k of dimension r.

The function N_i can be further expressed in terms of the characters of $k^\times$ (extended to functions on k as in Chapter 1, §4) as follows. Denote by H_i the subgroup $(k^\times)^{n_i} = (k^\times)^{d_i}$ of $k^\times$, which is cyclic of order $(q-1)/d_i$. Since $H_i^\perp$ is cyclic of order d_i containing characters χ of $k^\times$ such that $\chi^{d_i} = 1$, it is equal to $\{\chi \in \widehat{k^\times} : \chi^{d_i} = 1\}$. One finds, for $u \in k$,

$$\sum_{\chi \in H_i^\perp} \chi(u) = \begin{cases} 1 & \text{if } u = 0, \\ |H_i^\perp| = d_i & \text{if } u \in H_i = (k^\times)^{d_i} \\ 0 & \text{if } u \in k^\times \setminus H_i \quad (\text{since } H_i^{\perp\perp} = H_i), \end{cases}$$

in other words, $N_i = \sum_{\chi \in H_i^{\perp}} \chi = \sum_{\substack{\chi\in\widehat{k}^\times \\ \chi^{d_i}=1}} \chi$. Thus

$$N = \sum_{\substack{u_i \in k \\ \sum a_i u_i = 0}} \sum_{\substack{\chi_i \in \widehat{k}^\times \\ \chi_i^{d_i}=1}} \chi_0(u_0)\chi_1(u_1)\cdots\chi_r(u_r).$$

We proceed to evaluate the above sum by fixing a choice of $(\chi_0, \chi_1, \cdots, \chi_r)$ with $\chi_i^{d_i} = 1$ and summing over the u_i's. If all χ_i are trivial, then $\chi_i(u_i) = 1$ for all $u_i \in k$, thus the sum over $u_i \in k$ with $\sum a_i u_i = 0$ is the number of solutions to the linear equation $\sum a_i u_i = 0$, which is equal to q^r. Next, if some, but not all, χ_i are trivial, say, $\chi_{i_0} = 1$ then $\chi_{i_0}(u_{i_0}) = 1$ is independent of u_{i_0} so that the sum over all u_i satisfying $\sum a_i u_i = 0$ becomes $\prod_{\substack{j\neq i_0 \\ 0\le j\le r}} (\sum_{u_j\in k} \chi_j(u_j))$, which is equal to zero since some χ_j is nontrivial. At this point we have

$$\begin{aligned}
N &= q^r + \sum_{\substack{u_i \in k \\ \sum a_i u_i = 0}} \sum_{\substack{\chi_i \in \widehat{k}^\times \\ \chi_i^{d_i}=1, \chi_i \neq 1}} \chi_0(u_0)\chi_1(u_1)\cdots\chi_r(u_r) \\
&= q^r + \sum_{\chi_i^{d_i}=1, \chi_i\neq 1} \chi_0(a_0)^{-1}\cdots\chi_r(a_r)^{-1} \sum_{\substack{u_i \in k \\ \sum a_i u_i = 0}} \chi_0(a_0u_0)\cdots\chi_r(a_ru_r) \\
&= q^r + \sum_{\chi_i^{d_i}=1, \chi_i\neq 1} \chi_0(a_0)^{-1}\cdots\chi_r(a_r)^{-1} \sum_{\substack{y_i \in k^\times \\ \sum y_i = 0}} \chi_0(y_0)\cdots\chi_r(y_r) \text{ (letting } y_i = a_iu_i) \\
&= q^r + \sum_{\chi_i^{d_i}=1, \chi_i\neq 1} \chi_0(a_0)^{-1}\cdots\chi_r(a_r)^{-1} \sum_{\substack{y_0, v_i \in k^\times \\ \sum v_i + 1 = 0}} (\chi_0\chi_1\cdots\chi_r)(y_0)\chi_1(v_1)\cdots\chi_r(v_r) \\
&\quad \text{(letting } y_i = y_0v_i) \\
&= q^r + (q-1) \sum_{\substack{\chi_i^{d_i}=1, \chi_i\neq 1 \\ \chi_0\chi_1\cdots\chi_r = 1}} \chi_0(a_0)^{-1}\cdots\chi_r(a_r)^{-1} \sum_{\substack{v_i \in k^\times \\ v_1+\cdots+v_r+1=0}} \chi_1(v_1)\cdots\chi_r(v_r) \\
&= q^r + (q-1) \sum_{\substack{\chi_i^{d_i}=1, \chi_i\neq 1 \\ \chi_0\chi_1\cdots\chi_r = 1}} \chi_0(a_0)^{-1}\cdots\chi_r(a_r)^{-1} j(\chi_1, \cdots, \chi_r).
\end{aligned}$$

Here we used the fact that

$$\sum_{y_0 \in k^\times} (\chi_0\cdots\chi_r)(y_0) = \begin{cases} 0 & \text{if } \chi_0\cdots\chi_r \neq 1, \\ q-1 & \text{if } \chi_0\cdots\chi_r = 1. \end{cases}$$

As shown in Proposition 5 of Chapter 1, the Jacobi sum can be expressed as

$$j(\chi_1, \cdots, \chi_r) = \frac{1}{q} g(\chi_0, \psi)\cdots g(\chi_r, \psi)$$

for any nontrivial additive character ψ of k, where $\chi_0\chi_1\cdots\chi_r = 1$, and hence it has absolute value $q^{\frac{r-1}{2}}$. We have shown

Theorem 1. *The number N of solutions to $a_0x_0^{n_0}+\cdots+a_rx_r^{n_r}=0$ in k^{r+1} is*

$$N=q^r+(q-1)\sum_{\substack{\chi_i^{d_i}=1,\chi_i\neq 1\\ \chi_0\chi_1\cdots\chi_r=1}}\chi_0(a_0)^{-1}\cdots\chi_r(a_r)^{-1}j(\chi_1,\cdots,\chi_r)$$

$$=q^r+\frac{(q-1)}{q}\sum_{\substack{\chi_i^{d_i}=1,\chi_i\neq 1\\ \chi_0\chi_1\cdots\chi_r=1}}\chi_0(a_0)^{-1}\cdots\chi_r(a_r)^{-1}g(\chi_0,\psi)\cdots g(\chi_r,\psi),$$

where $d_i=\gcd(n_i,q-1)$. In particular, it satisfies

$$|N-q^r|\leq(q-1)q^{\frac{r-1}{2}}M,$$

where M counts the number of characters $(\chi_0,\cdots,\chi_r)\in(\widehat{k^\times})^{r+1}$ with $\chi_i\neq 1,\chi_i^{d_i}=1$ and $\chi_0\chi_1\cdots\chi_r=1$.

Next consider the nonhomogeneous equation $a_0x_0^{n_0}+\cdots+a_rx_r^{n_r}=b, b\in k^\times$. Dividing the above equation by $-b$ of necessary, we may assume that the equation is

$$a_0x_0^{n_0}+a_1x_1^{n_1}\cdots+a_rx_r^{n_r}+1=0$$

Denote by N_1 the number of solutions on k^{r+1} to the equation above, and N' the number of solutions to the equation below:

$$a_0x_0^{n_0}+a_1x_1^{n_1}+\cdots+a_rx_r^{n_r}+x_{r+1}^{q-1}=0.$$

Then, by Theorem 1 above,

$$N'=q^{r+1}+(q-1)\sum_{\substack{\chi_i^{d_i}=1,\chi_i\neq 1\\ \chi_0\chi_1\cdots\chi_{r+1}=1}}\chi_0(a_0)^{-1}\cdots\chi_r(a_r)^{-1}j(\chi_1,\cdots,\chi_{r+1}),$$

where $d_{r+1}=q-1$. Further, N' is related to N_1 by $N'=N+(q-1)N_1$. Here N counts the number of solutions with $x_{r+1}=0$ and $(q-1)N_1$ counts those with $x_{r+1}\neq 0$ since $x_{r+1}^{q-1}=1$ in this case. Hence

$$N_1=\frac{1}{q-1}(N'-N)=q^r+\sum_{\substack{\chi_i^{d_i}=1,\chi_i\neq 1\\ \chi_0\chi_1\cdots\chi_{r+1}=1}}\chi_0(a_0)^{-1}\cdots\chi_r(a_r)^{-1}j(\chi_1,\cdots,\chi_{r+1})$$

$$-\sum_{\substack{\chi_i^{d_i}=1,0\leq i\leq r\\ \chi_i\neq 1,\chi_0\chi_1\cdots\chi_r=1}}\chi_0(a_0)^{-1}\cdots\chi_r(a_r)^{-1}j(\chi_1,\cdots,\chi_r).$$

In the above expression, we may view the second summation corresponding to the first summation, but with $\chi_{r+1}=1$. Since $d_{r+1}=q-1$, the only condition on χ_{r+1} is to satisfy $\chi_0\cdots\chi_{r+1}=1$ for given $\chi_0,\cdots,\chi_r$ with $\chi_i^{d_i}=1$ and $\chi_i\neq 1$. Therefore the number of such $(\chi_0\cdots\chi_r)$'s is equal to $(d_0-1)\cdots(d_r-1)$. We have shown

Theorem 2. *The number N_1 of solutions to $a_0x_0^{n_0} + \cdots + a_rx_r^{n_r} = b,$, $a_i \in k^\times, b \in k^\times$, in k^{r+1} satisfies*

$$|N_1 - q^r| \leq (d_0 - 1)\cdots(d_r - 1)q^{r/2}.$$

where $d_i = \gcd(n_i, q-1)$.

A projective space of dimension r over k, denoted by $P^r(k)$, is obtained from $k^{r+1} - \{0\}$ by identifying collinear points. In other words, it consists of points $(x_0 : x_1 : \cdots : x_r)$ where $x_i \in k$, not all zero, and $(x_0 : x_1 : \cdots : x_r) = (tx_0 : tx_1 : \cdots : tx_r)$ for all $t \in k^\times$. Given a homogeneous equation $f(x_0, \cdots, x_r)$, it has the property that if $(b_0, b_1, \cdots, b_r)$ is a solution to $f = 0$, then so is $(tb_0, tb_1, \cdots, tb_r)$ for all $t \in k$. Thus we may consider the solutions to a homogeneous equation or to several homogeneous equations in a projective space, the collection of such solutions is called a projective variety.

Denote by $\overline{N}$ the number of points on the projective variety in $P^r(k)$ defined by

$$a_0x_0^m + a_1x_1^m + \cdots + a_rx_r^m = 0, \quad a_i \in k^\times.$$

Then $N = 1 + (q-1)\overline{N}$. In other words, by Theorem 1,

$$\overline{N} = 1 + q + \cdots + q^{r-1} + \sum_{\substack{\chi_i^d = 1, \chi_i \neq 1 \\ \chi_0 \cdots \chi_r = 1}} \chi_0(a_0)^{-1} \cdots \chi_r(a_r)^{-1} j(\chi_1, \cdots, \chi_r),$$

where $d = \gcd(m, q-1)$.

---

§2 Weil conjectures

For each integer $n \geq 1$, denote by k_n the degree n field extension of k in an algebraic closure $\overline{k}$, and by $\overline{N_n}$ the number of solutions to $a_0x_0^m + \ldots + a_rx_r^m = 0$ in $P^r(k_n)$. Let $d(n) = \gcd(m, q^n - 1)$. Then $|k_n| = q^n$ and

$$(1) \quad \overline{N_n} = 1 + q^n + \ldots + q^{n(r-1)} + \sum_{\substack{\chi_i \in \widehat{k_n^\times}, \chi_i^{d(n)} = 1, \\ \chi_i \neq 1, \chi_0 \cdots \chi_r = 1}} \chi_0(a_0)^{-1} \ldots \chi_r(a_r)^{-1} j(\chi_1, \ldots, \chi_r).$$

We study the characters occurring in the summation. Note that $d = \gcd(m, q-1)$ divides $d(n)$. A character χ of $k^\times$ with $\chi^d = 1$ yields a character $\chi \circ \mathrm{N}_{k_n/k}$ which has the same order as χ since $\mathrm{N}_{k_n/k}$ is surjective. On the other hand, there are exactly d characters of $k_n^\times$ with order dividing d. Therefore, the characters χ in $\widehat{k_n^\times}$ satisfying $\chi^d = 1$ are those arising from the characters of $k^\times$ of the same order by composing with the norm map. By abuse of language, sometimes we also denote

$\chi \circ \mathrm{N}_{k_n/k}$ by χ. Let ψ be a nontrivial additive character of k. Then, by Proposition 5 in Chapter 1, for $\chi_0, \ldots, \chi_r \in \widehat{k^\times}$ with $\chi_i^d = 1, \chi_i \neq 1, \chi_0 \ldots \chi_r = 1$, we have

$$\begin{aligned}
&\chi_0 \circ \mathrm{N}_{k_n/k}(a_0)^{-1} \ldots \chi_r \circ \mathrm{N}_{k_n/k}(a_r)^{-1} j(\chi_1 \circ \mathrm{N}_{k_n/k}, \ldots, \chi_r \circ \mathrm{N}_{k_n/k}) \\
&= (\chi_0(a_0) \ldots \chi_r(a_r))^{-n} \frac{1}{q^n} g(\chi_0 \circ \mathrm{N}_{k_n/k}, \psi \circ \mathrm{Tr}_{k_n/k}) \ldots g(\chi_r \circ \mathrm{N}_{k_n/k}, \psi \circ \mathrm{Tr}_{k_n/k}) \\
&= (\chi_0(a_0) \ldots \chi_r(a_r))^{-n} \frac{1}{q^n} g(\chi_0, \psi)^n \ldots g(\chi_r, \psi)^n (-1)^{(n+1)(r+1)} \\
&\text{(by Davenport-Hasse identity)} \\
&= (\chi_0(a_0)^{-1} \ldots \chi_r(a_r)^{-1} j(\chi_1, \ldots, \chi_r))^n (-1)^{(r+1)(n+1)}.
\end{aligned}$$

Likewise, if $k_n \supset k_{n'} \supset k$, then a term in (1) arising from characters $\chi_i \in \widehat{k_{n'}^\times}$, for $i = 0, \ldots, r$ is equal to

$$(\chi_0(a_0)^{-1} \ldots \chi_r(a_r)^{-1} j(\chi_1, \ldots, \chi_r))^{n/n'} (-1)^{(r+1)(\frac{n}{n'}+1)}.$$

Given $\chi_0, \ldots, \chi_r \in \widehat{k_n^\times}$, denote by $\mu(\chi_0, \ldots, \chi_r)$ the lowest degree of the field extension so that, up to composing with norm maps, $\chi_0, \ldots, \chi_r$ are the characters of $k^\times_{\mu(\chi_0, \ldots, \chi_r)}$. Then we may express (1) as

$$\begin{aligned}
&\overline{N_n} \\
&= 1 + q^n + \ldots + q^{n(r-1)} + \\
&\sum_{\substack{\chi_i^{d_n}=1, \chi_i \neq 1 \\ \chi_0 \cdots \chi_r = 1}} (\chi_0(a_0)^{-1} \ldots \chi_r(a_r)^{-1} j(\chi_1, \ldots, \chi_r))^{\frac{n}{\mu(\chi_0, \ldots, \chi_r)}} (-1)^{(r+1)(\frac{n}{\mu(\chi_0, \ldots, \chi_r)}+1)}.
\end{aligned}$$

Note that as n tends to infinity, there are only finitely many $d(n)$'s, in fact, they are factors of m', the largest factor of m prime to q. Hence, up to composing with norm maps, there are only finitely many multiplicative characters of finite extensions of k whose orders divide m'. Form the formal power series

(2)

$$\sum_{n=1}^{\infty} \overline{N_n} U^{n-1} = \sum_{i=0}^{r-1} \frac{q^i}{1 - q^i U} + (-1)^r \sum_{\substack{\chi_i^{m'}=1 \\ \chi_i \neq 1, \chi_0 \cdots \chi_r = 1}} \frac{-c(\chi_0, \ldots, \chi_r) U^{\mu(\chi_0, \ldots, \chi_r)-1}}{1 - c(\chi_0, \ldots, \chi_r) U^{\mu(\chi_0, \ldots, \chi_r)}},$$

where

$$\begin{aligned}
c(\chi_0, \ldots, \chi_r) &= (-1)^{r+1} \chi_0(a_0)^{-1} \ldots \chi_r(a_r)^{-1} j(\chi_1, \ldots, \chi_r) \\
&= (-1)^{r+1} \chi_0(a_0)^{-1} \ldots \chi_r(a_r)^{-1} \frac{1}{q^\mu} g(\chi_0, \psi \circ \mathrm{Tr}_{k_\mu/k}) \ldots g(\chi_r, \psi \circ \mathrm{Tr}_{k_u/k})
\end{aligned}$$

in which $\mu = \mu(\chi_0, \dots, \chi_r)$. Let τ be an automorphism in $\mathrm{Gal}(k_\mu/k)$. Then $\chi_i^\tau = \chi_i \circ \tau$ is a nontrivial character of the same order as χ_i, and $\chi_0^\tau \dots \chi_r^\tau = 1$. We have $\mu(\chi_0, \dots, \chi_r) = \mu(\chi_0^\tau, \dots, \chi_r^\tau)$. Further,

$$\begin{aligned} g(\chi_i^\tau, \psi \circ \mathrm{Tr}_{k_\mu/k}) &= \sum_{x \in k_\mu^\times} \chi_i^\tau(x) \psi \circ \mathrm{Tr}_{k_\mu/k}(x) \\ &= \sum_{x \in k_\mu^\times} \chi_i(x) \psi \circ \mathrm{Tr}_{k_\mu/k} \circ \tau^{-1}(x) \\ &= \sum_{x \in k_\mu^\times} \chi_i(x) \psi(\mathrm{Tr}_{k_\mu/k}(x)) \\ &= g(\chi_i, \psi \circ \mathrm{Tr}_{k_\mu/k}) \end{aligned}$$

and $\chi_i^\tau(a_i) = \chi_i(a_i)$ since $a_i \in k^\times$. This shows that $c(\chi_0^\tau, \dots, \chi_r^\tau) = c(\chi_0, \dots, \chi_r)$ for all $\tau \in Gal(k_\mu/k)$. Given $(\chi_0, \dots, \chi_r)$, there are $\mu(\chi_0, \dots, \chi_r)$ conjugates of $(\chi_0, \dots, \chi_r)$ and they all have the same c and μ. So we can rewrite (2) as

$$(3) \quad \sum_{n=1}^{\infty} \overline{N}_n U^{n-1} = \sum_{i=0}^{r-1} \frac{q^i}{1 - q^i U} + (-1)^r \sum_{(\chi_0, \dots, \chi_r) \in \Lambda_r} \frac{-\mu(\chi_0, \dots, \chi_r) c(\chi_0, \dots, \chi_r) U^{\mu(\chi_0, \dots, \chi_r) - 1}}{1 - c(\chi_0, \dots, \chi_r) U^{\mu(\chi_0, \dots, \chi_r)}},$$

where Λ_r is a set of representatives of the Galois orbits of the $(r+1)$-tuple of characters $(\chi_0, \dots, \chi_r)$, where χ_i are characters as above with $\chi_i^{m'} = 1, \chi_i \neq 1, \chi_0 \cdots \chi_r = 1$. Observe that in each quotient the numerator is ± 1 times the derivative of the denominator, hence we get

$$\sum_{n=1}^{\infty} \overline{N}_n U^{n-1} = \frac{d}{dU} \log Z(U) = \frac{Z'(U)}{Z(U)}$$

for some rational function $Z(U)$. We shall choose $Z(U)$ so that it is the quotient of two polynomials with constant term 1. It is called the zeta function attached to the variety defined by the equation $a_0 x_0^m +, \dots, +a_r x_r^m = 0$.

Exercise 1. Show that $(\chi_0^\tau, \dots, \chi_r^\tau) \neq (\chi_0, \dots, \chi_r)$ if $\tau \in Gal(k_\mu/k)$ is not the identity automorphism. Here $\mu = \mu(\chi_0, \dots, \chi_r)$.

Example 1. Let V_1 be the projective variety defined by $x_0^2 + x_1^2 + x_2^2 = 0$ over a finite field k with q elements. Suppose q is odd. We have $r = 2$ and $m = 2 = m'$. There is only one nontrivial character $\chi \in \widehat{k^\times}$ of order 2. So the only possible choice of χ_i is $\chi_i = \chi$ for $i = 0, 1, 2$. But then $\chi_0 \chi_1 \chi_2 = \chi^3 = \chi \neq 1$. When q is even, $m' = 1$. So in both cases the second summation in (3) is empty and

$$\sum_{n=1}^{\infty} \overline{N}_n U^{n-1} = \frac{1}{1-U} + \frac{q}{1-qU} = \frac{d}{dU} \log Z_{V_1}(U)$$

with

$$Z_{V_1}(U)=\frac{1}{(1-U)(1-qU)}.$$

Example 2. Let V_2 be the projective variety defined by $x_0^3+x_1^3+x_2^3=0$ in $P^2(k)$, where k has cardinality q. So $m=3, r=2, a_0=a_1=a_2=1$. Suppose $q\equiv 1 \pmod 3$ so that $m'=3$ divides $q-1$. Denote by $\eta,\overline{\eta}$ the characters of $k^\times$ of order 3. The only choices of (χ_0,χ_1,χ_2) with $\chi_i^3=1, \chi_i\neq 1$ and $\chi_0\chi_1\chi_2=1$ are (η,η,η) and $(\overline{\eta},\overline{\eta},\overline{\eta})$. Thus $\mu(\eta,\eta,\eta)=\mu(\overline{\eta},\overline{\eta},\overline{\eta})=1$ and $c(\eta,\eta,\eta)=-\frac{1}{q}g(\eta,\psi)^3$, $c(\overline{\eta},\overline{\eta},\overline{\eta})=-\frac{1}{q}g(\overline{\eta},\psi)^3$ for any nontrivial additive character ψ of k. We have

$$\begin{aligned}\sum_{n=1}^{\infty}\overline{N_n}U^{n-1}&=\frac{1}{1-U}+\frac{q}{1-qU}+\frac{\frac{1}{q}g(\eta,\psi)^3}{1+\frac{1}{q}g(\eta,\psi)^3U}+\frac{\frac{1}{q}g(\overline{\eta},\psi)^3}{1+\frac{1}{q}g(\overline{\eta},\psi)^3U}\\&=\frac{d}{dU}\log Z_{V_2}(U)\end{aligned}$$

and

$$Z_{V_2}(U)=\frac{(1+\frac{1}{q}g(\eta,\psi)^3U)(1+\frac{1}{q}g(\overline{\eta},\psi)^3U)}{(1-U)(1-qu)}.$$

Observe that the zeta functions in both examples satisfy a functional equation relating $Z_V(\frac{1}{qU})$ and $Z_V(U)$. Indeed, in Example 1 we have

$$Z_{V_1}(\frac{1}{qU})=\frac{1}{(qU^2)^{-1}}Z_{V_1}(U)=(qU^2)Z_{V_1}(U);$$

and in Example 2, write $g(\eta,\psi)=\varepsilon\sqrt{q}$ for some $\varepsilon\in S^1$, then $g(\overline{\eta},\psi)=\eta(-1)\overline{\varepsilon}\sqrt{q}=\overline{\varepsilon}\sqrt{q}$ since $\eta(-1)=\eta(-1)^3=1$. Therefore we have

$$Z_{V_2}(U)=\frac{(1+\varepsilon^3\sqrt{q}U)(1+\overline{\varepsilon^3}\sqrt{q}U)}{(1-U)(1-qU)}$$

and

$$\begin{aligned}Z_{V_2}(\frac{1}{qU})&=\frac{(\sqrt{q}U)^{-2}(\sqrt{q}U+\varepsilon^3)(\sqrt{q}U+\overline{\varepsilon^3})}{(qU^2)^{-1}(1-U)(1-qU)}\\&=\frac{(1+\varepsilon^3\sqrt{q}U)(1+\varepsilon^3\sqrt{q}U)}{(1-U)(1-qU)}\\&=Z_{V_2}(U).\end{aligned}$$

Now we examine the zeta function $Z_V(U)$ attached to the variety V defined by $a_0x_0^m+\ldots+a_rx_r^m=0$. We see from (3) that

$$Z_V(U)=\prod_{i=0}^{r-1}(1-q^iU)^{-1}\prod_{(\chi_0,\ldots,\chi_r)\in\Lambda_r}(1-c(\chi_0,\ldots,\chi_r)U^{\mu(\chi_0,\ldots,\chi_r)})^{(-1)^r}.$$

Notice that if $(\chi_0, \ldots, \chi_r)$ occurs in the above product, so does $(\overline{\chi_0}, \ldots, \overline{\chi_r})$. Moreover, we have $\mu(\overline{\chi_0}, \ldots, \overline{\chi_r}) = \mu(\chi_0, \ldots, \chi_r) := \mu$, and

$$\begin{aligned} c(\overline{\chi_0}, \ldots, \overline{\chi_r}) &= (-1)^{r+1}\overline{\chi_0}(a_0)^{-1} \ldots \overline{\chi_r}(a_r)^{-1} j(\overline{\chi_1}, \ldots, \overline{\chi_r}) \\ &= (-1)^{r+1}\overline{\chi_0(a_0) \ldots \chi_r(a_r) j(\chi_1, \ldots, \chi_r)} \\ &= \overline{c(\chi_0, \ldots, \chi_r)}. \end{aligned}$$

As $j(\chi_1, \ldots, \chi_r)$ has absolute value $q^{\mu(r-1)/2}$ by Proposition 5 in Chapter 1, we can write

$$c(\chi_0, \ldots, \chi_r) = \varepsilon(\chi_0, \ldots, \chi_r) q^{\mu(r-1)/2} \qquad \text{with} \qquad \varepsilon(\chi_0, \ldots, \chi_r) \in S^1.$$

Then one finds

$$\begin{aligned} & 1 - c(\chi_0, \ldots, \chi_r)(\frac{1}{q^{r-1}U})^{\mu} \\ &= -(q^{(r-1)/2}U)^{-\mu}\varepsilon(\chi_0, \ldots, \chi_r)(1 - \overline{\varepsilon(\chi_0, \ldots, \chi_r)} q^{\mu(r-1)/2}U^{\mu}) \\ &= -(q^{(r-1)/2}U)^{-\mu}\varepsilon(\chi_0, \ldots, \chi_r)(1 - c(\overline{\chi_0}, \ldots, \overline{\chi_r})U^{\mu}). \end{aligned}$$

If $(\overline{\chi_0}, \ldots, \overline{\chi_r})$ is not conjugate to $(\chi_0, \ldots, \chi_r)$, then $\varepsilon(\overline{\chi_0}, \ldots, \overline{\chi_r})\varepsilon(\chi_0, \ldots, \chi_r) = 1$; otherwise $\varepsilon(\chi_0, \ldots, \chi_r) = \pm 1$. This shows that

$$\begin{aligned} Z_V(\frac{1}{q^{r-1}U}) &= (-1)^r q^{r(r-1)/2} U^r \times \prod_{i=0}^{r-1} (1 - q^i U)^{-1} \\ &\quad \times \prod_{(\chi_0, \ldots, \chi_r) \in \Lambda_r} -\varepsilon(\chi_0, \ldots, \chi_r)^{(-1)^r} (q^{\frac{r-1}{2}} U)^{(-1)^{r+1}\mu(\chi_0, \ldots, \chi_r)} \\ &\quad (1 - c(\overline{\chi_0}, \ldots, \overline{\chi_r}) U^{\mu(\overline{\chi_0}, \ldots, \overline{\chi_r})})^{(-1)^r} \\ &= \pm (q^{\frac{r-1}{2}} U)^e Z_V(U), \end{aligned}$$

where the exponent e is the number of poles of Z_V minus the number of zeroes of Z_V. Observe that $r - 1$ is the dimension of the variety V.

From the above computation and the results for curves, Weil was led to the following far-reaching conjectures for nonsingular irreducible projective varieties, which suggested a deep connection between the arithmetic of algebraic varieties defined over finite fields and the topology of algebraic varieties defined over the complex numbers.

Let V be an irreducible nonsingular projective variety of dimension d defined over a finite field k of q elements. Denote by N_n the number of points of V over the degree n field extension of k. The zeta function Z_V attached to V is defined as

$$Z_V(U) = \exp(\sum_{n=1}^{\infty} N_n \frac{U^n}{n}),$$

which is a formal power series in U with rational coefficients. In 1949, Weil made the following 4 conjectural statements about Z_V:

(I) Rationality. $Z_V(U)$ is a rational function in U (with rational coefficients).

(II) Functional equation. There is an integer E, called the Euler-Poincaré characteristic of V, such that Z_V satisfies the functional equation

$$Z_V(\frac{1}{q^dU}) = \pm(q^{\frac{d}{2}}U)^E Z_V(U).$$

(III) Riemann hypothesis. There are polynomials $P_i(U), 0 \le i \le 2d$, with $P_0(U) = 1 - U$ and $P_{2d}(U) = 1 - q^dU$, such that

$$Z_V(U) = \frac{P_1(U)P_3(U)\dots P_{2d-1}(U)}{P_0(U)P_2(U)\dots P_{2d}(U)}$$

Further, each $P_i(U)$ is a polynomial with integral coefficients which can be written as

$$P_i(U) = \prod_{j=1}^{B_i}(1 - \alpha_{ij}U),$$

where α_{ij} are algebraic integers with $|\alpha_{ij}| = q^{i/2}$. Note that these conditions uniquely determine the polynomials P_i, if they exist.

(IV) Betti numbers. Define B_i, the degree of P_i, to be the ith Betti number of the variety V. Then the Euler-Poincaré characteristic E of V in (II) is equal to $\sum_{i=0}^{2d}(-1)^i B_i$. Further if V is obtained from a variety $\tilde{V}$ defined over the ring of integers of a number field by reduction modulo a prime ideal, then B_i is the rank of the ith cohomology group $H^i(\tilde{V}_h, \mathbf{Z})$, where $\tilde{V}_h$ is the complex projective variety defined by the same equations as $\tilde{V}$ with the usual topology.

The variety V_1 in Example 1 is a projective line in P^2, it has genus 0. Its Euler-Poincaré characteristic $E = 2 = 1 - 0 + 1$ and (I)-(IV) are satisfied. The variety V_2 in Example 2 for the case char $k \neq 2, 3$ is an elliptic curve, which has genus 1. We see that the Euler-Poincaré characteristic $E = 0 = 1 - 2 + 1$ and (I)-(IV) hold. The zeta functions for nonsingular projective curves over a finite field were introduced by F. K. Schmidt in 1931, who showed that for a projective nonsingular curve C of genus g defined over a field of q elements, its zeta function is of the form

$$Z_C(U) = \frac{P_1(U)}{(1 - U)(1 - qU)},$$

where $P_1(U)$ is a polynomial of degree $2g$ with integral coefficients and $P_1(0) = 1$, and further, $Z_C(U)$ satisfies the functional equation

$$Z_C(\frac{1}{qU}) = \pm(qU^2)^{1-g} Z_C(U).$$

The "Riemann hypothesis" for C is therefore that the zeroes of $P_1(U)$ have absolute value $q^{-1/2}$. This was first conjectured by E. Artin, who proved it for special cases. The case of $g = 1$ was proved by Hasse [11], and Weil proved it for curves of arbitrary genus in 1940 [15].

For the "Fermat hypersurface" V defined by $a_0x_0^m + \ldots + a_rx_r^m = 0$, it has dimension $d = r - 1$. When r is even, i.e., d is odd, we have

$$P_{2i}(U) = 1 - q^iU \quad \text{for} \quad 0 \le i \le d,$$

$$P_d(U) = \prod_{(\chi_0,\ldots,\chi_r)\in A_r} (1 - c(\chi_0,\ldots,\chi_r)U^{\mu(\chi_0,\ldots,\chi_r)}),$$

and other $P_i(U) = 1$. When r is odd, i.e., d is even, we have

$$P_{2i}(U) = 1 - q^iU \quad \text{for} \quad 0 \le i \le d,\ 2i \ne d,$$

$$P_d(U) = (1 - q^{d/2}U) \prod_{(\chi_0,\ldots,\chi_r)\in A_r} (1 - c(\chi_0,\ldots,\chi_r)U^{\mu(\chi_0,\ldots,\chi_r)}),$$

and all other $P_i(U) = 1$. Thus Z_V satisfies the Weil conjectures.

Exercise 2. Show that the Riemann hypothesis for a nonsingular projective curve C of genus g over a field k of q elements is equivalent to

$$|\overline{N_n} - q^n - 1| \le 2gq^{n/2} \quad \text{for all } n \ge 1.$$

Here $\overline{N_n}$ denotes the number of k_n rational points on C as usual.

Exercise 3. Let $V = P^d$ be the projective space of dimension d over a finite field k of q elements. Verify from the definition of the zeta function that

$$Z_V(U) = \frac{1}{(1-U)(1-qU)\ldots(1-q^dU)}.$$

Show that the Weil conjectures hold for $V = P^d$ over k.

Exercise 4. Determine the zeta function for the variety V defined by $x_0^3+x_1^3+x_2^3 = 0$ over k with $|k| = q \equiv 2(\text{mod}3)$ and q odd. Verify the Weil conjectures for Z_V.

It was Gauss who initiated the study of number of solutions N_p modulo p of a polynomial with integral coefficients. In particular, he wanted to know how N_p varies with respect to p. Suppose the polynomial defines an irreducible nonsingular projective variety of dimension d. Then Weil conjectures imply that

$$|N_p - (1 + p + \ldots + p^d)| \le bp^{d/2},$$

where b is the dth Betti number of the associcted projective variety over $\mathbf{C}$ defined by the same polynomial.

---

§3 Cohomological interpretation of the Weil conjectures

As pointed out by Weil himself, if one had a suitable cohomology theory for abstract varieties, analogous to the ordinary cohomology of varieties defined over C, then one could deduce his conjecture from various standard properties of the cohomology theory. Dwork [4] succeeded in proving the rationality and functional equation using p-adic analysis. Most other work on the Weil conjectures has centered around the search of a good cohomology theory which would give the Betti numbers as in (IV), and which should have its coefficients in a characteristic zero field so that the Lefschetz fixed point theorem would hold. There were several attempts. In 1963 Grothendieck developed the l-adic cohomology theory for abstract algebraic varieties using étale topology on algebraic varieties, from which he obtained another proof of rationality and functional equation of the zeta function [6-9]. The deepest part of the conjectures is the Riemann hypothesis, which Deligne succeeded in proving in 1973 using l-adic cohomology [1].

We explain briefly the connection between l-adic cohomology and the Weil conjectures. As in the previous section, let V be an irreducible nonsingular projective variety of dimension d defined over a finite field k of q elements. Denote by $\overline{V}$ the collection of points of V over an algebraic closure $\bar{k}$. Let l be a prime not dividing q. Endow the étale topology on $\overline{V}$. For each integer $r \geq 1$, there is an étale cohomology $H^i_{ét}(\overline{V}, \mathbf{Z}/l^r\mathbf{Z})$. The l-adic cohomology of $\overline{V}$ is defined by

$$H^i(\overline{V}, \mathbf{Q}_l) = (\varprojlim_{r} H^i_{ét}(\overline{V}, \mathbf{Z}/l^r\mathbf{Z})) \underset{\mathbf{Z}_l}{\otimes} \mathbf{Q}_l,$$

where $\mathbf{Z}_l$ is the ring of l-adic integers, i.e., $\mathbf{Z}_l = \{\sum_{i=0}^{\infty} a_i l^i : 0 \leq a_i \leq l-1\}$, which is the inverse limit of $\mathbf{Z}/l^r\mathbf{Z}$ as r tends to infinity; $\mathbf{Q}_l$ is the quotient field of $\mathbf{Z}_l$, which is also the completion of $\mathbf{Q}$ with respect to the l-adic metric on $\mathbf{Q}$. The l-adic cohomology has the following properties.

(a) The groups $H^i(\overline{V}, \mathbf{Q}_l)$ are finite-dimensional vector spaces over $\mathbf{Q}_l$ and they are zero except for $0 \leq i \leq 2d$.

(b) There is a cup-product structure

$$H^i(\overline{V}, \mathbf{Q}_l) \times H^j(\overline{V}, \mathbf{Q}_l) \to H^{i+j}(\overline{V}, \mathbf{Q}_l)$$

defined for all i, j.

(c) Poincaré duality. The top cohomology group $H^{2d}(\overline{V}, \mathbf{Q}_l)$ is 1-dimensional, and the cup-product defines a nondegenerate pairing

$$H^i(\overline{V}, \mathbf{Q}_l) \times H^{2d-i}(\overline{V}, \mathbf{Q}_l) \to H^{2d}(\overline{V}, \mathbf{Q}_l) \simeq \mathbf{Q}_l$$

for $0 \leq i \leq 2d$.

(d) Künneth formula. For two nonsingular varieties X, Y, there is a natural isomorphism of graded algebras

$$H^*(X, \mathbf{Q}_l) \otimes H^*(Y, \mathbf{Q}_l) \tilde{\to} H^*(X \times Y, \mathbf{Q}_l).$$

(e) Lefschetz fixed-point formula. Let $f : \overline{V} \to \overline{V}$ be a morphism with isolated fixed points, and each fixed point has multiplicity one. That is, the graph of f in $\overline{V} \times \overline{V}$ intersects the diagonal Δ in $\overline{V} \times \overline{V}$ transversally. Denote by $L(f, \overline{V})$ the number of fixed points of f. ($L(f, \overline{V})$ is finite since $\overline{V}$ is compact). Then

$$L(f, \overline{V}) = \sum_{i=0}^{2d} (-1)^i \operatorname{Tr}(f^{(i)}; H^i(\overline{V}, \mathbf{Q}_l)),$$

where $f^{(i)}$ is the induced pull-back map on H^i.

(f) Comparison theorem. If $\overline{V}$ comes from a nonsingular projective variety $\widetilde{V}$ defined over the ring of integers of a number field by reduction modulo a prime ideal, then

$$H^i(\overline{V}, \mathbf{Q}_l) \underset{\mathbf{Q}_l}{\otimes} \mathbf{C} \cong H^i(\widetilde{V}_h, \mathbf{C}),$$

where $\widetilde{V_h}$ is the associated complex variety with the classical topology.

(g) Cohomology class of a cycle. If Z is a subvariety of codimension i, then there is associated to Z a cohomology class $\eta(Z) \in H^{2i}(\overline{V}, \mathbf{Q}_l)$. This map extends by linearity to cycles. Rationally equivalent cycles have the same cohomology class. Intersection of cycles becomes cup-product of cohomology classes. Further, if P is a closed point of V, then $\eta(P) \in H^{2d}(\overline{V}, \mathbf{Q}_l)$ is nonzero.

These properties are enjoyed by the usual cohomology theory for an irreducible nonsingular projective variety over $\mathbf{C}$ developed mainly by Lefschetz and Hodge.

Now we discuss consequences of the above properties. The Frobenius morphism $\Phi : \overline{V} \to \overline{V}$ sends a point with coordinates (a_i) to the point with coordinates (a_i^q). A point P of $\overline{V}$ has coordinates over k_n if and only if it is fixed by Φ^n. Thus

$$\overline{N}_n = \text{the number of fixed points of } \Phi^n = L(\Phi^n, \overline{V}).$$

As V is nonsingular, we can calculate $\overline{N}_n$ by the Lefschetz fixed-point formula (e). It gives

$$\overline{N}_n = \sum_{i=0}^{2d} (-1)^i \operatorname{Tr}((\Phi^n)^{(i)}; H^i(\overline{V}, \mathbf{Q}_l)).$$

Substituting this into the definition of the zeta function attached to V, we get

$$Z_V(U) = \prod_{i=0}^{2d}\left[\exp\left(\sum_{n=1}^{\infty}\mathrm{Tr}((\Phi^n)^{(i)}; H^i(\overline{V}, \mathbf{Q}_l))\frac{U^n}{n}\right)\right]^{(-1)^i}$$
$$= \prod_{i=0}^{2d}\left[\exp\left(\sum_{n=1}^{\infty}\mathrm{Tr}((\Phi^{(i)})^n); H^i(\overline{V}, \mathbf{Q}_l))\frac{U^n}{n}\right)\right]^{(-1)^i}.$$

For each i, we may evaluate the exponential part using the following result on linear algebra.

Lemma 1. *Let f be an endomorphism of a finite-dimensional vector space W over a field K of characteristic zero. Then*

$$\exp\sum_{n=1}^{\infty}\mathrm{Tr}(f^n; W)\frac{U^n}{n} = \det(1 - fU; W)^{-1}$$

as formal power series in U with coefficients in K.

Proof. If W is one-dimensional, then f is multiplication by a scalar λ and

$$\exp\sum_{n=1}^{\infty}\mathrm{Tr}(f^n; W)\frac{U^n}{n} = \exp\sum_{n=1}^{\infty}\lambda^n\frac{U^n}{n} = \frac{1}{1-\lambda U} = \det(1 - fU; W)^{-1}.$$

For general case we prove by induction on $\dim W$. We may assume that K is algebraically closed so that f has an eigenvector, hence W contains a one-dimensional invariant subspace W'. We have a short exact sequence

$$0 \to W' \to W \to W/W' \to 0$$

and the product of left (resp. right) hand side on W' and on W/W' is equal to that on W. The desired result follows from induction.

It follows immediately from Lemma 1 that

Theorem 3. *The zeta function attached to an irreducible nonsingular projective variety V of* $\dim d$ *over k can be expressed as*

$$Z_V(U) = \frac{P_1(U)P_3(U)\ldots P_{2d-1}(U)}{P_0(U)P_2(U)\ldots P_{2d}(U)},$$

where $P_i(U) = \det(1 - \Phi^{(i)}U; H^i(V, \mathbf{Q}_l))$, and $\Phi^{(i)}$ is the map on $H^i(V, \mathbf{Q}_l)$ induced from the Frobenius morphism $\Phi : \overline{V} \to \overline{V}$.

We know that $Z_V(U)$ is a formal power series in U with coefficients in $\mathbf{Q}$, the above theorem says that it is also a rational function in U with coefficient in $\mathbf{Q}_l$, therefore it is a rational function in U with coefficients in $\mathbf{Q}$. However this does not mean that each individual $P_i(U)$ has rational coefficients, nor do we know that the P_i's in Theorem 3 are the P_i's stated in (III). On the other hand, since $\Phi^{(0)}$ acts on $H^0(\overline{V}, \mathbf{Q}_l)$ as the identity map, $P_0(U) = 1 - U$. Further, since the Frobenius morphism is a finite morphism of degree q^d, it induces multiplication by q^d on the top cohomology group $H^{2d}(\overline{V}, \mathbf{Q}_l)$. So $P_{2d}(U) = 1 - q^d U$. Let $B_i = \deg P_i = \dim H^i(\overline{V}, \mathbf{Q}_l)$.

Next we see that the functional equation (II) follows from the Poincaré duality (c). Indeed, the pairing on $H^i(\overline{V}, \mathbf{Q}_l) \times H^{2d-i}(\overline{V}, \mathbf{Q}_l) \to H^{2d}(\overline{V}, \mathbf{Q}_l)$ sends $(\Phi^{(i)}(v), (\Phi^{(2d-i)}(w))$ to $\Phi^{(i)}(v) \vee \Phi^{(2d-i)}(w) = \Phi^{(2d)}(v \vee w) = q^d(v \vee w)$ for all $v \in H^i(\overline{V}, \mathbf{Q}_l)$ and $w \in H^{2d-i}(\overline{V}, \mathbf{Q}_l)$. To proceed, recall the following fact from linear algebra.

Let A, B be vector spaces over a field K of dimension r such that there is a perfect pairing $<,>: A \times B \to K$. Suppose f and g are endomorphisms on A, B, respectively, such that there is a nonzero element $\lambda \in K$ satisfying

$$< fa, gb >= \lambda < a, b > \quad \text{for all } a \in A \text{ and } b \in B.$$

Then f, g are invertible. Further, ${}^t gf = \lambda I_A$. Therefore

$$\begin{aligned}
\det(1 -{}^t gU; B) &= \det(1 - gU; B) \\
&= \det(1 - \lambda f^{-1}U; A) \\
&= \det(-\lambda f^{-1}U; A)\det(1 - \frac{f}{\lambda U}; A) \\
&= \frac{(-\lambda U)^r}{\det(f; A)}\det(1 - \frac{f}{\lambda U}; A)
\end{aligned}$$

and $\det({}^t gf) = (\det g)(\det f) = \det \lambda I_A = \lambda^r$.

Applied to our situation above, we get

$$\begin{aligned}
P_{2d-i}(U) &= \det(1 - \Phi^{(2d-i)}U; H^{(2d-i)}(\overline{V}, \mathbf{Q}_l)) \\
&= \frac{(-1)^{B_i}(q^d U)^{B_i}}{\det(\Phi^{(i)}; H^{(i)}(\overline{V}, \mathbf{Q}_l))}\det(1 - \Phi^{(i)}/q^d U; H^{(i)}(\overline{V}, \mathbf{Q}_l)) \\
&= \frac{(-q^d U)^{B_i}}{\det(\Phi^{(i)}; H^{(i)}(\overline{V}, \mathbf{Q}_l))} P_i(\frac{1}{q^d U})
\end{aligned}$$

and $\det(\Phi^{(i)}; H^i(\overline{V}, \mathbf{Q}_l)) \det(\Phi^{(2d-i)}; H^{2d-i}(\overline{V}, \mathbf{Q}_l)) = q^{dB_i}$. In fact, the transpose of $q^{-(2d-i)/2}\Phi^{(2d-i)}$ is the inverse of $q^{-i/2}\Phi^{(i)}$. Combined with Theorem 3, this yields the functional equation (II) with

$$E = \sum_{i=0}^{2d} (-1)^i B_i.$$

Formally, we have shown that (I), (II), (IV) follows from the properties of l-adic cohomology once we have interpreted the zeta function as in Theorem 3. That this is the correct decomposition and that the Riemann hypothesis (III) holds were proved by Deligne [1] in 1973, using much deeper properties of l-adic cohomology. See also [5] and [13].

Theorem 4(Deligne). *The polynomials $P_i(U)$ in Theorem 3 have integral coefficients independent of l, and they can be written as*

$$P_i(U) = \sum_{j=1}^{B_i}(1 - \alpha_{ij}U),$$

where α_{ij} are algebraic integers with absolute value $q^{i/2}$.

While we shall not sketch Deligne's proof, we explain briefly why it is true. There is a cycle Z of codimension 1 such that $h = \eta(Z)$ is a nontrivial class in $H^2(\overline{V}, \mathbf{Q}_l)$ and $\Phi^{(2)}(h) = qh$. Then cup product with h $d-i$ times yields an isomorphism from $H^i(\overline{V}, \mathbf{Q}_l)$ to $H^{i+2(d-i)}(\overline{V}, \mathbf{Q}_l) = H^{2d-i}(\overline{V}, \mathbf{Q}_l)$. This together with the Poincaré duality gives rise to a nondegenerate pairing from $H^i(\overline{V}, \mathbf{Q}_l) \times H^i(\overline{V}, \mathbf{Q}_l)$ to $H^{2d}(\overline{V}, \mathbf{Q}_l)$. When V has an associated irreducible nonsingular complex projective variety $\widetilde{V_h}$, this is a nondegenerate pairing from $H^i(\widetilde{V_h}, \mathbf{C}) \times H^i(\widetilde{V_h}, \mathbf{C})$ to $H^{2d}(\widetilde{V_h}, \mathbf{C})$. When i is even, $H^i(\widetilde{V_h}, \mathbf{C})$ contains a $\Phi^{(i)}$-invariant real subspace $A^i(\widetilde{V_h})$ of dimension B_i over $\mathbf{R}$ such that on $A^i(\widetilde{V_h})$ this pairing is a nondegenerate scalar product and $q^{-i/2}\Phi^{(i)}$ is a unitary mapping for this scalar product. This shows that the eigenvalues of $\Phi^{(i)}$, i even, have the desired absolute value $q^{i/2}$. For i odd, we consider $\overline{V} \times \overline{V}$. The magnitude of eigenvalues of $\Phi^{(i)}$ on $H^i(\overline{V}, \mathbf{Q}_l)$ with i odd follows from that of induced Frobenius morphism on $H^{2i}(\overline{V} \times \overline{V}, \mathbf{Q}_l)$.

Remark. As shown above, the eigenvalues of $\Phi^{(i)}$ on $H^i(\overline{V}, \mathbf{Q}_l)$ are the α_{ij}'s occurring in $P_i(U) = \sum_{j=1}^{B_i}(1 - \alpha_{ij}U)$. In particular, the eigenvalues of $\Phi^{(i)}$ have absolute value $q^{i/2}$.

---

§4 Zeta functions as Euler products

Let k be a field and $P(T_1, \cdots, T_r)$ be an irreducible polynomial in $k[T_1, \cdots, T_r]$. The set of solutions of P in k^r is called an "affine algebraic hypersurface" in the affine space k^r. We may also consider solutions of P in any finite algebraic extensions of k. Denote by $\bar{k}$ an algebraic closure of k, and by V the set of solutions of P in $\bar{k}^r$. We say that V is defined over k. Given a point $x = (x_1, \cdots, x_r)$ of V, denote by $k(x)$ the field $k(x_1, \cdots, x_r)$ generated by the coordinates of x. It is a finite extension of k, and its degree over k is called the degree of x. Consider the homomorphism from $k[T_1, \cdots, T_r]$ to $\bar{k}$ sending T_i to x_i; its kernel $\mathcal{M}$ is a maximal ideal of $k[T_1, \cdots, T_r]$ containing P, and the quotient $k[T_1, \cdots, T_r]/\mathcal{M}$ is isomorphic to $k(x)$. Define $\deg \mathcal{M}$ to be $\deg x$. There are $\deg x = [k(x) : k]$ imbeddings of $k(x)$ into $\bar{k}$ over k. For each imbedding σ, the point $x^\sigma = (x_1^\sigma, \cdots, x_r^\sigma)$ also lies in V and the homomorphism from $k[T_1, \cdots, T_r]$ to $\bar{k}$ sending T_i to x_i^σ also has kernel $\mathcal{M}$ since polynomials in $\mathcal{M}$ have coefficients in k. Thus the ideal $\mathcal{M}$ corresponds to $\deg \mathcal{M}$ points of V. Conversely, given a maximal ideal $\mathcal{M}$ containing P, it corresponds to $\deg \mathcal{M}$ points of V as above. The orbit of x under all imbeddings of $k(x)$ over k into $\bar{k}$, that is, the set $\{x^\sigma : \sigma \in Gal(\bar{k}/k)\}$, is called a *closed point* of V, it is represented by the maximal ideal $\mathcal{M}$.

When k is a finite field, the zeta function of V is defined as

$$Z_V(u) = \prod_{\substack{\mathcal{M} \text{max ideal} \\ P \in \mathcal{M}}} (1 - u^{\deg \mathcal{M}})^{-1} = \prod_{\substack{x \text{ closed} \\ \text{point in } V}} (1 - u^{\deg x})^{-1}.$$

The points of V lying in k_n^r are those x with degree dividing n, so there are at most q^{nr} maximal ideals $\mathcal{M}$ containing P with $\deg \mathcal{M}$ dividing n. Here q is the cardinality of k. This shows that the infinite product converges absolutely for $|u|$ small. Further, for small u, we have

$$\frac{Z_V(u)'}{Z_V(u)} = \sum_{\substack{\mathcal{M} \max \\ P \in \mathcal{M}}} \frac{\deg \mathcal{M} \cdot u^{\deg \mathcal{M} - 1}}{1 - u^{\deg \mathcal{M}}} = \sum_{l=1}^{\infty} \sum_{\substack{\mathcal{M} \max \\ P \in \mathcal{M}}} \deg \mathcal{M} \cdot u^{l \deg \mathcal{M} - 1} = \sum_{\nu=1}^{\infty} N_\nu u^{\nu - 1},$$

where $N_\nu = \sum_{\deg \mathcal{M} | \nu} \deg \mathcal{M}$ is the number of points of V lying in k_ν^r.

In general, if V is a nonsingular projective variety defined over a finite field, it is a union of several affine varieties. The zeta function $Z_V(u)$ defined in §2 using the number of solutions over finite extensions of k has an Euler product:

$$Z_V(u) = \exp\left(\sum_{\nu=1}^{\infty} N_\nu \frac{u^\nu}{\nu}\right) = \prod_{\substack{x \text{ closed} \\ \text{point in } V}} (1 - u^{\deg x})^{-1}.$$

Example 3. Let C be the projective line over a finite field k. It has one "point at infinity" and the remaining points are on the affine line over k. The closed points on the affine line are parametrized by the monic irreducible polynomials in $k[T]$ or the maximal ideals of $k[T]$. We have

$$Z_C(u) = (1 - u)^{-1} \prod_{\substack{f \in k[T] \\ f \text{monic irred}}} (1 - u^{\deg f})^{-1}.$$

Example 4. Let C be a nonsingular projective curve defined by a homogeneous polynomial $P(T_0, T_1, T_2)$ with coefficients in a finite field k of q elements. The collection of solutions $\{(x_0 : x_1 : x_2)\}$ with $x_0 \neq 0$ can be identified with the points on the affine curve C defined by $P(T,Y) = P(1, \frac{T_1}{T_0}, \frac{T_2}{T_0})$, where $T = \frac{T_1}{T_0}, Y = \frac{T_2}{T_0}$. We may assume P is monic in Y. Denote by F the field $k(T)$ and by K the field $F[Y]/(P(T,Y))$, which is a finite extension of F, called the field of rational functions on $\mathcal{C}$. Let $\mathcal{O}$ be the integral closure of $k[T]$ in K. The maximal ideals $\mathcal{M}$ in $k[T,Y]$ containing P are the maximal ideals $\mathcal{P}$ of $\mathcal{O}$; further, if $\mathcal{P}$ corresponds to the closed point x on the affine curve C, then the residue field of $\mathcal{P}$, $\mathcal{O}/\mathcal{P}$, is isomorphic to $k(x)$. Define $\deg \mathcal{P} = \deg x = [\mathcal{O}/\mathcal{P} : k]$ so that the residue field $\mathcal{O}/\mathcal{P}$ has cardinality $q^{\deg \mathcal{P}}$, which is defined to be the norm $N\mathcal{P}$ of $\mathcal{P}$. The zeta function attached to C is

$$Z_C(u) = \prod_{\mathcal{P}} (1 - u^{\deg \mathcal{P}})^{-1}.$$

Call these prime ideals $\mathcal{P}$ "finite places" of $\mathcal{C}$ or K. The points in $\mathcal{C} \setminus C$ are solutions $\{(x_0 : x_1 : x_2)\}$ wtih $x_0 = 0$. They are regarded as "points at infinity" with respect to our choice of the affine subvariety. There are finitely many such points, they are represented by the maximal ideals $\mathcal{P}_\infty$ of the integral closure $\mathcal{O}'$ of $k[\frac{1}{T}]$ in K containing the principal ideal generated by $\frac{1}{T}$. These can be viewed as "infinite places". Again, $\deg \mathcal{P}_\infty$ is $[\mathcal{O}'/\mathcal{P}_\infty : k]$ and $N\mathcal{P}_\infty$is the cardinality of the residue field $\mathcal{O}'/\mathcal{P}_\infty$, that is, $q^{\deg \mathcal{P}_\infty}$.

The finite and infinite places together represent all closed points of $\mathcal{C}$, and we have

$$Z_{\mathcal{C}}(u) = \prod_{\mathcal{P} \text{ place of } K} (1 - u^{\deg \mathcal{P}})^{-1}.$$

The field $F = k(T)$ is called a rational function field of one variable over k. A function field K is a finite extension of $k(T)$; it is the field of rational functions on a nonsingular projective curve defined over k_ν, the maximal extension of k contained in K, called the field of constants of K.

References

[1] P. Deligne: La conjecture de Weil I, Inst. Hautes Etudes Sci. Publ. Math. 43 (1974), 273-307.

[2] P. Deligne: La conjecture de Weil II, Inst. Hautes Etudes Sci. Publ. Math. 52 (1980), 137-252.

[3] J. A. Dieudonné: On the history of the Weil conjectures. The Mathematical Intelligencer 10. Springer-Verlag, Berlin-Heidelberg-New York (1975).

[4] B. Dwork: On the rationality of the zeta function of an algebraic variety, Amer. J. Math. 82 (1960), 631-648.

[5] E. Freitag and R. Kiehl: Etale Cohomology and the Weil conjecture, Springer-Verlag, Berlin-Heidelberg-New York, 1988.

[6] A. Grothendieck (with M. Artin and J. L. Verdier): Théorie des topos et cohomologic étale des schemas (1963-64), Lecture Notes in Math. 269,270,305, Springer-Verlag, Berlin-Heidelberg-New York, 1972-73.

[7] A. Grothendieck: Formule de Lefschetz et rationalité des fonctions L, Seminaire Bourbaki 1964/65, Exposé 279. W.A. Benjamin, New York, 1966

[8] A. Grothendieck (by P. Deligne with J. F. Boutot, L. Illusie and J. L. Verdier) Cohomologie étale. Lecture Notes in Math. 569, Springer-Verlag, Berlin-Heidelberg-New York, 1977.

[9] A. Grothendieck: Cohomologie ℓ-adique et functions L (1965-66), Lecture Notes in Math. 589, Springer-Verlag, Berlin-Heidelberg-New York, 1977.

[10] R. Hartshorne: Algebraic Geometry, Springer-Verlag, Berlin-Heidelberg-New York, 1977.

[11] H. Hasse: Beweis des Analogons der Riemannschen Vermutung für die Artinschen and F. K. Schmidtschen Kongruenzzetafunktionen in gewissen elliptischen Fällen. Ges. d. Wiss. Nachrichten. Math. Phys. Klasse, 1933, Heft 3, 253-262.

[12] L. K. Hua and H. S. Vandiver: Characters over certain types of rings with applications to the theory of equations in finite fields, Proc. Nat. Acad. Sci. U.S.A, vol 35 (1949), 94-99.

[13] N. Katz: An overview of Deligne's proof of the Riemann hypothesis for varieties over finite fields, Proc. Symp. in Pure Math., Amer. Math. Soc. 28 (1976), 275-305.

[14] A. Weil: Numbers of solutions of equations in finite fields, Bulletin of Amer. Math. Soc. 55 (1949), 497-508.

[15] A. Weil: Sur les Courbes Algébriques et les Variétés Qui s'en Deduisent, Hermann, Paris (1948).

CHAPTER 3

Local and Global Fields

§1 Local fields

Take a function field $F = k(T)$ with the field of constants k being a finite field of q elements. Let x be a closed point of the projective line $\mathbf{P}^1$ over k. By studying the behavior of a nonzero rational function f at $x \in \mathbf{P}^1$ we mean to know if x is a zero or pole of f, and of which order. An easy way to get this is to use the irreducible polynomial $P(T)$ of x and to express $f(T) = g(T)/h(T)$ as a quotient of two polynomials $g(T)$ and $h(T)$ in $k[T]$. (Replace T by $\frac{1}{T}$ if x is the point at infinity.) Then the order of vanishing of f at x is the order of P dividing g minus that dividing h. Or even better, denote by $\mathcal{P}$ the ideal generated by $P(T)$ in $k[T]$, then the order of $P(T)$ dividing $g(T)$ is the highest power m such that $g(T)$ lies in $\mathcal{P}^m$. The last description extends to any function field K with field of constants k. Indeed, in this case a place of K is a maximal ideal $\mathcal{P}$ of the integral closure $\mathcal{O}$ of $k[T]$ or of $k[\frac{1}{T}]$ in K. To know the behavior of a nonzero rational function $f \in K$ at $\mathcal{P}$, express $f = g/h$, where g, h are in $\mathcal{O}$. This is possible because K is the quotient field of $\mathcal{O}$. Then the order of g at $\mathcal{P}$, denoted by $\mathrm{ord}_{\mathcal{P}} g$, is the highest power m such that $g \in \mathcal{P}^m$ and $\mathrm{ord}_{\mathcal{P}} f = \mathrm{ord}_{\mathcal{P}} g - \mathrm{ord}_{\mathcal{P}} h$. Using $\mathrm{ord}_{\mathcal{P}}$ we define, for each place $\mathcal{P}$ of K, a valuation $|\ |_{\mathcal{P}}$ on K by

$$\begin{cases} |0|_{\mathcal{P}} = 0, \\ |f|_{\mathcal{P}} = N\mathcal{P}^{-\mathrm{ord}_{\mathcal{P}} f} = \left(q^{-\deg \mathcal{P}}\right)^{\mathrm{ord}_{\mathcal{P}} f} = |\mathcal{O}/\mathcal{P}|^{-\mathrm{ord}_{\mathcal{P}} f} \quad \text{if } f \in K^\times. \end{cases}$$

Over the field of rationals $\mathbf{Q}$, the ring of integers $\mathbf{Z}$ plays the same role as $k[T]$ in $k(T)$. So each prime ideal (p) of $\mathbf{Z}$ is a finite place of $\mathbf{Q}$. If K is a finite (algebraic) field extension of $\mathbf{Q}$, called a number field, its ring of integers $\mathcal{O}$ is the integral closure of $\mathbf{Z}$ in K, and the maximal ideals $\mathcal{P}$ of $\mathcal{O}$ are the finite places of K. The residue field $\mathcal{O}/\mathcal{P}$ is finite, its cardinaity $N\mathcal{P}$ is called the norm of $\mathcal{P}$. We define $\mathrm{ord}_{\mathcal{P}}$ the same way as above, and a valuation $|\ |_{\mathcal{P}}$ on K by

$$\begin{cases} |0|_{\mathcal{P}} = 0, \\ |z|_{\mathcal{P}} = N\mathcal{P}^{-\mathrm{ord}_{\mathcal{P}} z} = |\mathcal{O}/\mathcal{P}|^{-\mathrm{ord}_{\mathcal{P}} z} \quad \text{if } z \in K^\times. \end{cases}$$

The valuation $|\ |_{\mathcal{P}}$ defined for a function field or a number field K above has the following properties. Let $x, y \in K$. Then
(1) $|x|_{\mathcal{P}} \geq 0$ and $|x|_{\mathcal{P}} = 0$ if and only if $x = 0$,
(2) $|xy|_{\mathcal{P}} = |x|_{\mathcal{P}}|y|_{\mathcal{P}}$,
(3) $|x+y|_{\mathcal{P}} \leq \max(|x|_{\mathcal{P}}, |y|_{\mathcal{P}})$.
Observe that property (3) implies
(4) triangular inequality : $|x+y|_{\mathcal{P}} \leq |x|_{\mathcal{P}} + |y|_{\mathcal{P}}$.
Therefore $|\ |_{\mathcal{P}}$ defines a metric on K. The completion of K with respect to this metric is denoted by $K_{\mathcal{P}}$, which is a field containing K as a dense subfield. The integers in $K_{\mathcal{P}}$ are those elements which don't have poles at $\mathcal{P}$; they form a ring because of properties (1)–(3), called the ring of integers $\mathcal{O}_{\mathcal{P}}$ of $K_{\mathcal{P}}$. In other words,

$$\mathcal{O}_{\mathcal{P}} = \{x \in K_{\mathcal{P}} : |x|_{\mathcal{P}} \leq 1\}.$$

The multiplicatively invertible elements in $\mathcal{O}_{\mathcal{P}}$ are called $\mathcal{P}$–adic units; they form a multiplicative group $\mathcal{U}_{\mathcal{P}}$, so

$$\mathcal{U}_{\mathcal{P}} = \{x \in K_{\mathcal{P}} : |x|_{\mathcal{P}} = 1\}.$$

Thus

$$\mathcal{P}_{\mathcal{P}} = \mathcal{O}_{\mathcal{P}} - \mathcal{U}_{\mathcal{P}} = \{x \in K_{\mathcal{P}} : |x|_{\mathcal{P}} < 1\}$$

is the unique maximal ideal of $\mathcal{O}_{\mathcal{P}}$, and the only nonzero ideals of $\mathcal{O}_{\mathcal{P}}$ are powers of $\mathcal{P}_{\mathcal{P}}$. The quotient $\mathcal{O}_{\mathcal{P}}/\mathcal{P}_{\mathcal{P}}$ is called the residue field of $K_{\mathcal{P}}$; it is isomorphic to $\mathcal{O}/\mathcal{P}$ before completion, thus is finite. The ideal $\mathcal{P}_{\mathcal{P}}$ is a principal ideal, generated by any element $\pi_{\mathcal{P}}$ with the largest valuation $(N\mathcal{P})^{-1}$, called a uniformizer at $\mathcal{P}$. Then $K_{\mathcal{P}}^{\times} = \mathcal{U}_{\mathcal{P}} \cdot \langle \pi_{\mathcal{P}} \rangle$. Further, let S be a set of representatives of the residue field in $\mathcal{O}_{\mathcal{P}}$. Then any element in $\mathcal{O}_{\mathcal{P}}$ is a Taylor series in powers of $\pi_{\mathcal{P}}$ with coefficients in S, the units are those with the constant term not in $\mathcal{P}_{\mathcal{P}}$, and an element in $K_{\mathcal{P}}$ is a Laurent series in powers of $\pi_{\mathcal{P}}$ with coefficients in S.

Example 1. $K = \mathbf{Q}$ and $\mathcal{P} = (p)$. We may choose $\pi_{\mathcal{P}} = p$ and $S = \{i : 0 \leq i \leq p-1\}$. Then

$$\mathcal{O}_{\mathcal{P}} = \mathbf{Z}_{\mathcal{P}} = \left\{ \sum_{i=0}^{\infty} a_i p^i : 0 \leq a_i \leq p-1 \right\},$$

$$\mathbf{Q} \cap \mathbf{Z}_{\mathcal{P}} = \left\{ \frac{m}{n} : m, n \in \mathbf{Z}, p \nmid n \right\},$$

$$\mathcal{P}_{\mathcal{P}} = p\mathbf{Z}_{\mathcal{P}} = \left\{ \sum_{i=1}^{\infty} a_i p^i : 0 \leq a_i \leq p-1 \right\},$$

$$\mathcal{U}_{\mathcal{P}} = \left\{ \sum_{i=0}^{\infty} a_i p^i : 0 \leq a_i \leq p-1 \text{ and } a_0 \neq 0 \right\},$$

$$\mathbf{Q}_{\mathcal{P}} = \left\{ \sum_{i > -\infty} a_i p^i : 0 \leq a_i \leq p-1 \right\}.$$

The residue field is $\mathbf{Z}_{\mathcal{P}}/p\mathbf{Z}_{\mathcal{P}} \cong \mathbf{Z}/p\mathbf{Z}$.

Exercise 1. Express -1 and $\frac{1}{r}$, $p \nmid r$, in Taylor series in powers of p with coefficients in S.

Example 2. $K = k(T)$ and $\mathcal{P}$ is a place of degree 1, say, at $a \in k$. We may choose $\pi_{\mathcal{P}} = T - a$ and $S = k$. The ring of integers consists of Taylor series in powers of $T - a$ with coefficients in k, $\mathcal{U}_{\mathcal{P}}$ consists of those nonvanishing at a, the residue field at $\mathcal{P}$ is isomorphic to k, and the field $K_{\mathcal{P}}$ is the field of formal power series $k((T-a))$, which is isomorphic to $k((T))$.

Observe the topology we put on $K_{\mathcal{P}}$ as a topological group. The ideals $\{\mathcal{P}_{\mathcal{P}}^n\}_{n \geq 0}$ form a neighborhood system of 0 so that, as an additive topological group, $K_{\mathcal{P}}$ is locally compact and totally disconnected (that is, each point is a connected component), and $\mathcal{O}_{\mathcal{P}}$ and hence $\mathcal{P}_{\mathcal{P}}^n$ for $n \geq 1$ are both open and compact. The group of units $\mathcal{U}_{\mathcal{P}}$ has a natural filtration

$$\mathcal{U}_{\mathcal{P}} \supset 1 + \mathcal{P}_{\mathcal{P}} \supset 1 + \mathcal{P}_{\mathcal{P}}^2 \supset \cdots ,$$

with the first quotient isomorphic to the multiplicative group of the residue field and the remaining successive quotients isomorphic to the residue field. The groups $\{1 + \mathcal{P}_{\mathcal{P}}^n\}_{n \geq 1}$ form a neighborhood system of 1 so that the multiplicative group $K_{\mathcal{P}}^{\times}$ is also locally compact, and $\mathcal{U}_{\mathcal{P}}$ is both open and compact. Note that the topology on $K_{\mathcal{P}}^{\times}$ is the topology induced from $K_{\mathcal{P}}$.

The valuation $|\ |_{\mathcal{P}}$ defined above does not have the archimedean property, it is called a nonarchimedean valuation, and the place $\mathcal{P}$ is called a nonarchimedean or finite place of K. When K is a function field, all valuations on K are nonarchimedean, but for a number field K, there are also archimedean valuations, which yield archimedean or infinite places of K. More precisely, suppose K is a number field of degree n over $\mathbf{Q}$. Then $K = \mathbf{Q}(\xi)$ for some element ξ in K. The irreducible polynomial $f(x)$ of ξ over $\mathbf{Q}$ has degree n, let $\xi_1, \cdots, \xi_{r_1}$ be its real roots and $\xi_{r_1+1}, \overline{\xi}_{r_1+1}, \xi_{r_1+2}, \overline{\xi}_{r_1+2}, \cdots, \xi_{r_1+r_2}, \overline{\xi}_{r_1+r_2}$ be its complex roots. Here $n = r_1 + 2r_2$ and $\overline{\xi}_j$ is the complex conjugation of ξ_j. The n embeddings of K into $\mathbf{C}$ over $\mathbf{Q}$ are given by $\sigma_i : \xi \longmapsto \xi_i$ for $i = 1, \cdots, r_1$, $\tau_j : \xi \longmapsto \xi_{i+j}$, $\overline{\tau}_j : \xi \longmapsto \overline{\xi}_{i+j}$ for $1 \leq j \leq r_2$. So $\sigma_i(K)$ is a subfield of $\mathbf{R}$ and $\tau_j(K)$, $\overline{\tau}_j(K)$ are subfields of $\mathbf{C}$. The usual absolute value on $\mathbf{R}$ or $\mathbf{C}$ restricted to the imbedded image of K defines a valuation on K, and the completion of K with respect to this valuation is $\mathbf{R}$ or $\mathbf{C}$, accordingly. On the other hand, τ_j and $\overline{\tau}_j$ yield the same valuation, and $\sigma_1, \cdots, \sigma_{r_1}, \tau_1, \cdots, \tau_{r_2}$ give rise to nonequivalent valuations on K, they are called the archimedean places of K since each valuation has the archimedean property. We say that K has r_1 real places and r_2 complex places.

A global field is either a number field or a function field of one variable over a finite field. If v is a place of a global field K, the completion of K at v, noted K_v, is called a local field. So local fields are $K_{\mathcal{P}}$, $\mathbf{R}$ and $\mathbf{C}$; of which $\mathbf{R}$ and $\mathbf{C}$ are archimedean local fields and the $K_{\mathcal{P}}$'s are nonarchimedean local fields. Note that they are all locally compact. The standard valuation $|\ |_v$ on K_v is defined as

$$|\ |_v = \begin{cases} |\ |_{\mathcal{P}} & \text{if } K_v = K_{\mathcal{P}}, \\ |\ |_{\mathbf{R}} & \text{if } K_v = \mathbf{R}, \\ |\ |_{\mathbf{C}}^2 & \text{if } K_v = \mathbf{C}. \end{cases}$$

In the last case, the valuation on $\mathbf{C}$ is squared to account for the fact that two complex imbeddings yield only one complex place. We point out that if $|\ |$ is a nonarchimedean valuation on a field K, and any positive power of $|\ |$ is also a nonarchimedean valuation on K and they define the same topology. These two valuations are said to be equivalent. A nonarchimedean place of a global field corresponds to an equivalence class of valuations, one of which is the standard $\mathcal{P}$-adic valuation. The reason for such a choice will be clear in the next section. It can be shown that any nontrivial valuation of a global field is equivalent to $|\ |_{\mathbf{R}}$ or $|\ |_{\mathbf{C}}$ or a standard $\mathcal{P}$-adic valuation.

The following lemma is basic.

Hensel's Lemma. *Let K be a nonarchimedean local field with ring of integers $\mathcal{O}$ and maximal ideal $\mathcal{P}$. Let $F(x)$ be a polynomial with coefficients in $\mathcal{O}$ such that modulo $\mathcal{P}, \overline{F(x)}$ factors as a product of two relatively prime polynomials $g(x)$ and $h(x)$ with coefficients in the residue field $\mathcal{O}/\mathcal{P}$. Then there exist two polynomials $G(x), H(x)$ in $\mathcal{O}[x]$ lifting $g(x)$ and $h(x)$ such that $F(x) = G(x)H(x)$ with $\deg G = \deg g$.*

Proof. We shall factor $F(x)$ by successive approximations. Let $G_0(x), H_0(x)$ be two polynomials in $\mathcal{O}[x]$ lifting $g(x), h(x)$, respectively, and such that $\deg G_0 = \deg g$, $\deg H_0 = \deg h$. Then $F(x) - G_0(x)H_0(x) = \pi F_1(x)$, where π is a uniformizer of $\mathcal{P}$ and $F_1(x) \in \mathcal{O}[x]$. Next we solve $F_1(x) \equiv g_1(x)H_0(x) + h_1(x)G_0(x) \bmod \mathcal{P}$, or equivalently, $\overline{F_1(x)} = \overline{g_1(x)}h(x) + \overline{h_1(x)}g(x)$ in $\mathcal{O}/\mathcal{P}[x]$. The last equation is solvable for $\overline{g_1}, \overline{h_1}$ in $\mathcal{O}/\mathcal{P}[x]$ since $g(x)$ and $h(x)$ are coprime. Choose $\overline{g_1}, \overline{h_1}$ so that $\deg \overline{g_1} < \deg g$, then $\deg \overline{h_1} \leq \deg F - \deg g$. Let g_1, h_1 be liftings of $\overline{g_1}, \overline{h_1}$ in $\mathcal{O}[x]$ with the same degrees as $\overline{g_1}, \overline{h_1}$, respectively. Put $G_1(x) = G_0(x) + \pi g_1(x)$ and $H_1(x) = H_0(x) + \pi h_1(x)$. Then $\deg G_1 = \deg g$, $\overline{G_1} = g$, $\overline{H_1} = h$ and $F(x) - G_1(x)H_1(x) = \pi^2 F_2(x)$ for some F_2 in $\mathcal{O}[x]$ with $\deg F_2 \leq \deg F$. Solve $F \equiv (G_1(x) + \pi^2 g_2(x))(H_1(x) + \pi^2 h_2(x)) \bmod \mathcal{P}^3$ and proceed as before. Continuing this process, finally we obtain two sequences of polynomials $\{G_0(x), G_1(x), \cdots\}$ and $\{H_0(x), H_1(x), \cdots\}$, with $G_{i+1}(x) \equiv G_i(x) \bmod \mathcal{P}^{i+1}$, $H_{i+1}(x) \equiv H_i(x) \bmod \mathcal{P}^{i+1}$, $\deg G_i = \deg g$ and $F \equiv G_i(x)H_i(x) \bmod \mathcal{P}^{i+1}$. Thus both are Cauchy sequences, converging to $G(x)$ and $H(x)$ with $\deg G = \deg G_0 = \deg g, \overline{G} = g, \overline{H} = h$. Since $F \equiv G_n(x)H_n(x) \bmod \mathcal{P}^{n+1}$ for $n \geq 1$, we see that $F(x) = G(x)H(x)$ when passing to limit. □

Exercise 2. Let K be a function field with the field of constants k. Let v be a place of K of degree n. Show that K is isomorphic to the field of formal power series $k_n((\pi_v))$ for any uniformizer π_v of K_v.

§2 Extensions of valuations

Let K be a nonarchimedean local field with a valuation $|\ |_K$. Let L be a finite separable extension of K with $[L:K]=n$. Then there are n embeddings $\sigma_1, \cdots, \sigma_n$ of L over K into an algebraic closure $\overline{K}$ of K. With these n embeddings we can define trace and norm of any element z in L via

$$\mathrm{Tr}_{L/K}(z) = \sigma_1(z) + \cdots + \sigma_n(z), \quad \mathrm{N}_{L/K}(z) = \sigma_1(z)\cdots\sigma_n(z).$$

It can be shown that $\mathrm{Tr}_{L/K}$ is a homomorphism from the additive group L to the additive group K, and $\mathrm{N}_{L/K}$ is a homomorphism from the multiplicative group $L^\times$ to the multiplicative group $K^\times$. Further, if M is a subfield of L containing K, then

$$\mathrm{Tr}_{L/K} = \mathrm{Tr}_{M/K} \circ \mathrm{Tr}_{L/M} \quad \text{and} \quad \mathrm{N}_{L/K} = \mathrm{N}_{M/K} \circ \mathrm{N}_{L/M}.$$

The relation between $\mathrm{Tr}_{L/K}(z), \mathrm{N}_{L/K}(z)$ and the coefficients of the irreducible polynomial of z is the same as in the finite field case, as stated in Theorem 5 and Exercise 2 of Chapter 1. The details are left as an exercise.

We would like to extend the valuation $|\ |_K$ of K to a valuation $|\ |_L$ of L. Suppose we could. If L is Galois over K, then all conjugates of z should have the same valuation. Thus $|z|_L^n = |\sigma_1(z)|_L \cdots |\sigma_n(z)|_L = |\mathrm{N}_{L/K}(z)|_L = |\mathrm{N}_{L/K}(z)|_K$. This shows that $|z|_L = |\mathrm{N}_{L/K}(z)|_K^{1/n}$, a formula which makes sense even when L is not Galois over K. This suggests the following

Theorem 1. *Let K be a nonarchimedean local field with a valuation $|\ |_K$. Let L be a separable field extension of K with degree n. Then $|\ |_K$ has a unique extension to a valuation $|\ |_L$ of L, given by*

$$|z|_L = |\mathrm{N}_{L/K}(z)|_K^{1/n} \quad \textit{for} \quad z \in L.$$

Further, L is complete with respect to $|\ |_L$.

Proof. Existence. Define $|\ |_L$ on L by $|z|_L = |\mathrm{N}_{L/K}(z)|_K^{1/n}$ for $z \in L$. Clearly, it satisfies (1) and (2). It remains to show that $|z+w|_L \le \max(|z|_L, |w|_L)$ for all $z, w \in L$. In view of its definition, it amounts to showing that if $z \in L$ is such that $|\mathrm{N}_{L/K}(z)|_K \le 1$, then $|\mathrm{N}_{L/K}(1+z)|_K \le 1$. Consider the irreducible polynomial $f(x) = x^m + a_{m-1}x^{m-1} + \cdots + a_0$ of z over K. As $\left(a_0(-1)^m\right)^{n/m} = \mathrm{N}_{L/K}(z)$, we get $|a_0|_K \le 1$. In other words, a_0 lies in the ring of integers $\mathcal{O}_K$ of K. Since $f(-1) = (-1)^m \mathrm{N}_{K(z)/K}(1+z) = (-1)^m + (-1)^{m-1}a_{m-1} + \cdots + a_0$, if we can show that all coefficients $a_i \in \mathcal{O}_K$, then so does $\mathrm{N}_{K(z)/K}(1+z)$ and hence $\mathrm{N}_{L/K}(1+z)$, implying $|\mathrm{N}_{L/K}(1+z)|_K \le 1$. Suppose otherwise. Let j be the largest index such that $|a_j|_K \ge |a_i|_K$ for $i = 0, \cdots, m-1$. Then $|a_j|_K > 1, m-1 \ge j > 0$, and $|a_i|_K < |a_j|_K$ for $i > j$. Denote by $\mathcal{P}_K$ the maximal ideal of $\mathcal{O}_K$. We have

$a_j^{-1} \in \mathcal{P}_K$, and $a_i a_j^{-1} \in \mathcal{P}_K$ for $m-1 \geq i > j$. The polynomial $a_j^{-1}f(x)$ lies in $\mathcal{O}_K[x]$, and modulo $\mathcal{P}_K$ it yields $\overline{a_j^{-1}f(x)} = g(x)h(x)$, where $h(x) = 1$, and $g(x) = \overline{a_j^{-1}f(x)} = x^j + \cdots$ is a polynomial in $\mathcal{O}_K/\mathcal{P}_K[x]$ of degree $j, m > j > 0$. Apply Hensel's Lemma : $a_j^{-1}f(x) = G(x)H(x)$ for two polynomials G, H in $\mathcal{O}_K[x]$ with $\deg G = j$ and $\deg H > 0$, contradicting the irreducibility of f. This proves that all $a_i \in \mathcal{O}_K$ and thus (3) holds.

Uniqueness. View L as an n-dimensional vector space over K. Any valuation $\| \ \|$ on L extending $| \ |_K$ defines a norm on L satisfying $\|\alpha x\| = |\alpha|_K \|x\|$ for $\alpha \in K$ and $x \in L$. Fix a basis $\{w_1, \cdots, w_n\}$ of L over K. For $x \in L$, write x as a K-linear combination of $w_1, \cdots, w_n$:

$$x = \alpha_1 w_1 + \cdots + \alpha_n w_n, \quad \alpha_i \in K.$$

Define $\|x\|_0 = \max_i |\alpha_i|_K$. Then it is a norm on L. If $\| \ \|$ is equivalent to $\| \ \|_0$ as norms, then any two valuations on L extending $| \ |_K$ are equivalent, showing the uniqueness of $\| \ \|_L$, and the completeness of L with respect to $\| \ \|_L$ since it is so with respect to $\| \ \|_0$. Our theorem will follow from

Theorem 2. *Let K be a local field with valuation $| \ |_K$. Let L be a finite-dimensional vector space over K. Then any norm $\| \ \|$ on L satisfying $\|\alpha x\| = |\alpha|_K \|x\|$ for $\alpha \in K$, $x \in L$, is equivalent to $\| \ \|_0$ defined above. Further, L is complete with respect to $\| \ \|$.*

Proof. We have to show the existence of two positive numbers μ and ν such that for all $x \in L$, $\|x\| \leq \mu \|x\|_0$ and $\|x\|_0 \leq \nu \|x\|$. One inequality is easy. From $x = \alpha_1 w_1 + \cdots + \alpha_n w_n$ and triangular inequality, we have

$$\begin{aligned} \|x\| &\leq \|\alpha_1 w_1\| + \cdots + \|\alpha_n w_n\| = |\alpha_1|_K \|w_1\| + \cdots + |\alpha_n|_K \|w_n\| \\ &\leq \mu \max_i |\alpha_i|_K = \mu \|x\|_0, \end{aligned}$$

where $\mu = \|w_1\| + \cdots + \|w_n\|$. To show the other inequality, define the functions $|x|_i = |\alpha_i|_K$ for $x = \alpha_1 w_1 + \cdots + \alpha_n w_n$ in L, $\alpha_i \in K$. It suffices to show the existence of positive numbers $\nu_1, \cdots, \nu_n$ such that $|x_i| \leq \nu_i \|x\|$ for $1 \leq i \leq n$. This is obviously true if $n = 1$. Hence we prove by induction on n. Let V be the subspace of L over K spanned by $w_1, \cdots, w_{n-1}$ and let $\| \ \|'$ be the restriction of $\| \ \|$ to V. By induction hypothesis $\nu_1', \cdots, \nu_{n-1}'$ exist, $\| \ \|'$ is equivalent to $\| \ \|_0'$ on V and V is complete. If ν_n exists, then $\nu_1, \cdots, \nu_{n-1}$ also exist. Indeed, write $x = x(V) + \alpha_n w_n$, where $x(V) = \alpha_1 w_1 + \cdots + \alpha_{n-1} w_{n-1}$. Then for $1 \leq i \leq n-1$,

$$\begin{aligned} |x|_i = |x(V)|_i \leq \nu_i' \|x(V)\|' &\leq \nu_i'(\|x\| + \|\alpha_n w_n\|) \leq \nu_i'(\|x\| + |x|_n \|w_n\|) \\ &\leq \nu_i'(1 + \nu_n \|w_n\|) \|x\| \end{aligned}$$

so that we may choose $\nu_i = \nu_i'(1 + \nu_n \|w_n\|)$ for $i = 1, \cdots, n-1$. Suppose ν_n does not exist. Then there exists a sequence $\{x_j\}_{j \geq 1}$ such that $|x_j|_n > j \|x_j\|$. Then

$x_j \notin V$, and since the inequality remains unchanged if x_j is replaced by a nonzero scalar multiple, we may assume

$$x_j = \alpha_{j1} w_1 + \cdots + \alpha_{jn-1} w_{n-1} + w_n, \quad \alpha_{ji} \in K.$$

This shows that $\|x_j\| < \frac{1}{j}$, hence $\lim x_j = 0$ and $\lim x_j - w_n = -w_n$ as $j \to \infty$. On the other hand, $x_j - w_n$ are elements of a closed subspace V, the sequence cannot have a limit outside V. Therefore ν_n and hence $\nu_1, \cdots, \nu_{n-1}$ exist, as desired. □

Remarks. (1) Denote by $\mathcal{O}_L$ the ring of integers of L with respect to $| \ |_L$. It was shown in the existence proof that every element in $\mathcal{O}_L$ is integral over $\mathcal{O}_K$. Conversely, if $x \in L$ is integral over $\mathcal{O}_K$, then its norm lies in $\mathcal{O}_K$ and hence it lies in $\mathcal{O}_L$ by definition of $| \ |_L$. Thus $\mathcal{O}_L$ consists of the elements in L integral over K. In other words, the two senses of "integrality" coincide.

(2) Theorem 1 also holds if K is an archimedean local field since

$$|z|_{\mathbf{C}} = |z\bar{z}|_{\mathbf{R}}^{1/2}, \quad z \in \mathbf{C}$$

is a valuation on $\mathbf{C}$ and the uniqueness proof remains valid on $\mathbf{C}$.

Let $K, L, | \ |_K, | \ |_L$ be as in Theorem 1. Denote by $\mathcal{O}_K, \mathcal{O}_L$ the ring of integers in K, L, respectively, and by $\mathcal{P}_K, \mathcal{P}_L$ the maximal ideal of $\mathcal{O}_K, \mathcal{O}_L$, respectively. As $\mathcal{O}_K \cap \mathcal{P}_L = \mathcal{P}_K$, the residue field $\mathcal{O}_L/\mathcal{P}_L$ of L is an extension of the residue field $\mathcal{O}_K/\mathcal{P}_K$ of K. Suppose $w_1, \cdots, w_f$ in $\mathcal{O}_L$ are such that, modulo $\mathcal{P}_L$, $\overline{w_1}, \cdots, \overline{w_f}$ are linearly independent over $\mathcal{O}_K/\mathcal{P}_K$. We claim that $w_1, \cdots, w_f$ are linearly independent over K. Suppose otherwise. Let $\sum_{i=1}^{f} a_i w_i = 0$ be a nontrivial relation. We may assume that all $a_i \in \mathcal{O}_K$ and not all in $\mathcal{P}_K$. Modulo $\mathcal{P}_L$, this gives rise to a nontrivial linear relation of $\overline{w_i}$'s over $\mathcal{O}_K/\mathcal{P}_K$, a contradiction. This shows that $\mathcal{O}_L/\mathcal{P}_L$ is a finite extension of $\mathcal{O}_K/\mathcal{P}_K$ with degree $f \le n = [L:K]$. We have also shown that any set of f elements in $\mathcal{O}_L$ which, modulo $\mathcal{P}_L$, form a basis of $\mathcal{O}_L/\mathcal{P}_L$ over $\mathcal{O}_K/\mathcal{P}_K$ are linearly independent over K. Next, choose a uniformizer π_K of K and π_L of L. As $\pi_K \in \mathcal{P}_L$, we have $\pi_K = u\pi_L^e$ for some positive integer e and some unit $u \in \mathcal{U}_L$, that is, $|\pi_K|_L = |\pi_L|_L^e$. Call e the ramification index of the extension L over K.

Theorem 3. *Let K, L be as in Theorem 1 and e and f be as defined above. Then*

$$[L:K] = n = ef.$$

Proof. Let $\mathcal{S}$ be any set of representatives of the residue field $\mathcal{O}_L/\mathcal{P}_L$ in $\mathcal{O}_L$. We know that any element in L is a Laurent series in powers of π_L with coefficients in $\mathcal{S}$. Since any power π_L^j can be written as $u_j \pi_L^i \pi_K^m$ for some unit u_j in $\mathcal{U}_L$, some integers i, m with $0 \le i \le e-1$, the same procedure shows that any element can be expressed as $\sum_{i=0}^{e-1}\sum_{m>-\infty} s_{im}\pi_L^i\pi_K^m$ with coefficients $s_{im}\in\mathcal{S}$. Let S be a set of representatives of the residue field $\mathcal{O}_K/\mathcal{P}_K$ in $\mathcal{O}_K$ and let $w_1,\cdots,w_f$ be elements in $\mathcal{O}_L$ such that modulo $\mathcal{P}_L$ they form a basis over $\mathcal{O}_K/\mathcal{P}_K$. Then we may choose $\mathcal{S}$ to be the set $\{\sum_{j=1}^{f} s_jw_j : s_j\in S\}$. This shows that $w_j\pi_L^i, 1\le j\le f$ and $0\le i\le e-1$, generate L over K.

It remains to show that $w_j\pi_L^i{}'s$ are linearly independent over K. Suppose otherwise. Let $\sum_{i,j} a_{ij}w_j\pi_L^i = 0$ be a nontrivial linear relation over K, where $0\le i\le e-1$ and $1\le j\le f$. We may assume that all a_{ij} are in $\mathcal{O}_K$ and some a_{ij} is a unit. Let i_0 be the smallest index m such that a_{mj} is a unit for some j. Then $a_{ij}\in\mathcal{P}_K$ for $i<i_0$ and all j so that $\sum_{\text{all } j \text{ and } i\neq i_0} a_{ij}w_j\pi_L^i\in\mathcal{P}_L^{i_0+1}$. Consequently, $\sum_{1\le j\le f} a_{i_0j}w_j\pi_L^{i_0}\in\mathcal{P}_L^{i_0+1}$, which implies that $\sum_{1\le j\le f} a_{i_0j}w_j\in\mathcal{P}_L$, or equivalently, modulo $\mathcal{P}_L$, $\sum_{1\le j\le f}\overline{a_{i_0j}w_j} = \bar{0}$ in $\mathcal{O}_L/\mathcal{P}_L$. This nontrivial relation over $\mathcal{O}_K/\mathcal{P}_K$ contradicts the linear independency of $\overline{w_j}'s$ over $\mathcal{O}_K/\mathcal{P}_K$. Therefore $w_j\pi_L^i{}'s, 1\le j\le f, 0\le i\le e-1$, from a basis of L over K. □

The extension L/K is said to be unramified if $e=1$, and totally ramified if $f=1$.

Exercise 3. Let K be a nonarchimedean local field and let L be a separable degree n field extension of K. Suppose the residue field $\mathcal{O}_K/\mathcal{P}_K$ of K has q elements, and $\mathcal{O}_L/\mathcal{P}_L$ has q^f elements. With the ramification index e, we have $n=ef$.

(1) Show that L contains q^f-1st roots of unity. (Hint. Use Hensel's lemma.)
(2) Let ξ be a primitive q^f-1st root of unity in L, and let $M=K(\xi)$. Show that M is an unramified extension of K of degree f and L is totally ramified over M. In fact, any unramified extension of K in L is contained in M so M is the maximal unramified extension of K in L.
(3) Let π_L be a uniformizer of $\mathcal{P}_L$. Let $f(x)=x^r+a_{r-1}x^{r-1}+\cdots+a_0$ be the irreducible polynomial of π_L over M. Show that $a_i\in\mathcal{P}_M, a_0\in\mathcal{P}_M-\mathcal{P}_M^2$, and $r=e$. Thus $L=M(\pi_L)$.

Denote by q the cardinality of the residue field $\mathcal{O}_K/\mathcal{P}_K$. Suppose that $|\ |_K$ is the standard valuation on K so that $|\pi_K|_K = q^{-1}$. Denote by $|\ |$ the standard valuation on L equivalent to the unique extension $|\ |_L$ of $|\ |_K$ to L. Thus

$$|\pi_L| = |\mathcal{O}_L/\mathcal{P}_L|^{-1} = q^{-f}.$$

We want to express $|\ |$. For this purpose, it sufficces to compute

$$|\pi_K| = |\pi_L|^e = q^{-ef} = q^{-n} = |\pi_K|_K^n$$

by Theorem 3. Hence $|\ | = |\ |_K^n$. Combined with Theorem 1, we have the following corollary.

Corollary 1. *Let K be a nonarchimedean local field with the standard valuation $| \ |_K$. Let L be a separable degree n field extension of K. Then the standard valuation $| \ |$ on L equivalent to the unique extension $| \ |_L$ of $| \ |_K$ to L is given by*

$$|x| = |\mathrm{N}_{L/K}\, x|_K \quad \textit{for all} \quad x \in L.$$

Theorem 1 says that a valuation of a local field has a unique extexsion in any finite extension of the field. Next theorem explains what happens when the base field is a global field.

Let K be a field and let L be a degree n field extension of K. Let $w_1 = 1, \cdots, w_n$ be a basis of L over K. Then

$$(*) \qquad w_i w_j = \sum_{k=1}^{n} a_{ijk} w_k \quad \text{with} \quad a_{ijk} \in K,$$

and this relation determines L up to isomorphism. Let A be a ring containing K. The tensor product $A \otimes_K L$ consists of $\sum_{i=1}^{n} c_i w_i, c_i \in A$. Algebraically, it is a ring with the componentwise addition and with multiplication table given by $(*)$. Note that both A and L are imbedded in $A \otimes_K L$. If there is a topology on A, then we put on $A \otimes_K L$ the topology coming from the product topology on A^n via the isomorphism $(c_1, \cdots, c_n) \longmapsto \sum_{i=1}^{n} c_i w_i$ from A^n to $A \otimes_K L$. One checks easily that both algebraic and topological structures on $A \otimes_K L$ are independent of the choice of a basis $\{w_1, \cdots, w_n\}$ of L over K.

Theorem 4. *Let K be a global field and let L be a finite separable extension of K of degree n. Let v be a place of K. Then L has at most n places dividing v, that is, $| \ |_v$ on K extends to at most n distinct (hence nonequivalent) valuations on K. Let $w_1, \cdots, w_r (r \leq n)$ be the places of L dividing v. Denote by K_v, L_{w_i} the completion of K at v and L at w_i, respectively. Then*

$$(5) \qquad K_v \otimes_K L \cong L_{w_1} \oplus \cdots \oplus L_{w_r}$$

algebraically and topologically, where the right hand side is given the product topology.

Proof. There is an element $\xi \in L$ such that $L = K(\xi)$. Let $f(x)$ be the irreducible polynomial of ξ over K. Factor $f(x) = f_1(x) \cdots f_r(x)$ as a product of irreducible polynomials $f_i(x)$ over K_v. As L is separable over K, the $f_i's$ are mutually coprime. Clearly, $r \leq n$. Algebraically, we have, by Chinese remainder theorem,

$$K_v \otimes_K L \cong K_v \otimes_K K[x]/(f(x)) \cong K_v[x]/(f(x)) \cong \prod_{i=1}^{r} K_v[x]/(f_i(x)).$$

Here each $K_v[x]/\big(f_i(x)\big)$ is a finite field extension, called L_i, of K_v. In the above isomorphisms, the element ξ is mapped to $x+\big(f(x)\big)$ in $K_v[x]/\big(f(x)\big)$, and then to $\Big(x+\big(f_1(x)\big),\cdots,x+\big(f_r(x)\big)\Big)$ in $\prod_{i=1}^{r} K_v[x]/\big(f_i(x)\big)$. Thus for each i, $1\le i\le r$, the map $\xi\longmapsto x+\big(f_i(x)\big)$ yields a nontrivial homomorphism from $K[\xi]$ into $K_v[x]/\big(f_i(x)\big)$. As $K[\xi]=L$ is a field, this means that the field L is imbedded in each finite extension $L_i=K_v[x]/\big(f_i(x)\big)$ of K_v. Denote by $|\ \ |_i$ the unique extension of $|\ \ |_v$ on K_v to L_i. As K is dense in K_v, $K\underset{K}{\otimes}L$ is dense in $K_v\underset{K}{\otimes}L$, hence L is dense in each L_i. So $|\ \ |_i$ restricted to L corresponds to a place w_i of L and L_i is the completion L_{w_i}.

We have to show that the valuations $|\ \ |_i$ are distinct, and these are the only valuations of L extending $|\ \ |_v$. Let $|\ \ |$ be a valuation of L extending $|\ \ |_v$. Then $|\ \ |$ extends by continuity to a real-valued function, also denoted by $|\ \ |$, on $K_v\underset{K}{\otimes}L$; it satisfies

$$|\alpha\beta|=|\alpha||\beta|$$

and

$$\begin{cases}|\alpha+\beta|\le\max(|\alpha|,|\beta|) & \text{if } |\ \ |_v \text{ is nonarchimedean,}\\ \text{or }\ |\alpha+\beta|\le|\alpha|+|\beta| & \text{if } |\ \ |_v \text{ is archimedean,}\end{cases}$$

by continuity. Consider the restriction of $|\ \ |$ to L_i. If there is an element $\alpha\in L_i$ such that $|\alpha|\ne 0$, then for all nonzero β in L_i, the relation $|\alpha|=|\beta||\alpha\beta^{-1}|\ne 0$ implies $|\beta|\ne 0$, so $|\ \ |$ on L_i is a valuation of L_i extending $|\ \ |_v$, hence is equal to $|\ \ |_i$ by Theorem 1. As $|\ \ |$ is nonzero on $K_v\underset{K}{\otimes}L$, it is nonzero on some L_i. This shows that all extensions of $|\ \ |_v$ on L are among $|\ \ |_1,\cdots,|\ \ |_r$. Further, if $|\ \ |_i=|\ \ |_j$ with $i\ne j$, then choosing $|\ \ |=|\ \ |_i=|\ \ |_j$ would result in $|\ \ |$ nonzero on L_i and on L_j. Let $\alpha\in L_i^\times$ and $\beta\in L_j^\times$. Viewed as elements in $K_v\underset{K}{\otimes}L$, we have $\alpha\beta=(0\cdots0\ \underset{\substack{\uparrow\\ i\text{th place}}}{\alpha}\ 0\cdots0)\cdot(0\cdots0\ \underset{\substack{\uparrow\\ j\text{th place}}}{\beta}\ 0\cdots0)=(0\cdots0)=0$ in $K_v\underset{v}{\otimes}L$. But $0=|0|=|\alpha\beta|=|\alpha||\beta|\ne 0$ is a contradiction. This proves the algebraic part of the theorem.

It remains to show that (5) is also a topological homomorphism. For $x=(x_1,\cdots,x_r)\in L_{w_1}\oplus\cdots\oplus L_{w_r}$, define $\|x\|_0=\max\limits_{1\le i\le r}|x_i|_i$. Then $\|\ \ \|_0$ is a norm on $L_{w_1}\oplus\cdots\oplus L_{w_r}$ viewed as a vector space over K_v, and it induces the product topology. On the other hand, by Theorem 2, any two norms on finite-dimensional vector spaces over K_v are equivalent, so $\|\ \ \|_0$ induces the tensor product topology on $K_v\underset{K}{\otimes}L$. □

Corollary 2. *Let K be a global field and L a finite-dimensional separable extension of K. Let v be a place of K and $w_1,\cdots,w_r$ be the places of L dividing v. Let $\xi\in L$. Then*

$$\mathrm{N}_{L/K}(\xi)=\prod_{i=1}^{r}\mathrm{N}_{L_{w_i}/K_v}(\xi)\quad\textit{and}\quad \mathrm{Tr}_{L/K}(\xi)=\sum_{i=1}^{r}\mathrm{Tr}_{L_{w_i}/K_v}(\xi).$$

Proof. As shown in the above proof, $\sum_{i=1}^{r}[L_{w_i} : K_v] = n = [L : K]$, hence Corollary holds for $\xi \in K$. Next assume $K(\xi) = L$. Let $f(x)$ and $f_i(x)$ be as in the proof above; we have $\mathrm{N}_{L/K}(\xi) = (-1)^n f(0)$ and $\mathrm{N}_{L_{w_i}/K_v}(\xi) = (-1)^{[L_{w_i}:K_v]} f_i(0)$, and $\mathrm{Tr}_{L/K}(\xi) = -$ coefficient of x^{n-1} in f and $\mathrm{Tr}_{L_{w_i}/K_v}(\xi) = -$ coefficient of $x^{[L_{w_i}:K_v]-1}$ in $f_i(x)$. As $f(x) = f_1(x)\cdots f_r(x)$, the global norm and trace of ξ are related to local norm and trace of ξ as stated. Finally, suppose $M = K(\xi)$ is an intermediate field. Let $v_1, \cdots, v_s$ be the places of M dividing v. Then $w_1, \cdots, w_r$ are the places of L dividing one of $v_1, \cdots, v_s$. Fix a place v_i of M. We have

$$\begin{aligned}\prod_{w_j|v_i} \mathrm{N}_{L_{w_j}/K_v}(\xi) &= \prod_{w_j|v_i} \mathrm{N}_{M_{v_i}/K_v} \circ \mathrm{N}_{L_{w_j}/M_{v_i}}(\xi)\\ &= \mathrm{N}_{M_{v_i}/K_v}(\xi)^{\sum_{w_j|v_i}[L_{w_j}:M_{v_i}]}\\ &= \mathrm{N}_{M_{v_i}/K_v}(\xi)^{[L:M]}.\end{aligned}$$

Therefore

$$\begin{aligned}\prod_{w_j} \mathrm{N}_{L_{w_j}/K_v}(\xi) &= \prod_{v_i|v}\prod_{w_j|v_i} \mathrm{N}_{L_{w_j}/K_v}(\xi) = \prod_{v_i|v} \mathrm{N}_{M_{v_i}/K_v}(\xi)^{[L:M]}\\ &= \mathrm{N}_{M/K}(\xi)^{[L:M]} = \mathrm{N}_{L/K}(\xi).\end{aligned}$$

Similar proof shows $\mathrm{Tr}_{L/K}(\xi) = \sum_{i=1}^{r} \mathrm{Tr}_{L_{w_i}/K_v}(\xi)$. □

Corollary 3. *Let K be a global field and L a finite-dimensional separable extension of K. Let v be a place of K and $w_1, \cdots, w_r$ be the places of L dividing v. Denote by $|\ |_v$ and $|\ |_w$ the standard valuation on K_v, L_w, respectively. Then for $\xi \in L$, we have $|\mathrm{N}_{L/K}(\xi)|_v = \prod_{i=1}^{r} |\xi|_{w_i}$.*

Proof. When v is nonarchimedean, this follows from Corollaries 1 and 2; when v is archimedean, this follows from Corollary 2 and the definitions of standard archimedean valuations. □

Theorem 5. *Let K be a global field and $\xi \in K^\times$. Then $|\xi|_w = 1$ for all except possibly finitely many places w of K.*

Proof. This is obvious if $K = \mathbf{Q}$ or a rational function field of one variable over a finite field. If K is a function field, but not a rational function field, then there is a rational function subfield F such that K is a finite separable extension of F. If K is a number field, we take F to be the field $\mathbf{Q}$ of rationals. Let $f(x) = x^n + a_{n-1}x^{n-1} + \cdots + a_0$ be the irreducible polynomial of ξ over F. As each a_i is integral at all except for finitely many places of F, there is a finite set S of places of F so that at each place v of F outside S, all a_i are in $\mathcal{O}_v$. As $a_0 \neq 0$, enlarging S if necessary, we may assume that $|a_0|_v = 1$ for all places v of F outside S. Let

$\mathcal{S}$ consist of places of K dividing places in S. Then $\mathcal{S}$ is finite by Theorem 4. For each place w of K outside $\mathcal{S}$, w divides a unique place v of F outside S; we see that the element ξ is integral over $\mathcal{O}_v$, hence it lies in $\mathcal{O}_w$ by Remark (1) following Theorem 1. Moreover, by Corollaries 1 and 3, we have $|\xi|_w = |N_{K_w/F_v}(\xi)|_v \leq 1$ and $\prod_{w|v} |\xi|_w = |N_{K/F}(\xi)|_v = |a_0|_v^{[K:F(\xi)]} = 1$, which imply $|\xi|_w = 1$ for all w outside $\mathcal{S}$. This proves the theorem. □

In the funcion field case, the theorem above says that a nonzero rational function on a nonsingular projective curve defined over a finite field has zeros and poles at finitely many closed points.

Theorem 6. *(Product formula) Let K be a global field and $\xi \in K^\times$. Then*

$$\prod_{\substack{w \text{ places} \\ \text{of } K}} |\xi|_w = 1.$$

The infinite product $\prod_w |\xi|_w$ is in fact a finite product by Theorem 5, hence is well-defined.

Proof. Case 1. $K = \mathbf{Q}$. We may assume $\xi = n$ is a nonzero integer. Factor n as a product of prime powers : $n = \pm p_1^{e_1} \cdots p_r^{e_r}, e_i > 0, p_1, \cdots, p_r$ distinct primes. This shows that $|n|_{p_i} = p_i^{-e_i}$ for $i = 1, \cdots, r$, $|n|_p = 1$ for $p \neq p_1, \cdots, p_r$, and $|n|_{\mathbf{R}} = n$. Thus $\prod_{v \text{ places of } \mathbf{Q}} |n|_v = 1$.

Case 2. $K = k(T)$ is a rational function field over a finite field k of q elements. We may assume $\xi = g(T)$ is a polynomial in $k[T]$. Factor $g(T)$ as a product of monic irreducible polynomials : $g(T) = a g_1(T)^{e_1} \cdots g_r(T)^{e_r}$, where $a \in k^\times, e_i > 0, g_1, \cdots, g_r$ distinct. Denote by v_i the place of K with $g_i(T)$ as a uniformizer. Then $|g(T)|_{v_i} = q^{-f_i e_i}$, where $f_i = \deg g_i$, $|g(T)|_v = 1$ for $v \neq v_1, \cdots, v_r, \infty$, and $|g(T)|_\infty = q^{\deg g}$ since g has a pole at ∞ with order equal to $\deg g = \sum_{i=1}^r f_i e_i$, and ∞ is a place of degree 1. This proves that $\prod_{v \text{ places of } K} |\xi|_v = 1$.

Case 3. K is a finite separable extension of $F = \mathbf{Q}$ or $k(T)$. Let $\xi \in K^\times$. Then

$$\begin{aligned}\prod_{\substack{w \text{ places} \\ \text{of } K}} |\xi|_w &= \prod_{\substack{v \text{ places} \\ \text{of } F}} \prod_{w|v} |\xi|_w \\ &= \prod_v |N_{K/F}(\xi)|_v \quad \text{by Corollary 3} \\ &= 1\end{aligned}$$

since $N_{K/F}(\xi) \in F^\times$ and theorem holds for F. □

In the function field case, the above theorem says that for a nonzero rational function ξ on a nonsingular projective curve C the total degree of zeros of ξ is equal to the total degree of poles of ξ counting multiplicity. Here the degree is counted as follows: if ξ vanishes at a closed point x of degree f with order e, then the degree of zero of ξ at x is fe.

Corollary 4. *If K is a function field with the field of constants k and $\xi \in K^\times$ lies in $\mathcal{O}_w$ for all places w of K, then $\xi \in k^\times$.*

Proof. Since ξ does not have poles, so it cannot have zeros. In other words, it is a nonzero constant. □

---

§3 Adèles and idèles

Given an algebraic group G defined over a global field K, we would like to study the connection between its global points $G(K)$ and its local points $G(K_v)$. A natural object to think of is the infinite profuct $\prod_v G(K_v)$ over all places v of K. Unfortunately, it does not have good topological properties. Instead, we should take only the "restricted product"of $G(K_v)'s$. Thus we begin by explaining the general concept of restricted product.

Let $\{G_v\}_{v\in\Sigma}$ be a family of locally compact topological groups parametrized by the set Σ; let Σ_0 be a finite subset of Σ. For each $v \in \Sigma - \Sigma_0$, we fix a compact open subgroup H_v of G_v. To each finite subset S of Σ containing Σ_0, define $G_S = \prod_{v\in S} G_v \prod_{v\in\Sigma-S} H_v$ and endow G_S with the product topology. Note that G_S is still locally compact. If S_1 and S_2 are two such sets with $S_1 \supset S_2$, then $G_{S_1} \supset G_{S_2}$ and the topology on G_{S_2} is the induced topology from G_{S_1}. The restricted product of $\{G_v\}$ with respect to $\{H_v\}$ is the direct limit G of G_S, as S runs through all finite subsets of Σ containing Σ_0, under the inclusion map. A set U is open in G if and only if $U \cap G_S$ is open in G_S for all finite subset S containing Σ_0. Thus an open neighborhood of the identity element in G contains $\prod_{v\in S} U_v \prod_{v\in\Sigma-S} H_v$ for some finite set $S \supset \Sigma_0$ and U_v open in G_v containing 1. Therefore G is a locally compact topological group, independent of the choice of Σ_0.

Take a global field K. Let $\Sigma = \Sigma_K$ be the set of places of K, and $\Sigma_0 = \Sigma_\infty$ be the set of archimedean places of K (which is empty if K is a function field). First consider the case $G = G_a$, the additive group. Then $G_v = G(K_v) = K_v$ for $v \in \Sigma$, and for $v \in \Sigma - \Sigma_0$, let $H_v = G(\mathcal{O}_v) = \mathcal{O}_v$. The restricted product of $\{K_v\}_{v\in\Sigma}$ with respect to $\{\mathcal{O}_v\}_{v\in\Sigma-\Sigma_0}$ is a ring under componentwise addition and multiplication, called the ring of adèles of K, denoted by A_K. So

$$A_K = \{(x_v) \in \prod_{v\in\Sigma_K} K_v : x_v \in \mathcal{O}_v \text{ for almost all } v\}.$$

Theorem 5 implies that any element of K lies in $\mathcal{O}_v$ for almost all places v of K, hence we may imbed K diagonally in A_K.

Consider a separable field extension L of K of degree n. Let $\omega_1, \cdots, \omega_n$ be a basis of L over K. Let v be a place of K and $w_1, \cdots, w_r$ be the places of L dividing v. Theorem 4 says that $K_v \underset{K}{\otimes} L = K_v\omega_1 \oplus \cdots \oplus K_v\omega_n \cong L_{w_1} \oplus \cdots \oplus L_{w_r}$ algebraically and topologically. We want to study the relationship between $\mathcal{O}_v\omega_1 \oplus \cdots \oplus \mathcal{O}_v\omega_n$ and $\mathcal{O}_{w_1} \oplus \cdots \oplus \mathcal{O}_{w_r}$. Clearly, $\mathcal{O}_{w_1} \oplus \cdots \oplus \mathcal{O}_{w_r}$ consists of the elements in $K_v \underset{K}{\otimes} L$ integral over $\mathcal{O}_v$. So except for finitely many v's, we have $\omega_1, \cdots, \omega_n$ integral over $\mathcal{O}_v$ by Theorem 5, and hence $\mathcal{O}_v\omega_1 \oplus \cdots \oplus \mathcal{O}_v\omega_n$ is contained in $\mathcal{O}_{w_1} \oplus \cdots \oplus \mathcal{O}_{w_r}$. Conversely, let β be an element in $K_v \underset{K}{\otimes} L$ integral over $\mathcal{O}_v$. Write $\beta = \alpha_1\omega_1 + \cdots + \alpha_n\omega_n$ with $\alpha_i \in K_v$. Let $\sigma_1, \cdots, \sigma_n$ be the n imbeddings of L over K into an algebraic closure $\overline{K}$ of K. For $x = x_1\omega_1 + \cdots + x_n\omega_n \in K_v \underset{K}{\otimes} L$ with $x_i \in K_v$, denote by $x^{(j)} = x_1\sigma_j(\omega_1) + \cdots + x_n\sigma_j(\omega_n) \in K_v \underset{K}{\otimes} \overline{K}$ for $j = 1, \cdots, n$. Then $\sum_{j=1}^{n} x^{(j)} = x_1 \operatorname{Tr}_{L/K}\omega_1 + \cdots + x_n \operatorname{Tr}_{L/K}\omega_n = \sum_{j=1}^{n} x_j \sum_{i=1}^{r} \operatorname{Tr}_{L_{w_i}/K_v}(\omega_j) = \sum_{i=1}^{r} \operatorname{Tr}_{L_{w_i}/K_v}(x)$ by Corollary 2. We define $\operatorname{Tr}_{L/K}(x)$ to be $\sum_{j=1}^{n} x^{(j)} = \sum_{i=1}^{r} \operatorname{Tr}_{L_{w_i}/K_v}(x)$. Thus if x is integral over $\mathcal{O}_v$, then $\operatorname{Tr}_{L/K}(x) \in \mathcal{O}_v$. The relation $\beta = \alpha_1\omega_1 + \cdots + \alpha_n\omega_n$ yields the following matrix equation

$$\begin{pmatrix} \omega_1^{(1)} & \cdots & \omega_n^{(1)} \\ \vdots & \ddots & \vdots \\ \omega_1^{(n)} & \cdots & \omega_n^{(n)} \end{pmatrix} \begin{pmatrix} \alpha_1 \\ \vdots \\ \alpha_n \end{pmatrix} = \begin{pmatrix} \beta^{(1)} \\ \vdots \\ \beta^{(n)} \end{pmatrix}$$

As $\omega_1, \cdots, \omega_n$ are K-linearly independent, the matrix $W = \left(\omega_j^{(i)}\right)$ has nonzero determinant. By Cramer's rule, we get

$$\alpha_i = \frac{\det B_i}{\det W},$$

where B_i is the matrix obtained from W by replacing the ith column by $\begin{pmatrix} \beta^{(1)} \\ \vdots \\ \beta^{(n)} \end{pmatrix}$.

As it stands, we can't say much about α_i, but we know a lot more about

$$\alpha_i^2 = \frac{\det({}^tB_iB_i)}{\det({}^tWW)}$$

since ${}^tWW = \left(\operatorname{Tr}_{L/K}\omega_i\omega_j\right)$ has entries in $\mathcal{O}_v$, and so does ${}^tB_iB_i$. Further, $d = \det\left({}^tWW\right)$ is a nonzero element in K, by Theorem 5, we have that $|d|_v = 1$ for almost all places v of K, and hence α_i^2 and thus α_i lies in $\mathcal{O}_v$ for almost all places v of K. We have shown $\beta \in \mathcal{O}_v\omega_1 \oplus \cdots \oplus \mathcal{O}_v\omega_n$ for almost all v. This proves

Lemma 1. *Let L be a separable field extension of a global field K of degree n. Let $\omega_1, \cdots, \omega_n$ be a basis of L over K. Then for almost all places v of K, we have*

$$\mathcal{O}_v\omega_1 \oplus \cdots \oplus \mathcal{O}_v\omega_n = \mathcal{O}_{w_1} \oplus \cdots \oplus \mathcal{O}_{w_r},$$

where $w_1, \cdots, w_r$ are the places of L dividing v.

Theorem 4 and Lemma 1 together yield

Lemma 2. *Let L be a separable field extension of a global field K of degree n. Then $A_K \underset{K}{\otimes} L$ is isomorphic to A_L algebraically and topologically, and $L = K \underset{K}{\otimes} L$ is imbedded in A_L diagonally.*

Now we are ready to prove

Theorem 7. *Let K be a global field. Then K is a discrete subgroup of A_K with A_K/K compact under the quotient topology.*

Proof. Let F be $\mathbf{Q}$ or a rational function field $k(T)$ over a finite field k such that K is a separable finite extension of F of degree n, say. Let $w_1, \cdots, w_n$ be a basis of K over F. Then as an additive group, $A_K = A_F \underset{F}{\otimes} K = A_F w_1 \oplus \cdots \oplus A_F w_n$. Suppose we can show that F is a discrete subgroup of A_F and that there is a compact set Ω in A_F such that $A_F = F + \Omega$. Then K is a discrete subgroup in $A_F \underset{F}{\otimes} K = A_K$. Moreover, letting $\widetilde{\Omega} = \Omega w_1 + \cdots + \Omega w_n$, we get a compact subset $\widetilde{\Omega}$ of A_K such that $A_K = K + \widetilde{\Omega}$, thus $A_K/K \cong \widetilde{\Omega}/K \cap \widetilde{\Omega}$ is compact. Hence it suffices to prove Theorem 7 for the cases $K = \mathbf{Q}$ and $k(T)$.

Case 1. $K = k(T)$. Let $\Omega = \prod_{v \in \Sigma_K} \mathcal{O}_v$, it is open and compact in A_K. For a place v, choose the uniformizer π_v at v to be the monic irreducible polynomial in T vanishing at v if $v \neq \infty$, and $\pi_v = \frac{1}{T}$ if v is the place at infinity ∞. Choose the set S_v of representatives of the residue field at v to be polynomials in $k[T]$ with degree less than $\deg v$. Then every element x_v in K_v can be written as $x_v = \sum_{i > -\infty} s_i \pi_v^i$ with $s_i \in S_v$. Denote by $\langle x_v \rangle = \sum_{i<0} s_i \pi_v^i$, called the polar part of x_v. Given an element $x = (x_v)_{v \in \Sigma_K}$ in A_K, we have $\langle x_v \rangle$ nonzero for only finitely many places, and $\langle x_v \rangle \in \mathcal{O}_w$ for all places $w \neq v$. Let $\gamma = \sum_v \langle x_v \rangle$. Then $\gamma \in K$ and $x - \gamma$ is integral everywhere. This shows that $A_K = K + \Omega$. Further, by Corollary 4, $K \cap \Omega = k$ is finite so K is discrete in A_K and $A_K/K \cong \Omega/k$ is compact.

Case 2. $K = \mathbf{Q}$. Let $\Omega = \left[-\frac{1}{2}, \frac{1}{2}\right] \prod_p \mathbf{Z}_p$. At a finite place p of $\mathbf{Q}$, choose the uniformizer $\pi_p = p$ and the set S_p of representatives of the residue field at p to be $\{0, 1, \cdots, p-1\}$. For $x_p \in \mathbf{Q}_p$, define the polar part $\langle x_p \rangle$ the same way as above. Then for any $x = (x_v) \in A_{\mathbf{Q}}$, we have $\langle x_p \rangle = 0$ for almost all p so that $\gamma = \sum_p \langle x_p \rangle$

is a rational number, and $x-\gamma \in \mathbf{Z}_p$ for all p. Finally, there is an integer $n \in \mathbf{Z}$ such that the archimedean component of $x-\gamma-n$ lies in $\left[-\frac{1}{2}, \frac{1}{2}\right]$. This shows that $A_{\mathbf{Q}} = \mathbf{Q} + \Omega$. Observe that $\left(-\frac{1}{2}, \frac{1}{2}\right) \prod_p \mathbf{Z}_p$ is an open neighborhood of 0 not containing any other rational number. Thus $\mathbf{Q}$ is imbedded in $A_{\mathbf{Q}}$ as a discrete subgroup. Finally, $A_{\mathbf{Q}}/\mathbf{Q} \cong \left(\mathbf{R} \cdot \prod_p \mathbf{Z}_p\right)/\mathbf{Z} \cong (\mathbf{R}/\mathbf{Z}) \prod_p \mathbf{Z}_p$ is compact. □

Next take $G = G_m$, the multiplicative group, the same $\Sigma = \Sigma_K$ and $\Sigma_0 = \Sigma_\infty$ as before. Then for $v \in \Sigma_K$, $G_v = G(K_v) = K_v^\times$, and for $v \in \Sigma - \Sigma_0$, let $H_v = G(\mathcal{O}_v) = \mathcal{U}_v$. The restricted product of $\{K_v^\times\}$ with respect to $\{\mathcal{U}_v\}$ is called the group of idèles, denoted by I_K. Thus

$$I_K = \{(x_v) \in \prod_{v \in \Sigma_K} K_v^\times : x_v \in \mathcal{U}_v \text{ for almost all } v\}.$$

Note that I_K is the group of units in A_K algebraically, but the topology on I_K is finer than the induced topology.

Exercise 4. Show that the automorphism $x \longmapsto x^{-1}$ on I_K is not continuous under the induced topology on I_K. In fact, the topology we put on I_K is the weakest topology such that I_K is a topological group.

Theorem 5 implies that $K^\times$ may be imbedded in I_K diagonally. The quotient group $I_K/K^\times$ is called the idèle class group of K. Using the standard valuation $|\ |_v$ on each completion K_v of K at v, we define a map $|\ |$, called the norm map, from I_K to the multiplicative group of positive reals by

$$|x| = \prod_v |x_v|_v \quad \text{for} \quad x = (x_v) \in I_K.$$

Since x_v is a unit for almost all v, the above product is in fact a finite product. Note that $|\ |$ is a continuous homomorphism. Its kernel I_K^1 consists of idèles of norm 1. For a number field K, the image of $|\ |$ is all positive reals and we have

$$I_K = I_K^1 \times \left(\mathbf{R}_{>0}^\times\right)_w \cong I_K^1 \times \mathbf{R}_{>0}^\times,$$

where w is an archimedean place of K and $\left(\mathbf{R}_{>0}^\times\right)_w$ is a subgroup of $K_w^\times$; whereas for a function field K with q elements in its field of constants, the image of $|\ |$ is an infinite cyclic subgroup of $\mathbf{R}_{>0}$ generated by a power of q, say, q^n, and we have

$$I_K = I_K^1 \times \langle x \rangle \cong I_K^1 \times \mathbf{Z},$$

where x is an idèle with $|x| = q^n$. The product formula (Theorem 6) says that $K^\times$ is a subgroup of I_K^1. We know that K is a discrete subgroup of A_K by Theorem 7, and the topology on I_K is finer than the induced topology, hence $K^\times$ is a discrete subgroup of I_K, and therefore of I_K^1.

Theorem 8. *Let K be a global field. Then $K^\times$ is a discrete subgroup of I_K^1 with $I_K^1/K^\times$ compact under the quotient topology.*

It remains to prove the second assertion. For this, we shall need a geometric lemma. Given an idèle $a = (a_v)$, define a "parallelotope of size a" to be the set

$$P_a = \{x = (x_v) \in A_K : |x_v|_v \le |a_v|_v \text{ for all places } v \text{ of } K\}.$$

Then $P_a = aP_1$, which is a compact subset of A_K. Since A_K/K is compact, denote by $\overline{\mu}_K$ the Haar measure on A_K/K with total volume 1, and by μ_K the Haar measure on A_K which gives counting measure on K and induces $\overline{\mu}_K$ on A_K/K. Then P_a has a finite volume under μ_K and $\mu_K(P_a) = |a|\mu_K(P_1)$. The following lemma says that if the volume of P_a is sufficiently large, then P_a contains at least two elements in K. This is a special case of Minkowski's theorem stated in adèlic language.

Lemma 3. *There is a constant $c > 0$ such that for every idèle a with $|a| > \frac{1}{c}$, we have $P_a \cap K^\times \ne \phi$.*

Proof. Choose an idèle $b = (b_v)$ of K such that $b_v = 1$ if v is nonarchimedean and $b_v = \frac{1}{4}$ if v is archimedean, then the difference set $P_b \smallsetminus P_b \subset P_1$. Suppose $P_a \cap K^\times$ is empty. Then P_{ab} meets any translation of K in A_K at most at one point, for if y_1, y_2 are two distinct elements in $(x+K) \cap P_{ab}$, then $\alpha = y_1 - x, \beta = y_2 - x$ are two distinct elements in K and $\alpha - \beta$ lies in $P_{ab} \smallsetminus P_{ab} \subset P_a$, contradicting our assumption. Denote by $f, \overline{f}$ the characteristic function of $P_{ab}, (P_{ab} + K)/K$, respectively. We have

$$\begin{aligned}\mu_K(P_{ab}) &= \int_{A_K} f(x)d\mu_K(x) = \int_{A_K/K} \sum_{\alpha \in K} f(x+\alpha)d\overline{\mu}_K(\overline{x}) \\ &= \int_{A_K/K} \overline{f}(\overline{x})d\overline{\mu}_K(\overline{x}) \quad \text{since card}\left((x+K)\cap P_{ab}\right) \le 1 \text{ for all } x \in A_K, \\ &\le \int_{A_K/K} d\overline{\mu}_K(\overline{x}) = 1.\end{aligned}$$

Therefore if we set $c = \mu_K(P_b)$, then we have $c|a| \le 1$. Hence for idèles a with $|a| > \frac{1}{c}, P_a \cap K^\times$ is nonempty. □

Now we prove Theorem 8. Let a_0 be an idèle such that $|a_0| > \frac{1}{c}$. Then for any $a \in I_K^1$, we have $|a^{-1}a_0| = |a_0| > \frac{1}{c}$, so that there exists a nonzero element $\alpha \in K^\times$ lying in $P_{a^{-1}a_0}$, that is, $\alpha a \in P_{a_0}$, by Lemma 3. Further, since $\alpha a \in I_K^1$, there exists $\beta \in K^\times$ such that $\beta(\alpha a)^{-1} \in P_{a_0}$. Combining the last two statements, we get $\beta \in P_{a_0^2}$. On the other hand, $P_{a_0^2} \cap K$ is both compact and discrete, hence is finite, say,

$$P_{a_0^2} \cap K = \{0, \gamma_1, \cdots, \gamma_s\}.$$

We have shown that $a\alpha \in P_{a_0}$ and $(a\alpha)^{-1} \in \bigcup_{i=1}^{s} \gamma_i^{-1} P_{a_0}$. Set

$$B = P_{a_0} \cup \bigcup_{i=1}^{s} \gamma_i^{-1} P_{a_0}$$

$$\text{and} \qquad B^* = \{x \in I_K : x, x^{-1} \in B\}.$$

There is a finite set S of places of K such that at the places v of K outside S, $(a_0)_v$ and $(\gamma_i)_v$ are units for $i = 1, \cdots, s$, thus B^* is the product of a compact subset of $\prod_{v \in S} K_v^\times$ and $\prod_{v \notin S} \mathcal{U}_v$, hence is a compact subset of I_K. We have shown that for any $a \in I_K^1$, there is an $\alpha \in K^\times$ such that $a\alpha \in B^*$. This proves that $I_K^1/K^\times$ is a quotient of $I_K^1 \cap B^*$, hence is compact. □

We derive some consequences of Theorem 8. First consider the case where K is a function field. The free abelian group generated by places of K is called the group $\mathcal{D}iv(K)$ of divisors of K. The degree of a divisor $\mathcal{D} = \sum_v n_v v$, where $n_v \in \mathbf{Z}$ and $n_v = 0$ for almost all v, is defined as $\deg \mathcal{D} = \sum_v n_v \deg v$. Thus the degree map $\deg : \mathcal{D}iv(K) \longrightarrow \mathbf{Z}$ is a homomorphism with kernel equal to $\mathcal{D}iv^0(K)$, the subgroup of degree zero divisors. We shall prove in Chapter 5, §4 (Theorem 5) that the image of the degree map is $\mathbf{Z}$. For $\xi \in K^\times$, the divisor of ξ is $\operatorname{div}\xi = \sum_v n_v v$, where n_v is the order of ξ at v. We know from Theorems 5 and 6 that $n_v = 0$ for almost all v and $\deg(\operatorname{div}\xi) = \sum n_v \deg v = 0$. Hence $\operatorname{div}\xi$ lies in $\mathcal{D}iv^0(K)$, called a principal divisor. Denote by $\mathcal{P}rin(K)$ the group of principal divisors. The quotient $\mathcal{D}iv^0(K)/\mathcal{P}rin(K)$ is called the Jacobian of K, denoted by $\mathrm{Jac}(K)$.

For an idèle $x = (x_v)$ in I_K, associate $\operatorname{div} x = \sum_v (\operatorname{ord}_v x_v) v$. As x_v is a unit for almost all v, we have $\operatorname{div} x \in \mathcal{D}iv(K)$. Endow the discrete topology on $\mathcal{D}iv(K)$. Then the map $\operatorname{div} : I_K \longrightarrow \mathcal{D}iv(K)$ is a continuous homomorphism, which is surjective with kernel equal to $\prod_v \mathcal{U}_v$. Observe that for $x \in I_K$, we have $|x| = \prod_v |x_v|_v = \prod_v q^{-\operatorname{ord}_v x_v \deg v} = q^{-\deg(\operatorname{div} x)}$, where q is the cardinality of the field of constants in K. The subgroup I_K^1 of norm 1 idèles is mapped to the subgroup $\mathcal{D}iv^0(K)$ surjectively, and $K^\times$ is mapped onto $\mathcal{P}rin(K)$. Thus div induces the isomorphism $I_K^1/K^\times \prod_v \mathcal{U}_v \cong \mathcal{D}iv^0(K)/\mathcal{P}rin(K) = \mathrm{Jac}(K)$ algebraically and topologically. As $I_K^1/K^\times$ is compact, we find that $\mathrm{Jac}(K)$ is both compact and discrete, thus is finite. This proves

Corollary 5. *Let K be a function field. Then $Jac(K) = \mathcal{D}iv^0(K)/\mathcal{P}rin(K) \cong I_K^1/K^\times \prod_v \mathcal{U}_v$ is finite.*

Next consider the case where K is a number field. For each nonarchimedean place v of K, write $\mathcal{M}_v$ for the maximal ideal of the ring of integers $\mathcal{O}_K$ of K

corresponding to the place v. Denote by $\mathcal{F}(K)$ th free abelian group generated by $\mathcal{M}_v$'s, and by $\mathcal{P}rin(K)$ the abelian group generated by nonzero principal ideals. Since every nonzero ideal in $\mathcal{O}_K$ is a product of maximal ideals, unique up to order, $\mathcal{P}rin(K)$ is a subgroup of $\mathcal{F}(K)$. The quotient $\mathcal{F}(K)/\mathcal{P}rin(K)$ is called the ideal class group $Cl(K)$ of K.

For an idèle $x = (x_v) \in I_K^1$, let $\mathcal{I}(x)$ denote the ideal $\prod_{v \text{ nonarch}} \mathcal{M}_v^{\mathrm{ord}_v x_v}$. Since x_v has order zero for almost all v, $\mathcal{I}(x)$ is an element in $\mathcal{F}(K)$. The map $\mathcal{I} : I_K^1 \longrightarrow \mathcal{F}(K)$ sending x to $\mathcal{I}(x)$ is a group homomorphism. Given an element $\prod_{v \text{ nonarch}} \mathcal{M}_v^{n_v}$ in $\mathcal{F}(K)$, we construct an idèle $x = (x_v) \in I_K^1$ such that $\mathcal{I}(x) = \prod \mathcal{M}_v^{n_v}$ as follows. At each nonarchimedean place v, choose a uniformizer π_v and let $x_v = \pi_v^{n_v}$. Then $x_v = 1$ for almost all v. Fix an archimedean place w. For all archimedean places $v \neq w$, let $x_v = 1$, and let x_w be an element in K_w such that $\prod_{v \text{ all places}} |x_v|_v = 1$. Thus $x = (x_v)$ has the required property. Endow $\mathcal{F}(K)$ with the discrete topology. Then $\mathcal{I}$ is a continuous surjective homomorphism with kernel equal to $I_\infty^1 \prod_{v \text{ nonarch}} \mathcal{U}_v$, where I_∞^1 consists of $x = (x_v)$ in $\prod_{v\in\Sigma_\infty} K_v^\times$ with $\prod_{v\in\Sigma_\infty} |x_v|_v = 1$. Further, $\mathcal{I}$ maps $K^\times$ onto $\mathcal{P}rin(K)$, hence it induces an isomorphism $I_K^1/K^\times \cdot I_\infty^1 \prod_{v \text{ nonarch}} \mathcal{U}_v \cong \mathcal{F}(K)/\mathcal{P}rin(K)$ algebraically and topologically. As $I_K^1/K^\times$ is compact, this shows that $\mathcal{F}(K)/\mathcal{P}rin(K)$ is discrete and compact, hence is finite. We have shown

Corollary 6. *(Minkowski) Let K be a number field. Then*

$$Cl(K) = \mathcal{F}(K)/\mathcal{P}rin(K) \cong I_K^1/K^\times \cdot I_\infty^1 \prod_{v \text{ nonarch}} \mathcal{U}_v$$

is finite.

Elements in $K^\times \cap I_\infty^1 \prod_{v \text{ nonarch}} \mathcal{U}_v$ are called units of the number field K. Dirichlet showed that the group of units is the product of the (finite) group of roots of unity in K and a free abelian group of rank $r_1 + r_2 - 1$. Here r_1, r_2 are the number of real and complex places of K, respectively.

For a general algebraic group G defined over a global field K, the group of adelic points $G(A_K)$ is the restricted product of $\{G(K_v)\}_{v\in\Sigma_K}$ with respect to $\{G(\mathcal{O}_v)\}_{v\in\Sigma_K-\Sigma_\infty}$.

References

[1] E. Artin and J. Tate : Class field theory, Harvard (1961).

[2] J.W.S. Cassels and A. Fröhlich : Algebraic Number Theory, Thompson, Washington (1967), republished by Academic Press, London.

[3] S. Iyanaga : The Theory of Numbers, North–Holland, Amsterdam–Oxford (1969).

[4] A. Weil : Basic Number Theory, Springer–Verlag, New York–Heidelberg–Berlin (1973).

CHAPTER 4

The Riemann–Roch Theorem

§1 Characters of a restricted product

Recall that a character of a topological group G is a continuous homomorphism from G to S^1. For latter uses, sometimes we allow such a homomorphism to take values in $\mathbf{C}^\times$, and call it a quasi-character. The group of characters of G, denoted by $\widehat{G}$, is called the topological dual of G; it is endowed with the compact open topology so that $\widehat{G}$ is again a topological group. We study the dual of the group of G which is the restricted product of a family of locally compact commutative topological groups $G_v, v \in \Sigma$, with respect to chosen open compact subgroups H_v of $G_v, v \in \Sigma - \Sigma_0$, where Σ_0 is a finite subset of the index set Σ. Each G_v is imbedded in G as a subgroup by letting the components outside v being the identity elements of the groups in question.

Let χ be a (quasi –) character of G. Denote by χ_v the restriction of χ on G_v. Then χ_v is a (quasi –) character of G_v.

Proposition 1. *χ_v is trivial on H_v for almost all v, and, for $a = (a_v) \in G$, we have*

$$\chi(a) = \prod_v \chi_v(a_v).$$

Observe that the above is a finite product since $a_v \in H_v$ for almost all v and χ_v is trivial on H_v for almost all v.

Proof. Let U be an open neighborhood of 1 in $\mathbf{C}^\times$ containing no subgroup other than $\{1\}$. Then $N = \chi^{-1}(U)$ is an open neighborhood of the identity of G. It follows from the definition of the topology put on G that N contains $\prod_{v \in S} N_v \prod_{v \in \Sigma - S} H_v$ for some finite subset S of Σ, where N_v is an open neighborhood of the identity in G_v for $v \in S$. This proves that χ_v is trivial on H_v for all v outside S since $\chi_v(H_v)$ is a subgroup contained in U. Given $a = (a_v) \in G$, let S' be a finite subset of

$\sum$ containing S such that for v outside $S', a_v \in H_v$ and $\chi_v(H_v) = 1$. Then for $v \in \Sigma - S', \chi_v(a_v) = 1$ and

$$\chi(a) = \chi\left(\prod_{v\in S'} a_v \prod_{v\in\Sigma-S'} a_v\right) = \chi\left(\prod_{v\in S'} a_v\right) = \prod_{v\in S'} \chi(a_v) = \prod_{v\in\Sigma} \chi(a_v). \quad \square$$

Proposition 2. *Given, for each $v \in \Sigma$, a (quasi –) character χ_v of G_v such that $\chi_v(H_v) = 1$ for almost all v, then $\chi(a) := \prod_v \chi_v(a_v)$ for $a = (a_v) \in G$ defines a (quasi –) character χ of G.*

Proof. Clearly , χ is a homomorphism. Let S be a finite subset of Σ such that $\chi_v(H_v) = 1$ for v outside S. Suppose S has cardinality s. Given an open neighborhood U of 1 in $\mathbf{C}^\times$, let V be an open neighborhood of 1 such that $V^s \subset U$. For each $v \in S, \chi_v$ is continuous, let N_v be an open neighborhood of the identity of G_v such that $\chi_v(N_v) \subset V$. Then $\chi\left(\prod_{v\in S} N_v \prod_{v\in\Sigma-S} H_v\right) = \prod_{v\in S} \chi_v(N_v) \subset V^s \subset U$ and $\prod_{v\in S} N_v \prod_{v\in\Sigma-S} H_v$ is an open neighborhood of the identity of G. This proves that χ is continuous, hence is a (quasi –) character. $\square$

In view of Propositions 1 and 2, we write $\chi = \prod_v \chi_v$ and keep in mind that $\chi_v(H_v) = 1$ for almost all v.

Denote by $\widehat{G_v}$ the dual group of G_v, it is locally compact. For $v \in \Sigma - \Sigma_0$, we have $H_v^\perp = \{\chi_v \in \widehat{G_v} : \chi_v(H_v) = 1\} = \widehat{G_v/H_v}$. As H_v is open, G_v/H_v is discrete so that its dual $H_v^\perp$ is compact. As H_v is compact, its dual $\widehat{H}_v$ is discrete. The restriction map from $\widehat{G}_v$ to $\widehat{H}_v$ has kernel $H_v^\perp$, so $\widehat{G}_v/H_v^\perp$, being isomorphic to a subgroup of $\widehat{H}_v$, is discrete. This shows that $H_v^\perp$ is a compact and open subgroup of $\widehat{G}_v$.

Theorem 1. *The restricted product of $\widehat{G}_v, v \in \Sigma$, with respect to $H_v^\perp, v \in \Sigma - \Sigma_0$, is isomorphic to the character group $\widehat{G}$ of G algebraically and topologically.*

Proof. That $(\chi_v) \mapsto \chi = \prod_v \chi_v$ is an algebraic isomorphism follows from Propositions 1 and 2. We have only to check that the topology is the same. Indeed, $\chi = \prod_v \chi_v \in \widehat{G}$ is close to the trivial character of G if and only if $\chi(B)$ is close to 1 for a large compact subset B of G which we may assume of the form $B = \prod_{v\in S} B_v \prod_{v\in\Sigma-S} H_v$, where S is a finite subset of Σ and for $v \in S, B_v$ is a compact subset of G_v. The latter holds if and only if $\chi_v(B_v)$ is close to 1 for $v \in S$ and $\chi_v(H_v) = 1$ for $v \in \Sigma - S$ (since $\chi_v(H_v)$ is a subgroup close 1 and $\{1\}$ is the only such kind of group in $\mathbf{C}^\times$), which means that for $v \in S$, χ_v is close to the trivial character in $\widehat{G}_v$, and for $v \in \Sigma - S$, $\chi_v \in H_v^\perp$; in other words, (χ_v) is close to the identity in the restricted product of $\{\widehat{G}_v\}$ with respect to $\{H_v^\perp\}$. □

---

§2 Standard additive characters

For the remainder of this chapter, K denotes a function field of one variable with the field of constants k containing q elements. We shall pick standard additive characters of K locally and globally. Fix a nontrivial additive character φ of k (cf. Chapter 1, §4).

We start with the case $K = k(T)$. Denote by ∞ the "infinite place" of K with uniformizer $\frac{1}{T}$. Every element x in K_∞ can be written as $x = \sum\limits_{i<\infty} a_i T^i$ with $a_i \in k$. The standard additive character ψ_∞ of K_∞ is defined by

$$\psi_\infty(x) = \psi_\infty(a_n T^n + a_{n-1}T^{n-1} + \cdots + a_0 + a_{-1}T^{-1} + \cdots) = \varphi(-a_{-1}).$$

Next we extend ψ_∞ to a character ψ' of $K_\infty \prod\limits_{v\neq\infty} \mathcal{O}_v$ such that ψ' is trivial on $\prod\limits_{v\neq\infty} \mathcal{O}_v$ and ψ' on K_∞ is equal to ψ_∞. As $K_\infty \prod\limits_{v\neq\infty} \mathcal{O}_v \cap K = k[T]$ and ψ' is trivial on $k[T]$, we may further extend ψ' to a character ψ on $K + K_\infty \prod\limits_{v\neq\infty} \mathcal{O}_v$, trivial on K. As shown in the proof of Theorem 7 in Chapter 3, $A_K = K + K_\infty \prod\limits_{v\neq\infty} \mathcal{O}_v$, thus we have constructed the standard character ψ on A_K, trivial on K, whose restriction to K_∞ is ψ_∞. For $v \neq \infty$ denote by ψ_v the restriction of ψ to K_v, it is the standard character of K_v.

Proposition 3. *Let $K = k(T)$. For $v \neq \infty, \psi_v$ is a nontrivial additive character of K_v. In fact, ψ_v is trivial on $\mathcal{O}_v$ but not on $\mathcal{P}_v^{-1}$. Further, let $\pi_v = P_v(T)$ be the monic irreducible polynomial in $k[T]$ vanishing at v. Then every $x \in K_v$ can be written in a unique way as $x = \sum\limits_{i>-\infty} s_i \pi_v^i$, where $s_i \in k[T]$ with $\deg s_i < \deg P_v = \deg v$, and $\psi_v(x) = \psi_v\left(s_{-1}\pi_v^{-1}\right) = \varphi\left(a_{(\deg v)-1}\right)$, where*

$$s_{-1} = a_0 + a_1 T + \cdots + a_{(\deg v)-1} T^{(\deg v)-1}.$$

Proof. It follows from the definition of ψ that ψ_v is trivial on $\mathcal{O}_v$. If we can show that ψ_v is as described, then ψ_v is not trivial on $\mathcal{P}_v^{-1}$ since φ is a nontrivial character of k. It suffices to compute the value of ψ_v on $x = s_{-n}\pi_v^{-n} + \cdots + s_{-1}\pi_v^{-1} = f(T)/P_v(T)^n$, where $s_i \in k[T]$ has degree less than $\deg P_v$, thus $\deg f(T) < \deg P_v(T)^n =: \beta$. As observed before, $f(T)/P_v(T)^n \in \mathcal{O}_w$ for all places $w \neq v, \infty$. Hence

$$1 = \psi\left(\frac{f(T)}{P_v(T)^n}\right) = \psi_\infty\left(\frac{f(T)}{P_v(T)^n}\right)\psi_v\left(\frac{f(T)}{P_v(T)^n}\right)$$

implies

$$\psi_v(x) = \psi_\infty \left(-\frac{f(T)}{P_v(T)^n} \right).$$

To determine the above value, write $f(T) = T^{\beta-1}\left(a_{\beta-1} + a_{\beta-2}T^{-1} + \cdots\right)$ and $P_v(T)^n = T^\beta\left(1 + b_{\beta-1} + \cdots\right)$ so that $-f(T)/P_v(T)^n = -T^{-1}\left(a_{\beta-1} + x\right)$ for some element x in $\mathcal{P}_\infty$. It then follows from the definition of ψ_∞ that

$$\psi_v(x) = \psi_\infty\left(-f(T)/P_v(T)^n\right) = \varphi\left(a_{\beta-1}\right).$$

Finally we observe that $a_{\beta-1}$ is the cofficient of $T^{(\deg v-1)}$ in s_{-1}. □

Next consider the case where K is a finite degree separable extension of $F = k(T)$. We have already defined the standard local and global additive characters ψ_v, ψ attached to F. Let w be a place of K and let v be the place of F divisible by w. Define ψ_w to be $\psi_v \circ \mathrm{Tr}_{K_w/F_v}$. Since $\mathrm{Tr}_{K_w/F_v}(\mathcal{O}_w)$ is an $\mathcal{O}_v$ – module contained in $\mathcal{O}_v$, it is equal to $\mathcal{P}_v^n$ for some integer $n \geq 0$. This shows that ψ_w is trivial on $\mathcal{O}_w$ if w does not divide ∞; and on the other hand, $\mathrm{Tr}_{K_w/F_v}\left(\pi_v^{-n-1}\mathcal{O}_w\right) = \mathcal{P}_v^{-1}$ on which ψ_v is nontrivial, hence each ψ_w is a nontrivial character of K_w. Futher, the ψ_w's define a character $\psi = \prod_w \psi_w$ of A_K by Proposition 2. Corollory 2 of Chapter 3 implies that ψ is trivial on K. It is the standard character of A_K trivial on K.

Let w be a place of K and v be the place of F divisible by w. The set of elements x in K_w such that $\mathrm{Tr}_{K_w/F_v}(x\mathcal{O}_w) \subset \mathcal{O}_v$ is an $\mathcal{O}_w$ -module and is not equal to K_w as we have observed, hence it is equal to $\mathcal{P}_w^{-d_w}$ for some integer $d_w \geq 0$. The ideal $\mathcal{D}_w = \mathcal{P}_w^{dw}$ is called the *different* of K_w over F_v. For any nontrivial character η_w of K_w, the largest integer n such tha η_w is trivial on $\mathcal{P}_w^{-n}$ is called the *order* of η_w. Hence for w not dividing ∞, the order of the standard character ψ_w is d_w; the order of ψ_∞ on F_∞ is -1.

Implicitly contained in the triviality on K of the standard global additive character ψ is the so-called "Residue Theorem" which we now explain. Recall that the field of constants k of K has q elements. Let w be a place of K of degree n so that the residue field of K_w is a degree n field extension of k. The residue field of K_w consists of the roots of $x^{q^n} = x$ in an algebraic closure of k. Since $x^{q^n} - x$ has q^n distinct roots in the residue field, by Hensel's Lemma, the equation has q^n distinct solutions in K_w, and they constitute the degree n extension of k in K_w, denote it by k_n. So k_n is a set of representatives of the residue field of K_w in $\mathcal{O}_w$, and hence K_w is isomorphic to the field of formal power series $k_n((\pi_w))$ for any uniformizer π_w in K_w. For $\alpha \in K_w$, write $\alpha = \sum_{i > -\infty} a_i\pi_w^i$ with $a_i \in k_n$. Define $d\alpha = \left(\sum_{i>-\infty} ia_i\pi_w^{i-1}\right) d\pi_w$, called a local k-differential. It can be easily shown that such differentials enjoy the usual formal properties of differentials. The K_w – module generated by such differentials is a one-dimensional vector space over K_w with $d\pi_w$ as a basis. Given a differential $\omega = fd\pi_w$ with $f = \sum_{i>-\infty} a_i\pi_w^i, a_i \in k_n$, it can be shown that the coefficient a_{-1} is independent of the choice of the uniformizer π_w, and it is defined as the residue of ω at w:

$$\mathrm{Res}_w\ \omega = a_{-1}.$$

Global (meromorphic) k–differentials are the K– module generated by df's with $f \in K$. It is also a one-dimensional vector space over K, generated by df for any $f \in K$ with $df \neq 0$. Given a global k–differential $\omega = gdf, gf \in K$, it is also a local k–differential at every place ω of$K$. As g, f are integral at almost all places of K, there are only finitely many places where ω is nonholomorphic. Write X for the projective irreducible nonsingular curve defined over k such that K is the field of rational functions on X. Then there is a finite extension k_n of k such that all poles of ω are among the points on X over k_n, i.e., $X(k_n)$. Similar to what happens to a meromorphic differential on the one point compactification of the complex plane, one has

Theorem 2 (Residue Theorem). *For every global k–differential ω of $K = k(X)$, we have*

$$\sum_{P \in X(\bar{k})} \mathrm{Res}_P\ \omega = 0.$$

Our purpose is to see that the residue theorem follows from the triviality of ψ on K. First study the case $K = k(T)$ so that $X(\bar{k}) = \mathbf{P}^1(\bar{k})$. We may write ω as $f(T)dT$. Express $f(T)$ as a sum of its partial fractions:

$$f(T) = \sum_{i=0}^{n} a_i T^i + \sum_{v \neq \infty} \frac{g_v(T)}{P_v(T)^{n_v}}, \quad a_i \in k,$$

where $P_v(T)$ is the monic irreducible polynomial vanishing at the closed point v, and $g_v(T)$ is a polynomial in $k[T]$ of degree less than $n_v \deg P_v, n_v = 0$ for almost all v. Since residue is linear over k, it suffices to prove the theorem for $f(T) = T^n, n \geq 0$, and $f(T) = \frac{g_v(T)}{P_v(T)^n}$. If $f(T) = T^n$, ω is integral at all places $v \neq \infty$; pick $\pi_\infty = \frac{1}{T}$ so that $T = \pi_\infty^{-1}$, $dT = -1 \cdot \pi_\infty^{-2} d\pi_\infty$ and $f(T)dT = \pi_\infty^{-n} \cdot (-\pi_\infty^{-2})\, d\pi_\infty$ which obviously has residue zero. If $f(T) = \frac{g_v(T)}{P_v(T)^n}$, ω is integral at all places except v and ∞. Suppose degree $P_v(T) = r$. Let $\alpha_1, \cdots, \alpha_r$ be its roots in $k_r \subset \bar{k}$. Write $f(T) = \sum_{i=1}^{n} \frac{s_i}{P_v(T)^i}$ as in Proposition 3. Further, there are constants β_{ij} in $k_r, 1 \leq i \leq n, 1 \leq j \leq r$ such that

$$f(T) = \sum_{i=1}^{n} \frac{s_i}{P_v(T)^i} = \sum_{i=1}^{n} \sum_{j=1}^{n} \frac{\beta_{ij}}{(T - \alpha_j)^i}.$$

The residue of $f(T)dT$ at α_j is β_{1j} and $\sum_{j=1}^{r} \beta_{1j}$ is the coefficient of T^{r-1} in s_{-1}. At ∞, we have $f(T) = T^{-1}(b_{-1} + b_0 T^{-1} + \cdots)$ so that $f(T)dT = (-b_{-1}\pi_\infty^{-1} +$

$b_0 + b_1\pi_\infty^{-1} + \cdots)d\pi_\infty$ and the residue of $f(T)dT$ at ∞ is $-b_{-1}$. In view of the standard character ψ of A_K/K, we find $1 = \psi(f(T)) = \psi_\infty(f(T))\psi_v(f(T)) = \varphi\left(\sum_{P\in X(\bar{k})} \operatorname{Res}_P f(T)dT\right)$. For any $\alpha \in k$, the character ψ^α is also trivial on K, this shows that $\varphi^\alpha\left(\sum_P \operatorname{Res}_P f(T)dT\right) = 1$ for all $\alpha \in k$, hence $\sum_{P\in X(\bar{k})} \operatorname{Res}_P f(T)dT = 0$.

When K is a separable finite degree field extension of $k(T)$, we write ω as fdT with $f \in K$. Suppose the poles of ω are in $X(k_n)$. Extending the field of constants if necessary, we may assume that the poles of ω are in $X(k)$. The key point is to show that for each place v of $k(T)$ such that ω has a pole at a place w of K dividing v,

$$(1) \qquad \sum_{\substack{w \text{ place of } K \\ w|v}} \operatorname{Res}_w(\omega) = \operatorname{Res}_v(\operatorname{Tr}_{K/k(T)}\omega) := \operatorname{Res}_v(\operatorname{Tr}_{K/k(T)} f)dT.$$

The theorem then follows from the result for $k(T)$. □

Remark. The residue theorem can be also formulated in terms of the residues at closed points of X over k, that is, in terms of places of K. Indeed, let ω be a k-differential of K. The residue of ω at a place v of K is an element in the residue field of v, which we denote by k_v. It is an extension of k of degree equal to $\deg v$.

Theorem 2′ (Residue Theorem). *Let ω be a k-differential of K. Then*

$$\sum_{v \text{ places of } K} \operatorname{Tr}_{k_v/k}(\operatorname{Res}_v \omega) = 0.$$

When $K = k(T)$, we may assume $\omega = f(T)dT$ with $f(T) = \frac{g_v(T)}{P_v(T)^n}, v \neq \infty$, as before. Then $k_v = k_r$ is a degree r extension of k. The place v splits completely in the field $L = k_r(T)$ into r places $\alpha_1, \cdots, \alpha_r$ of degree 1 in L. Suppose at v, ω can be written as $\left(\cdots + a_{-1}P_v(T)^{-1} + a_0 + \cdots\right)dP_v(T) = \beta dP_v(T)$ with $a_i \in k_r$ so that $\operatorname{Res}_v \omega = a_{-1}$. Since each α_i is a simple root of $P_v(T)$, we may choose $P_v(T)$ as a uniformizer of L_{α_i}. Recall that the field K_v is diagonally imbedded in $K_v \otimes_K L = L_{\alpha_1} \oplus \cdots \oplus L_{\alpha_r}$, and the images of $\beta \in K_v$ in the L_{α_i}'s are β under the action of elements of $Gal(k_r/k)$, which acts on the coefficients a_i and leaves $P_v(T)$ invariant. This shows that $\sum_{\alpha_i} \operatorname{Res}_{\alpha_i} \omega$ is the sum of conjugates of a_{-1} under $Gal(k_r/k)$, in other words, it is equal to $\operatorname{Tr}_{k_r/k}(a_{-1}) = \operatorname{Tr}_{k_r/k}(\operatorname{Res}_v \omega)$. Hence the new statement holds for $K = k(T)$.

When K is a finite separable extension of $F = k(T)$, given a k-differential ω of K and a place v of F, we have

$$\sum_{\substack{w \text{ places of } K \\ w|v}} \operatorname{Tr}_{k_w/k}(\operatorname{Res}_w \omega) = \sum_{w|v} \operatorname{Tr}_{k_v/k} \operatorname{Tr}_{k_w/k_v}(\operatorname{Res}_w \omega).$$

One can show that

$$\text{(2)} \qquad \mathrm{Tr}_{k_w/k_v}(\mathrm{Res}_w\ \omega) = \mathrm{Res}_v(\mathrm{Tr}_{K_w/F_v}\ \omega),$$

thus

$$\sum_{\substack{w \text{ places of } K \\ w|v}} \mathrm{Tr}_{k_w/k}(\mathrm{Res}_w\ \omega) = \mathrm{Tr}_{k_v/k}\left(\sum_{w|v} \mathrm{Res}_v(\mathrm{Tr}_{K_w/F_v}\ \omega)\right)$$
$$= \mathrm{Tr}_{k_v/k}\,\mathrm{Res}_v\left(\sum_{w|v} \mathrm{Tr}_{K_w/F_v}\ \omega\right)$$
$$= \mathrm{Tr}_{k_v/k}\,\mathrm{Res}_v\left(\mathrm{Tr}_{K/F}\ \omega\right).$$

Therefore

$$\sum_{w} \mathrm{Tr}_{k_w/k}(\mathrm{Res}_w\ \omega) = \sum_{v \text{ places of } K}\ \sum_{w|v} \mathrm{Tr}_{k_w/k}(\mathrm{Res}_w\ \omega)$$
$$= \sum_{v} \mathrm{Tr}_{k_v/k}\,\mathrm{Res}_v\left(\mathrm{Tr}_{K/F}\ \omega\right) = 0$$

since $\mathrm{Tr}_{K/F}\ \omega$ is a k-differential of F.

The reader is referred to [2] and [3] for more detailed discussions of differentials and the residue theorem. Note that (1) is a special case of (2).

Exercise 1. Show that there is an additive character ψ of $A_{\mathbf{Q}}$ trivial on $\mathbf{Q}$ such that ψ is trivial on $\prod_p \mathbf{Z}_p$ and the restriction of ψ to $\mathbf{R} = \mathbf{Q}_\infty$ is $\psi_\infty(x) = e^{-2\pi i x}$. Describe the restriction ψ_p of ψ to $\mathbf{Q}_p$ for any prime p.

---

§3 Duality

Let v be a place of K. Using the standard additive character ψ_v derived in the previous section, we obtain an isomorphism between K_v and its dual $\widehat{K}_v$.

Theorem 3. *The map $\theta : x \mapsto \varphi_v^x$ gives rise to an isomorphism algebraically and topologically between K_v and its topological dual $\widehat{K}_v$.*

Recall that ψ_v^x is the additive character on K_v mapping y to $\psi_v(xy)$; it is a character because multiplication by x is a continuous homomorphism on the additive group of K_v.

Proof. As $\psi_v^{x+y} = \psi_v^x \psi_v^y$, the map θ is an algebraic homomorphism. Further, as ψ_v is nontrivial and multiplication by x on K_v is surjective if $x \neq 0, \psi_v^x$ is the trivial character only when $x = 0$. This shows that θ is 1-1. Denote by H the image of θ, that is , $H = \{\psi_v^x : x \in K_v\}$, by $\overline{H}$ its closure in $\widehat{K}_v$. Then

$$\overline{H}^{\perp} = \{y \in K_v; \psi_v^x(y) = 1 \quad \text{for all} \quad x \in K_v\}.$$

Since ψ_v is nontrivial, $\psi_v(yK_v) = 1$ implies $y = 0$. This shows that $\overline{H}^{\perp} = 0$, thus $\overline{H} = \widehat{K}_v$, H is dense in $\widehat{K}_v$. Next we prove that θ is continuous. Start with an open neighborhood of the trivial character in H, say, consisting of characters sending a compact set $B \subset K_v$ into an open neighborhood V of 1 in $\mathbf{C}$, we may assume that $B = \{x \in K_v : |x|_v \leq M\}$. As ψ_v is continuous, there is a positive number δ such that ψ_v sends all elements x in K_v with $|x|_v \leq \delta$ into V. Let U be the open neighborhood of $0 \in K_v$ consisting of elements x in K_v with $|x|_v < \delta/M$. Then elements in UB have valuation $< \delta$ and thus are mapped to V by ψ_v. This shows that $\theta(U)$ is contained in the prescribed neighborhood of the trivial character. To show the continuity of θ^{-1}, given an open neighborhood U of 0 in K_v, we have to find a compact set $B \subset K_v$ and an open neighborhood V of 1 in $\mathbf{C}$ such that all characters in H sending B to V are contained in $\theta(U)$. As ψ_v is nontrivial, there is an element x_0 in K_v such that $\psi_v(x_0) \neq 1$. Let U be the open disc in $\mathbf{C}$ centered at 1 with radius $|\psi_v(x_0) - 1|$. We may assume that U consists of elements in K_v with valuation less than ε. Let $|x_0|_v = \delta$. Choose B to be the ball in K_v with center 0 and radius $\delta\varepsilon^{-1}$. Suppose $\psi_v^x(B) \subset V$, i.e., $\psi_v(xB) \subset V$. If $x \notin U$, then $|x_v| \geq \varepsilon$ and xB contains x_0, which contradicts $\psi_v(x_0) \notin V$. We have shown that θ is bi-continuous. Hence H is locally compact. In particular, H is closed in $\widehat{K}_v$, hence is equal to $\widehat{K}_v$. This completes the proof of Theorem 3. □

Take the special case $K = k(T)$. Theorem 1 says that the global dual $\widehat{A}_K$ is isomorphic to the restricted product of $\{\widehat{K}_v\}$ with respect to $\{\mathcal{O}_v^{\perp}\}$. By Theorem 3, we may identify $\widehat{K}_v$ with K_v using the standard additive character ψ_v, and under this identification, $\mathcal{O}_v^{\perp} = \{\psi_v^x : \psi_v^x(\mathcal{O}_v) = \psi_v(x\mathcal{O}_v) = 1\}$ is identified with $\mathcal{O}_v$ for $v \neq \infty$ and with $\mathcal{P}_\infty^2$ for $v = \infty$ since ψ_v for $v \neq \infty$ is trivial on $\mathcal{O}_v$ but not on $\mathcal{P}_v^{-1}$ by Proposition 3 and ψ_∞ is trivial on $\mathcal{P}_\infty^2$ but not on $\mathcal{P}_\infty$ by definition. Therefore $\widehat{A}_K$ is isomorphic to A_K algebraically and topologically by means of the global additive character $\psi = \prod_v \psi_v$. More precisely, $\widehat{A_K} = \{\psi^x : x \in A_K\}$ and $\psi^x \mapsto x$ is the isomorphism.

Next assume K is a finite degree separable field extension of $F = k(T)$. By Lemma 2 in Chapter 3, the additive topological group A_K is isomorphic to $A_F \otimes_F K$, which is isomorphic to $\underbrace{A_F \oplus \cdots \oplus A_F}_{n \text{ times}} = A_F^n$, where $n = [K : F]$. Thus $\widehat{A}_K$ is isomorphic to $\widehat{A_F^n} = \widehat{A}_F^n$, which in turn is isomorphic to A_F^n, and hence to A_K algebraically and topologically. On the other hand, by Theorem 1, $\widehat{A}_K$ is isomorphic to the restricted product of $\{\widehat{K}_w\}$ with respect to $\{\mathcal{O}_w^{\perp}\}$. Using the standard additive character ψ_w at a place w of K, we may, by Theorem 3, identify $\widehat{K}_w$ with K_w and $\mathcal{O}_w^{\perp}$ with $\mathcal{D}_w^{-1} = \mathcal{P}_w^{-d_w}$, the inverse of the different of K_w/F_v, where w divides the place v of F and $v \neq \infty$. This implies that A_K is isomorphic to the restricted product of $\{K_w\}$ with respect to $\{\mathcal{D}_w^{-1}\}$. In particular, $\prod_{w|\infty} \mathcal{O}_w \prod_{w\nmid\infty} \mathcal{D}_w^{-1}$ is open and compact in A_K. In view of the topology on A_K, this implies that $d_w \geq 0$ for almost all w (in order to be open) and $d_w \leq 0$ for almost all w (in order to be compact), thus $d_w = 0$ for almost all w. This result in turn implies that $x \mapsto \psi^x$ yields an isomorphism from A_K to $\widehat{A_K}$. We summarize the above discussion in

Theorem 4. *The map $x \mapsto \psi^x$ gives rise to an isomorphism from A_K to $\widehat{A_K}$ algebraically and topologically. Further, suppose that K is a separable finite degree extension of the rational function field $F = k(T)$. Then for almost all places v of F, the different of K_w over F_v is trivial. Here w is any place of K dividing v.*

Corollary 1. $K^\perp = \{\psi^\alpha : \alpha \in K\}$. *In other words, under the identification of $\widehat{A_K}$ with A_K via $\psi, K^\perp$ is identified with K.*

Proof. Clearly, for $\alpha \in K, \psi^\alpha$ is trivial on K. So $K^\perp$ is a subgroup of $\widehat{A_K}$ containing K. As A_K/K is compact, $K^\perp = \widehat{A_K/K}$ is discrete. Now $K^\perp/K$, being a discrete subgroup of the compact group A_K/K, is finite. On the other hand, $K^\perp$ is a vector space over K and K is infinite, the only way to have $K^\perp/K$ finite is that $K^\perp$ is 1-dimensional over K, i.e., $K^\perp = K$. □

Remark. Any choice of a nontrivial local (resp. global) additive character will establish an isomorphism between K_v (resp. A_K) and its topological dual. But in the global duality if we want $K^\perp$ to be K itself, then the global additive character has to be trivial on K.

---

§4 The Riemann-Roch Theorem

Recall that for an idèle $a = (a_v)$ of K, we defined the parallelotope of size a to be

$$P_a = \{(x_v) \in A_K : |x_v|_v \leq |a_v|_v \quad \text{for all places } v \text{ of } K\} = aP_1 = a \cdot \prod_v \mathcal{O}_v.$$

As P_a is an open and compact subgroup of A_K and K is a discrete subgroup of A_K, the intersection $\Lambda_a = P_a \cap K$ is a finite subgroup. Further, Λ_a is invariant under multiplication by k. Hence Λ_a ia a finite-dimensional vector space over k; denote its dimension by $\lambda(a)$ so that the cardinality of Λ_a is $q^{\lambda(a)}$.

Identify the dual of A_K with A_K via the standard additive character ψ as in Theorem 4. Consider the dual $P_a^\perp = \{x \in A_K : \psi(xP_a) = 1\} = a^{-1} \cdot P_1^\perp = a^{-1} \prod_v \mathcal{O}_v^\perp$. We have shown in the previous section that $\mathcal{O}_v^\perp = \mathcal{O}_v$ for almost all v and for each $v, \mathcal{O}_v^\perp = \mathcal{P}_v^{n_v}$ for some integer n_v, there exists an idèle c such that

$P_1^\perp = cP_1$. Hence $P_a^\perp P_1$ is also a parallelotope, and $P_a^\perp \cap K^\perp = P_a^\perp \cap K = \Lambda_{ca^{-1}}$ has cardinality $q^{\lambda(ca^{-1})}$. On the other hand, $P_a^\perp \cap K^\perp = (P_a + K)^\perp = (A_K/(P_a+K))^\wedge$, which is isomorphic to $A_K/(P_a + K)$. Denote by μ a Haar measure on A_K and by $\overline{\mu}$ the induced measure on A_K/K. Then we have

$$q^{\lambda(ca^{-1})} = |P_a^\perp \cap K^\perp| = |A_K/(P_a + K)| = \overline{\mu}(A_K/K)/\overline{\mu}((P_a + K)/K).$$

The natural map from P_a to $(P_a + K)/K$ is surjective with kernel $P_a \cap K = \Lambda_a$. Hence

$$\overline{\mu}((P_a + K)/K) = \mu(P_a)/|\Lambda_a| = |a|\mu(P_1)/q^{\lambda(a)} = q^{-\deg(\text{div } a)-\lambda(a)}\mu(P_1).$$

Choose μ to be the measure $\mu = \prod_v \mu_v$, where each μ_v is the Haar measure on K_v with $\mu_v(\mathcal{O}_v) = 1$. Then $\mu(P_1) = 1$. So far, we have shown that

$$q^{\lambda(ca^{-1})} = q^{\lambda(a)+\deg(\text{div } a)}\overline{\mu}(A_K/K).$$

This implies that $\overline{\mu}(A_K/K)$ is an integral power of q; denote the exponent by $g-1$. Letting $a = 1$, we determine the value g. We have $\lambda(1) = 1, \deg(\text{div } a) = 0$, and

$$q^{\lambda(c)} = q^1 \cdot q^{g-1} = q^g,$$

i.e., $g = \lambda(c) \geq 0$. We have proven

Theorem 5. *There is a nonnegative integer g and an idèle c of K such that for all $a \in I_K$, we have*

$$\lambda(a) = \lambda(ca^{-1}) - \deg(\text{div } a) - g + 1.$$

We translate the above theorem into the usual form of the Riemann-Roch theorem. A divisor $\mathcal{D} = \sum n_v v$ of K is said to be effective if all coefficients $n_v \geq 0$, in this case we write $\mathcal{D} \geq 0$. We take a closer look of Λ_a:

$$\begin{aligned}\Lambda_a &= \{\alpha \in K^\times : \alpha \in P_a\} = \{\alpha \in K : \alpha a^{-1} \in P_1\} \\ &= \{\alpha \in K^\times : \text{div } \alpha + \text{div } a^{-1} \geq 0\} \cup \{0\}.\end{aligned}$$

In algebraic geometry, given a divisor $\mathcal{D} \in \mathcal{D}iv(K)$, it is customary to denote

$$L(\mathcal{D}) = \{\alpha \in K^\times : \text{div } \alpha + \mathcal{D} \geq 0\} \cup \{0\}.$$

Let a be an idèle such that $\text{div } a^{-1} = \mathcal{D}$, then we have $\Lambda_a = L(\mathcal{D}), \Lambda_{ca^{-1}} = L(\mathcal{K}-\mathcal{D})$, where $\mathcal{K} = \text{div } c^{-1}$. Write $\lambda(\mathcal{D})$ for the dimension of $L(\mathcal{D})$ over k. Then Theorem 5 can be written as

Theorem 5′ (Riemann-Roch Theorem). *There is a nonnegative integer g and a divisor $\mathcal{K} \in \mathcal{D}iv(K)$ such that for all divisor $\mathcal{D} \in \mathcal{D}iv(K), \lambda(\mathcal{D})$ is finite and*

$$\lambda(\mathcal{D}) = \lambda(\mathcal{K} - \mathcal{D}) + \deg \mathcal{D} - g + 1.$$

The integer g is called the *genus* of the curve X such that k is the field of rational functions on X. Also call g the *genus* of K. The divisor $\mathcal{K}$ arises from the idèle c which depends on the choice of the additive character ψ used to identify $\widehat{A}_K$ with A_K. The only constraint on ψ is that it should be nontrivial on A_K and trivial on K. As shown in Corollary 1, if ψ is such a character, then other choices are of the form ψ^α with $\alpha \in K^\times$. A different choice of ψ amounts to multiplying c by α^{-1} and hence replacing $\mathcal{K}$ by $\mathcal{K} + \operatorname{div} \alpha$ for some $\alpha \in K^\times$. Call $\mathcal{K} + \mathcal{P}rin(K)$ the class of canonical divisors, which is uniquely determined by K.

Corollary 2. $\lambda(\mathcal{K})$ *and* $\deg \mathcal{K} = 2g - 2$.

Proof. We have already seen that $\lambda(\mathcal{K}) = g$. To obtain $\deg \mathcal{K}$, simply let $\mathcal{D} = \mathcal{K}$ in Theorem 5′. □

Corollary 3. *For any divisor $\mathcal{D}$ in $\mathcal{D}iv(K)$ of degree $> 2g - 2$, we have*

$$\lambda(\mathcal{D}) = \deg \mathcal{D} - g + 1.$$

Proof. If $\deg \mathcal{D} > 2g - 2$, then $\deg(\mathcal{K} - \mathcal{D}) < 0$. For any $\alpha \in K, \operatorname{div} \alpha + \mathcal{K} - \mathcal{D}$ has negative degree, hence cannot be effective. This shows that $\lambda(\mathcal{K} - \mathcal{D}) = 0$, thus $\lambda(\mathcal{D}) = \deg \mathcal{D} - g + 1$. □

The idèlic equivalence of Corollary 3 is

Corollary 4. *For any idèle $a = (a_v) \in I_K$ with $\deg(\operatorname{div} a) < 2 - 2g$, we have $A_K = K + P_a = K + \prod\limits_v a_v \mathcal{O}_v$.*

Proof. This is because $\lambda(ca^{-1}) = 0$, which means $\overline{\mu}(A_K/K) = \overline{\mu}((P_a + K)/K)$, which in turn implies $A_K = P_a + K$. □

We end this section by mentioning an implication of the Riemann-Roch Theorem on holomorphic differentials of K. Given a nonzero global k- differential ω of K, at each place v of K, express ω in terms of $\alpha d\pi_v$, where $\alpha \in k_v((\pi_v))$ is a formal Laurent series in π_v with coefficients in k_v isomorphic to the residue field at v. Define the order of α at v to be the order of ω at v, which is independent of the choice of the uniformizer π_v. Since $\omega = f dg$ for two functions $f, g \in K$, ω has nonzero order only at finitely many places, the sum $\sum\limits_v (\operatorname{ord}_v \omega) v$ is a divisor in

$\mathcal{D}iv(K)$, called the divisor of ω and written as $\operatorname{div} w$. When $\operatorname{div}\omega$ is an effective divisor, we say that ω is a holomorphic differential.

Let $K = k(T)$ be the rational function field over k, and let ψ be the standard additive character of A_K. We have seen in §3 that $P_1^{\perp} = \prod_v \mathcal{O}_v^{\perp} = \mathcal{P}_\infty^2 \prod_{v\neq\infty} \mathcal{O}_v$, thus we may choose the idèle c to be $c_\infty = \pi_\infty^2$ and $c_v = 1$ for $v \neq \infty$ so that $P_1^{\perp} = P_c$. Then $\mathcal{K} = \operatorname{div} c^{-1} = -2\infty$ is a canonical divisor of K. Next consider the divisor $\omega = dT$. We show that $\operatorname{div}\omega = -2\infty = \mathcal{K}$. Indeed, at a place $v \neq \infty$, let π_v be the monic irreducible polynomial in $k[T]$ vanishing at v. Express $T = a_0 + a_1\pi_v + a_2\pi_v^2 + \cdots$ with $a_i \in k_v$, then $dT = (a_1 + 2a_2\pi_v + \cdots)d\pi_v$. Claim $a_1 \neq 0$. Suppose otherwise, then $T \equiv a_0 \pmod{\pi_v^2}$. Let $r = \deg v$. We have $T^{q^r-1} \equiv 1 \pmod{\pi_v^2}$, which implies that π_v^2 divides $T^{q^r-1} - 1$, contradicting the separability of the equation $T^{q^r-1} - 1$ over k. This proves $\operatorname{ord}_v \omega = 0$ for $v \neq \infty$. At ∞, choose $\pi_\infty = \frac{1}{T}$, we have seen before that $dT = -\pi_\infty^{-2} d\pi_\infty$. Therefore $\operatorname{div}\omega = -2\infty$, as desired.

Next assume K is a separable finite degree field extension of $F = k(T)$. Let ω be a place of K and v the place of F divisible by w. Denote by e_w the ramification index of the extension K_w over F_v. This means that $\pi_v = u_w\pi_w^{e_w}$ for some unit u_w in K_w. In particular, if $e_w = 1$, that is, K_w over F_v is unramified, we may choose $\pi_w = \pi_v$. Let $u_w = u_0 + u_1\pi_w + u_2\pi_w^2 + \cdots$ be the power series expansion of u_w in $k_w((\pi_w))$. When $v \neq \infty$, we have

$$\begin{aligned} T &= a_0 + a_1\pi_v + a_2\pi_v^2 + \cdots \\ &= a_0 + a_1u_w\pi_w^{e_w} + a_2u_w^2\pi_w^{2e_w} + \cdots \\ &= a_0 + a_1u_0\pi_w^{e_w} + (a_1u_1\pi_w^{e_w+1} + a_1u_2\pi_w^{e_w+2} + \cdots) \\ &\quad + a_2(u_0 + u_1\pi_w + \cdots)^2\pi_w^{2e_w} + \cdots \end{aligned}$$

and when $v = \infty$, we have

$$T = -\pi_\infty^{-1} = -u'_\infty\pi_w^{-e_w} = -(u'_0\pi_w^{-e_w} + u'_1\pi_w^{1-e_w} + \cdots),$$

where $u'_w = u'_0 + u'_1\pi_w + \cdots \in k_w((\pi_w))$ is the inverse of u_w. We see immediately that if K_w is unramified over F_v, then $\operatorname{ord}_v dT = \operatorname{ord}_w dT$; if K_w is tamely ramified over F_v, that is, $e_w > 1$ but prime to the characteristic p of k, then $\operatorname{ord}_w dT = (e_w - 1)$ or $(-e_w - 1)$ according to $v \neq \infty$ or $v = \infty$. Finally, if K_w is wildly ramified over F_v, that is, $e_w > 1$ and divisible by p, then $\operatorname{ord}_w dT \geq e_w$ if $v \neq \infty$ and $\operatorname{ord}_w dT \geq -e_w$ if $v = \infty$.

Recall the definition of the different $\mathcal{D}_w = \mathcal{P}_w^{d_w}$ of K_w over F_v, it is given by

$$\mathcal{P}_w^{-d_w} = \{x \in K_w : \operatorname{Tr}_{K_w/F_v}(x\mathcal{O}_w) \subset \mathcal{O}_v\} \supseteq \mathcal{O}_w.$$

One can show that K_w is unramified over F_v if and only if $\mathcal{D}_w = \mathcal{O}_w$. In view of Theorem 4, we know that K_w is unramified over F_v for almost all w. Further, if K_w is tamely ramified over F_v, then $d_w = e_w - 1$; and if K_w is wildly ramified over F_v, we have $d_w \geq e_w$. A deeper study of the ramification theory shows that as a differential in K, the divisor of dT is $\operatorname{div} c^{-1}$, where $P_c = \prod_w \mathcal{O}_w^{\perp}$ using the standard character defined in §2. In other words, $\operatorname{div} dT = \mathcal{K}$.

Theorem 6. *Given a canonical divisor $\mathcal{K}$ of K, there is a k-differential ω of K such that $\operatorname{div}\omega = \mathcal{K}$. Further, the dimension of holomorphic k-differentials of K is the genus g of K.*

Proof. We have shown that div dT is a canonical divisor of K. As any two canonical divisors of K differ by a principal divisor, we can multiply dT by a suitable function f in K such that $\omega = f dT$ has the desired divisor. Futher, the space of k-differentials of K is one - dimensional over K, any differential ω' can be written as $\omega' = g\omega$ for some $g \in K$. Then ω' is holomorphic if and only if $\operatorname{div}\omega' = \operatorname{div} g + \operatorname{div}\omega = \operatorname{div} g + \mathcal{K} \geq 0$, which is equivalent to g belonging to the space $L(\mathcal{K})$. By Corollary 2, $L(\mathcal{K})$ has dimension g. This completes the proof of the theorem. □

We may regard the genus defined in the Riemann-Roch theorem as algebraic genus of the curve C with $K = k(C)$; Theorem 6 says that the algebraic genus of C is equal to the geometric genus.

---

§5 Counting points on curves over finite fields

Let C be a nonsingular projective curve of genus g defined over a finite field k with q elements. Denote by p the characteristic of k and by $\overline{N}_n$ the number of k_n-rational points on $C(\bar{k})$. We have discussed in Chapter 2 that the formal power series $\sum_{n\geq 1} \overline{N}_n U^{n-1} = Z'_C(V)/Z_C(V)$, where $Z_C(V)$ is the zeta function of C. According to the Weil conjectures, $Z_C(V)$ has the form

$$Z_C(V) = \frac{P_1(U)}{(1-U)(1-qU)},$$

where $P_1(U) = \prod_{i=1}^{2g}(1-\omega_i U)$ is a polynomial of degree $2g$ satisfying the functional equation $P_1(U) = \pm(q^{\frac{1}{2}}U)^{2g} P_1(\frac{1}{qU})$ and $|\omega_i| = q^{\frac{1}{2}}$ for $i = 1, \cdots, 2g$ (Riemann hypothesis for C). We shall prove the Riemann hypothesis in this section as a consequence of the Riemann-Roch theorem while leaving the proof of the remaining statements to the next chapter. As the proof of the analytic behavior of $P_1(U)$ does not make use of the size of ω_i, we shall actually prove $|\omega_i| = q^{\frac{1}{2}}$ assuming the functional equation.

It was an exercise in Chapter 2 that $|\omega_i| = q^{\frac{1}{2}}$ for $i = 1, \cdots, 2g$ is equivalent to

$$|\overline{N}_n - q^n - 1| \leq 2gq^{n/2} \qquad \text{for} \quad n \geq 1.$$

Observe that the above inequality can be relaxed:

Lemma 1. *$|\omega_i| = q^{\frac{1}{2}}$ for $i = 1, \cdots, 2g$ if and only if there exists a number $M > 0$ such that $|\overline{N}_{2n} - q^{2n} - 1| \leq Mq^{2n/2}$ for n sufficiently large.*

Proof. The necessity is clear. We prove sufficiency. Call $a_n = \overline{N}_n - q^n - 1$. From the definition of $Z_C(U)$ it follows that $-a_n = \omega_1^n + \cdots + \omega_{2g}^n$ for $n \geq 1$. Therefore $\sum_{n=1}^{\infty} a_{2n}U^{2n} = \sum_{i=1}^{2g} \frac{\omega_i^2 U^2}{1-\omega_i^2 U^2}$. The estimate on a_{2n} for n large implies that the power series converges absolutely to a holomorphic function on $|U| < q^{-1/2}$, which implies that the rational function has poles outside the disc $|U| < q^{-1/2}$, i.e., $|\omega_i| \leq q^{1/2}$ for $i = 1, \cdots, 2g$. This in turn implies $|\omega_i| \geq q^{1/2}$ by the functional equation. Therefore $|\omega_i| = q^{1/2}$ for $i = 1, \cdots, 2g$. □

Thus we may assume that q is an even power of p and q is large. It suffices to show that, as a function of q,

$$\overline{N}_1 = q + O(q^{1/2}).$$

To prove this, we follow Stepanov's idea. Let x_0 be a k-rational point of C. (Replace k by a finite extension, if necessary, to guarantee its existence). If we can construct a rational function f on C which vanishes at all other k-rational points of C of order $\geq m$, then from $m(\overline{N}_1 - 1) \leq$ #of zeros of f = # of poles of f, we conclude that

$$\overline{N}_1 \leq 1 + \frac{1}{m}(\#\text{ of poles of } f).$$

If, further, f does not have too many poles, then we get a useful upper bound for $\overline{N}_1$, which essentially is of the same strength as $\overline{N}_1 = q+O(q^{1/2})$. This construction was given by Stepanov and by W. Schmidt separately, but the proofs were rather involved. Here we adopt Bombieri's much simpler method [1].

The k-rational points on C are the fixed points of the Frobenius morphism $\varphi : C(\overline{k}) \to C(\overline{k})$. For convenience, we consider rational functions on C over $\overline{k}$ although we only need to consinder those over a finite extension of k. The Riemann-Roch theorem holds for such rational functions with the same statements. Given a rational function f on C, $f \circ \varphi$ is a q-th power of a rational function, and $\text{div}\,(f \circ \varphi) = q \cdot \varphi^{-1}(\text{div}\, f)$. (For example, if $f(x) = x - a$, then $f \circ \varphi(x) = x^q - a = (x - a^{1/q})^q$.) Write

$L_m = L(mx_0)$

$=$ the space over $\overline{k}$ of the $\overline{k}$ - rational functions f on $C(\overline{k})$ such that $\text{div}\, f + mx_0 \geq 0$

$= \{$the k-rational functions $C(\overline{k})$ such that $\text{div}\, f + mx_0 \geq 0\} \otimes_k \overline{k}$.

Since $\varphi(x_0) = x_0$, we have $L_m \circ \varphi \subseteq L_{qm}$; also $\dim L_m \circ \varphi = \dim L_m$ since φ is surjective. If A, B are subspaces of L_m, L_n, respectively, denote by AB the subspace of L_{m+n} spanned by fg with $f \in A$, $g \in B$. Let $L_m^{(p^\mu)} = \{f^{p^\mu} : f \in L_m\} \subset L_{mp^\mu}$. Note that $\dim L_m^{(p^\mu)} = \dim L_m$.

Lemma 2. *If $\ell p^\mu < q$, the natural homomorphism $L_\ell^{(p^\mu)} \underset{\bar{k}}{\otimes} (L_m \circ \varphi) \to L_\ell^{(p^\mu)}(L_m \circ \varphi)$ is an isomorphism. Consequently,* $\dim L_\ell^{(p^\mu)}(L_m \circ \varphi) = (\dim L_\ell)(\dim L_m)$.

Proof. Observe that if $f, g \in L_{i+1} \setminus L_i$, then there is a constant $\alpha \in \bar{k}$ such that $f - \alpha g \in L_i$. This shows that $\dim L_{i+1} \leq \dim L_i + 1$. (And the Riemann-Roch theorem says that $\dim L_{i+1} = \dim L_i + 1$ if $i > 2g - 2$.) Hence there is a basis $s_1, s_2, \cdots, s_r$, of L_m such that $\operatorname{ord}_{x_0} s_1 < \operatorname{ord}_{x_0} s_2 < \cdots < \operatorname{ord}_{x_0} s_r$. In order to prove the lemma, it suffices to show that if $f_i \in L_\ell$ for $i = 1, \cdots, r$ are such that $\sum_{i=1}^{r} f_i^{p^\mu} \cdot s_i \circ \varphi = 0$, then $f_1 = \cdots = f_r = 0$. Suppose otherwise, say, $\sum_{i=j}^{r} f_i^{p^\mu} \cdot s_i \circ \varphi = 0$ and $f_j \neq 0$. We find

$$\begin{aligned}\operatorname{ord}_{x_0} f_j^{p^\mu} \cdot s_j \circ \varphi &= \operatorname{ord}_{x_0} \left(- \sum_{i=j+1}^{r} f_i^{p^\mu} \cdot s_i \circ \varphi \right) \\ &\geq \min_{j+1 \leq i \leq r} \operatorname{ord}_{x_0} \left(f_i^{p^\mu} \cdot s_i \circ \varphi \right) \\ &\geq -\ell p^\mu + q \operatorname{ord}_{x_0} s_{j+1}\end{aligned}$$

since $\operatorname{ord}_{x_0} f_i^{p^\mu} = p^\mu \operatorname{ord}_{x_0} f_i \geq -\ell p^\mu$ and $\operatorname{ord}_{x_0}(s_i \circ \varphi) = q \, ord_{x_0} s_i$, and s_{j+1} has lowest order at x_0 among $s_{j+1}, \cdots, s_r$. This then implies

$$p^\mu \operatorname{ord}_{x_0} f_j \geq -\ell p^\mu + q(\operatorname{ord}_{x_0} s_{j+1} - \operatorname{ord}_{x_0} s_j) \geq -\ell p^\mu + q > 0$$

by assumption. Hence $f_j = 0$, contradicting our assumption. □

Now we are ready to prove the following upper bound of $\overline{N}_1$.

Theorem 7. *Assume $q = p^\alpha$ with α an even integer and $q > (g+1)^4$. We have*

$$\overline{N}_1 < q + (2g+1)q^{1/2} + 1.$$

Proof. Assume $\ell p^\mu < q$, where ℓ and μ are to be specified later. By Lemma 2, the map $\delta : L_\ell^{(p^\mu)}(L_m \circ \varphi) \to L_\ell^{(p^\mu)} L_m \subseteq L_{\ell p^\mu + m}$ given by $\sum_{i=1}^{r} f_i^{p^\mu}(s_i \circ \varphi) \mapsto \sum_{i=1}^{r} f_i^{p^\mu} s_i$ is a well-defined homomorphism. By the Riemann-Roch theorem, $\dim L_i \geq i + 1 - g$, and the equality holds if $i > 2g - 2$. Thus, by Lemma 2,

$$\begin{aligned}\dim \ker \delta &\geq (\dim L_\ell)(\dim L_m) - \dim L_{\ell p^\mu + m} \\ &\geq (\ell + 1 - g)(m + 1 - g) - (\ell p^\mu + m + 1 - g) \qquad \text{if } \ell, m \geq g.\end{aligned}$$

Suppose $\ker \delta \neq \{0\}$. Let $f = \sum_{i=1}^{r} f_i^{p^\mu}(s_i \circ \varphi)$ be a nonzero element in $\ker \delta$. If $x \neq x_0$ is a k-rational point of C, then $\varphi(x) = x$ and

$$f(x) = \sum_{i=1}^{r} f_i(x)^{p^\mu} s_i(\varphi(x)) = \sum_{i=1}^{r} f_i(x)^{p^\mu} s_i(x) = (\delta f)(x) = 0.$$

This shows that f vanishes at all k-rational points of C except x_0. Further, since $s_i \circ \varphi$ is a q-th power, so f is a p^μ-th power, and hence f has at least $(\overline{N}_1 - 1)p^\mu$ zeros. On the other hand, f lies in $L_\ell^{(p^\mu)}(L_m \circ \varphi) \subseteq L_{\ell p^\mu + qm}$, hence it has at most $\ell p^\mu + qm$ poles.

We conclude that if $\ell p^\mu < q, \ell, m \geq g$ and $\dim \ker \delta > 0$, i.e., if $(\ell + 1 - g)(m + 1 - g) > (\ell p^\mu + m + 1 - g) > (\ell p^\mu + m + 1 - g)$, then

$$\overline{N}_1 \leq 1 + \frac{1}{p^\mu}(\ell p^\mu + qm) = \ell + 1 + qm/p^\mu.$$

Under the condition $q = p^\alpha, \alpha$ even, and $q > (g+1)^4$, we may choose

$$\mu = \alpha/2, \qquad m = p^\mu + 2g = \sqrt{q} + 2g, \qquad \ell = \left[\frac{g}{g+1}\sqrt{q}\right] + g + 1,$$

and obtain the desired conclusion. □

Next we show how to use Theorem 7 to derive $\overline{N}_1 = q + O(q^{1/2})$. The field $K = k(C)$ is a separable finite degree extension of a rational function field $F = k(T)$. Let L be the Galois closure of K in an algebraic closure of F. Denot by C' the projective nonsingular curve over k such that $L = k(C')$ is the field of k-rational functions on C'. Geometrically, we have $C' \to C \to \mathbf{P}^1$, where C' is a Galois cover of $\mathbf{P}^1$ with Galois group $G = Gal(L/K)$, and C' is Galois over C with Galois group $H \subseteq G$. We are concerned with the k-rational points on C, they correspond to the degree 1 places of K. Let w be such a place (or point) and let v be the place of F divisible by w. Since $k = k_w \supseteq k_v$, this shows that v is a place of F of degree 1. Geometrically, this means that the k-rational points of C are over the k-rational points of $\mathbf{P}^1$. Denote by S the preimage in C' of the k-rational points on $\mathbf{P}^1$. Given $x \in \mathbf{P}^1(k)$, the Galois group G acts on the fibre of x in C' transitively, the Frobenius morphism φ on C' also acts on the fibre of x. Given y in the fibre of x, we have $\varphi y = \eta(y)$ for some $\eta \in G$. If the cover is unramified at x, then there are $|G|$ points on the fibre of x and η is unique. Denote by $S(\eta) = \{y \in S : \varphi(y) = \eta(y)\}$, for $\eta \in G$. Note that when η is the identity element in $G, S(\eta)$ is the set of k-rational points on C'. Using the same argument as in the proof of Theorem 7, but with δ replaced by $\delta_\eta : L_\ell^{(p^\mu)}(L_m \circ \varphi) \to L_\ell^{(p^\mu)}(L_m \circ \eta)$, we obtain the same estimate:

$$|S(\eta)| \leq q + (2g' + 1)\sqrt{q} + 1$$

where g' is the genus of C'. On the other hand, by Hurwitz theorem,

$$\sum_{\eta \in G} |S(\eta)| = |G||\mathbf{P}^1(k)| + O(1).$$

Here $O(1)$ takes care of the ramification points. Since $|\mathbf{P}^1(k)| = q + 1$, comparison of the last two inequalities yields

$$|S(\eta)| = q + O(\sqrt{q}) \quad \text{for each} \quad \eta \in G.$$

The points of C are the H-orbits on C' and the k-rational points on C are those orbits Hx such that $Hx = H\varphi(x)$, i.e., $\varphi(x) = \eta(x)$ for some $\eta \in H$. As remarked above, these points x are in S. Hence $\bigcup_{\eta \in H} S(\eta)$ are the points in C' lying over the k-rational points in C. By Hurwitz theorem again,

$$\sum_{\eta \in H} |S(\eta)| = |H|\overline{N}_1 + O(1),$$

hence we get

$$\overline{N}_1 = q + O(q^{1/2}).$$

We have shown

Theorem 8. *Let C be a nonsingular projective curve of genus g defined over a finite field k with q elements. Suppose that the zeta function $Z_C(U)$ has the form*

$$Z_C(U) = \frac{P_1(U)}{(1-U)(1-qU)}$$

where $P_1(U) = \prod_{i=1}^{2g}(1 - w_i U)$ and $Z_C(U)$ satisfies the functional equation

$$Z_C(U) = \pm \left(q^{\frac{1}{2}}U\right)^{2g-2} Z_C\left(\frac{1}{qU}\right).$$

Then $|w_i| = q^{1/2}$ for $i = 1, \cdots, 2g$, that is, the "Riemann hypothesis" for C holds.

References

[1] E. Bombieri: Counting points on curves over finite fields [d'après S. A. Stepanov], Lecture notes in Math. 383, p.p. 234-241, Springer-Verlag, Berlin-Heidelberg-New York (1974).

[2] S. Iyanaga: The Theory of Numbers. North-Holland, Amsterdam - Oxford (1969).

[3] J.-P. Serre: Algebraic Groups and Class Fields, Springer-Verlag, GTM 117, New York(1988).

[4] J. Tate: Fourier analysis in number fields and Hecke's zeta-functions. Thesis, Princeton University 1950. Published in Algebraic Number Theory, J.W.S. Cassels and A. Fröhlich etal, Thompson, Washington D.C. (1967), republished by Academic Press, London.

[5] A. Weil: Basic Number Theory, Springer-Verlag, Berlin-Heidelberg-New York (1973).

CHAPTER 5

Zeta and L–functions

§1 *L*–functions of idèle class characters

Throughout this chapter we fix a function field K of genus g with the field of constants k containing q elements. A quasi-character χ of the group of idèles I_K is a continuous homomorphism from I_K to $\mathbf{C}^\times$. It is called an idèle class quasi-character if it is trivial on $K^\times$. For example, the $|\ |$ defined in Chapter 3, §3 via $|x| = \prod_v |x_v|_v$ for $x = (x_v) \in I_K$ is an idèle class quasi-character because of the product formula (Chapter 3, Theorem 6). Thus for any complex number s, $|\ |^s$ is also an idèle class quasi-character. The relation between quasi-characters and characters of $I_K/K^\times$ is explained in

Proposition 1. *Let χ be a quasi-character of the idèle class group $I_K/K^\times$. Then there exists a complex number s_0 such that $\chi_1 = \chi|\ |^{-s_0}$ is a character of finite order of the idèle class group $I_K/K^\times$.*

Proof. Recall the degree map from I_K to $\mathbf{Z}$ which sends an idèle x to deg(div x). The quasi-character $|\ |$ can also be described as

$$|x| = q^{-\deg \operatorname{div} x} \quad \text{for } x \in I_K.$$

The image of I_K under the degree map is an infinite cyclic subgroup generated by r, say. (In fact, $r = 1$ as we shall see in §4.) Let x_0 be an idèle with $\deg\operatorname{div} x_0 = r$. There exists a complex number s_0 such that $\chi(x_0) = |x_0|^{s_0}$. Then $\chi_1 = \chi|\ |^{-s_0}$ is a quasi-character of $I_K/K^\times$ and it is trivial on the infinite cyclic group $\langle x_0\rangle$ generated by x_0. As observed before, the kernel of the degree map is I_K^1, in other words, $I_K = I_K^1 \cdot \langle x_0\rangle$, the value group of χ_1 on I_K is $\chi(I_K^1/K^\times)$. But $I_K^1/K^\times$ is compact by Theorem 8 in Chapter 3, this implies that the image of χ_1 is contained in S^1, and χ_1 is a character of $I_K/K^\times$ of finite order. □

We have seen in Chapter 4, §1 that if χ is a quasi-character of I_K with its restriction to $K_v^\times$ being χ_v, then χ_v is trivial on $\mathcal{U}_v$ for almost all v. If χ_v is not trivial on $\mathcal{U}_v$, then it is trivial on a neighborhood $\mathcal{P}_v^{n_v}$ of 1. The smallest positive n_v is called the exponent of the conductor of χ_v, and we say that χ and χ_v are ramified at v. When χ_v is trivial on $\mathcal{U}_v$, put $n_v = 0$, and we say that χ and χ_v are unramified at v. Thus $f(\chi) = \sum_v n_v v$ is a divisor of K, called the *conductor* of χ.

Exercise 1. Describe all unramified characters of $I_K/K^\times$. Show that they form a group isomorphic to $\mathrm{Jac}(K) \times S^1$.

At a place v where χ is unramified, χ_v is determined by its value at a uniformizer π_v, and $\chi_v(\pi_v)$ is independent of the choice of π_v. Define

$$L(s, \chi_v) = \left(1 - \chi_v(\pi_v)(Nv)^{-s}\right)^{-1},$$

where $Nv = q^{\deg v}$ is the cardinality of the residue field of K_v. At the places v where χ is ramified, put $L(s, \chi_v) = 1$. Let

$$L(s, \chi) = \prod_v L(s, \chi_v),$$

it is called the L–function attached to the quasi-character χ. Observe that

$$L(s, \chi|\ |^{s_0}) = L(s + s_0, \chi),$$

hence we shall assume that χ is an idèle class character and study the analytic bahavior of $L(s, \chi)$.

Notice that when χ is the trivial character χ_0 of $I_K/K^\times$,

$$L(s, \chi_0) = \prod_v (1 - Nv^{-s})^{-1} =: \zeta_K(s) = Z_C(q^{-s}),$$

where $Z_C(U)$ is the zeta function attached to the projective nonsingular curve C so that K is the field of k–rational functions on C, as defined in Chapter 2, §4. There we have seen that $L(s, \chi_0)$ converges absolutely for Re $s \gg 0$. Hence for any character χ of $I_K/K^\times$, $L(s, \chi)$ converges absolutely to a holomorphic function on a right half plane which tends to 1 as Re $s \longrightarrow \infty$. The purpose of this chapter is to prove the following two theorems.

Theorem 1. *Let C be a nonsingular irreducible projective curve of genus g defined over a finite field k of q elements. Then the zeta function $Z_C(U)$ of C defined by the Euler product*

$$Z_C(U) = \prod_{\substack{k\text{–closed points} \\ v \text{ of } C}} (1 - U^{\deg v})^{-1},$$

which converges absolutely to a holomorphic function for $|U| < q^{-1}$, is in fact a rational function in U of the form

$$Z_C(U) = \frac{P_1(U)}{(1 - U)(1 - qU)},$$

where $P_1(U)$ is a polynomial of degree $2g$ with coefficients in $\mathbf{Z}$ such that

$$P_1(U) = (q^{\frac{1}{2}}U)^{2g} P_1(\frac{1}{qU}),$$

$P_1(0) = 1$ *and* $P_1(1) = h = |Jac(K)|$.

Thus Theorem 1 above and Theorem 8 in Chapter 4 together prove Weil conjectures for curves.

Theorem 2. *Let χ be a character of the idèle class group $I_K/K^\times$ and $\chi \neq |\ |^t$ for any $t \in \mathbf{C}$. Then the L-function defined above by Euler product converges absolutely to a holomorphic function for* $\mathrm{Re}\ s > 1$. *Further, $L(s,\chi) = P(q^{-s},\chi)$, where $P(U,\chi)$ is a polynomial of degree $2g-2+\deg f(\chi)$ satisfying $P(0,\chi) = 1$ and the functional equation*

$$P(U,\chi) = c(\chi)(q^{\frac{1}{2}}U)^{2g-2+\deg f(\chi)} P(\frac{1}{qU}, \chi^{-1}),$$

or equivalently,

$$L(s,\chi) = c(\chi)\, q^{(\frac{1}{2}-s)(2g-2+\deg f(\chi))} L(1-s, \chi^{-1}).$$

Here $f(\chi)$ is the conductor of χ and $c(\chi)$ is a constant.

Exercise 2. Show that the zeta function of an elliptic curve E over k has the form

$$Z_E(u) = \frac{1 + au + qu^2}{(1-u)(1-qu)}$$

where a is an integer with $|a| \leq 2\sqrt{q}$. (Recall that an elliptic curve has genus 1.)

The group of quasi-characters of the idèle class group, denoted by $\mathcal{A}(I_K)$, can be endowed with an analytic structure so that the connected component of a quasi-character χ consists of $\chi|\ |^s$, $s \in \mathbf{C}/\frac{2\pi i}{\log q}\mathbf{Z}$, which is a cylinder. The connected components of $\mathcal{A}(I_K)$ are parametrized by the characters of $I_K^1/K^\times$. The L-function above can be regarded as a holomorphic function defined on a certain right half-cylinder of each connected component of $\mathcal{A}(I_K)$. Theorems 1 and 2 say that it has a meromorphic continuation to the whole manifold $\mathcal{A}(I_K)$, which is holomorphic everywhere except on the connected component of the trivial character χ_0, where it has simple poles at χ_0 and $\chi_0|\ |$. Further L satisfies the functional equation

$$L(\chi) = \tilde{c}(\chi)L(|\ |\chi^{-1}),$$

where $\tilde{c}(\chi)$ is an exponential function on the component of χ.

§2 Fourier transforms

To prove the above two theorems, we shall follow the philosophy in Tate's thesis [1], using Fourier analysis on the adèlic groups. Denote by dx the additive Haar measure on A_K such that the induced measure $\overline{dx}$ on A_K/K has total volume 1. Fix the standard additive character ψ on A_K as defined in §2 of the previous chapter, and identify the dual $\widehat{A}_K$ of A_K with A_K using ψ. We shall consider Fourier transforms on the space $\mathcal{S}(A_K)$ of Schwartz functions on A_K, these are locally constant functions on A_K with compact support. Clearly, $\mathcal{S}(A_K)$ is generated by the characteristic functions of translations of the parallelotopes P_a of size a, for $a \in I_K$. With the chosen dx we define the Fourier transform $\widehat{\Phi}$ of $\Phi \in \mathcal{S}(A_K)$ to be

$$\widehat{\Phi}(\eta) = \int_{A_K} \Phi(x)\eta(x)dx \quad \text{for } \eta \in \widehat{A}_K,$$

or using the identification $\widehat{A}_K \cong A_K$, we may regard $\widehat{\Phi}$ as a function on A_K :

$$\widehat{\Phi}(y) = \widehat{\Phi}(\psi^y) = \int_{A_K} \Phi(x)\psi^y(x)dx = \int_{A_K} \Phi(x)\psi(yx)dx, \quad y \in A_K.$$

Proposition 2. *(i) The Fourier transform of a Schwartz function is again a Schwartz function.*

(ii) Given $\Phi \in \mathcal{S}(A_K)$, we have $\hat{\hat{\Phi}}(x) = \Phi(-x)$ for $x \in A_K$.

Proof. (i) We may assume $\Phi \in \mathcal{S}(A_K)$ is the characteristic function of $x_0 + P_a$ for some $x_0 \in A_K$ and $a \in I_K$. Then

$$\widehat{\Phi}(y) = \int_{A_K} \Phi(x)\psi(yx)dx = \int_{P_a} \Phi(x_0 + x)\psi(y(x_0 + x))dx = \psi(yx_0)\int_{P_a} \psi(yx)dx.$$

Since

$$\int_{P_a} \psi(yx)dx = \begin{cases} \mathrm{vol}(P_a) & \text{if } y \in P_a^{\perp}, \\ 0 & \text{if } y \notin P_a^{\perp}, \end{cases}$$

and ψ is locally constant, we see immediately that $\widehat{\Phi}$ is a Schwartz function.

(ii) Given $\Phi \in \mathcal{S}(A_K)$, we know from (i) that $\widehat{\Phi} \in \mathcal{S}(A_K)$. Suppose that $\Phi, \widehat{\Phi}$ are supported within parallelotopes P_a, P_b, respectively. Let $z \in A_K$. By definition,

$$\begin{aligned}\hat{\hat{\Phi}}(z) &= \int_{P_b} \widehat{\Phi}(y)\psi(yz)dy = \int_{P_b}\int_{P_a} \Phi(x)\psi(xy)dx\psi(yz)dy \\ &= \int_{P_a} \Phi(x)\int_{P_b} \psi((x+z)y)dydx.\end{aligned}$$

We may assume that P_b is large so that $P_b^\perp$ is contained in P_a. If $z \notin P_a$, then $x + z \notin P_b^\perp$ and the integral over P_b is zero so that $\hat{\hat{\Phi}}(z) = 0$. If $z \in P_a$, then the integral over P_b is zero unless $x \in -z + P_b^\perp$, in which case the integral over P_b is the volume of P_b. So for $z \in P_a$,

$$\hat{\hat{\Phi}}(z) = \text{vol}(P_b) \int_{-z+P_b^\perp} \Phi(x)dx.$$

We can further assume that P_b is so large that Φ is constant on $-z + P_b^\perp$. This shows that, for $z \in P_a$,

$$\hat{\hat{\Phi}}(z) = \text{vol}(P_b)\ \text{vol}(P_b^\perp)\Phi(-z) = \Phi(-z)\,\text{vol}(P_1)\ \text{vol}(P_1^\perp).$$

It was shown in §4 of the previous chapter that $\text{vol}(A_K/K) = q^{g-1}\,\text{vol}(P_1)$. We choose dx with $\text{vol}(A_K/K) = 1$, hence $\text{vol}(P_1) = q^{1-g}$. Then

$$\text{vol}(P_1^\perp) = \text{vol}(P_c) = |c|\,\text{vol}(P_1) = q^{2g-2}q^{1-g} = q^{g-1}$$

since $\deg(\text{div}\, c^{-1}) = \deg \mathcal{K} = 2g - 2$ by Corollary 2 in Chapter 4, §4. This proves that $\hat{\hat{\Phi}}(z) = \Phi(-z)$ for $z \in A_K$. □

Because of property (ii), we say that the measure dx is self-dual, it is called the Tamagawa measure on A_K. Denote by Φ_a the characteristic function of P_a. It follows from the proof above that $\widehat{\Phi}_a = |a|q^{1-g}$ times the characteristic function of $P_a^\perp$. Write $P_a^\perp = P_{ca^{-1}}$. We may express the characteristic function of P_c as $\Phi_1(c^{-1}x)$. Noting that $|c^{-1}| = q^{2-2g}$, we have shown

Corollary 1. *Let Φ_a denote the characteristic function of P_a. Then*

$$\widehat{\Phi}_a = |a||c^{-1}|^{1/2}\Phi_{ca^{-1}},$$

where c is any idèle such that $P_1^\perp = P_c$. In particular, $\widehat{\Phi}_1(x) = |c^{-1}|^{1/2}\Phi_1(c^{-1}x)$.

c^{-1} is called a *differental idèle* of K.

From the proof of Proposition 2 we also have

Corollary 2. *Let Φ be the characteristic function of $x_0 + P_a$. Then $\widehat{\Phi} = \psi^{x_0}\widehat{\Phi}_a = |a||c^{-1}|^{1/2}\psi^{x_0}\Phi_{ca^{-1}}$, where c is as in Corollary 1.*

Given a Schwartz function Φ on A_K, the sum $\sum_{\alpha\in K} \Phi(x+\alpha)$ is finite for any $x \in A_K$ since Φ has compact support and K is discrete, thus it defines a function

$$h_\Phi(x) = \sum_{\alpha\in K} \Phi(x+\alpha), \quad x \in A_K/K,$$

which is locally constant. The dual of A_K/K is $\widehat{A_K/K} = K^\perp = \{\psi^\beta : \beta \in K\} \cong K$. The Fourier transform of h_Φ is a function $\widehat{h}_\Phi$ on $\widehat{A_K/K}$ defined by

$$\widehat{h}_\Phi(\beta) = h_\Phi(\psi^\beta) = \int_{A_K/K} h_\Phi(x)\psi(\beta x)\overline{dx}.$$

Here $\overline{dx}$ is the measure on A_K/K induced from dx on A_K. Replacing h_Φ by its definition and using the fact that ψ is trivial on K, we find

$$\widehat{h}_\Phi(\beta) = \int_{A_K/K} \sum_{\alpha\in K} \Phi(x+\alpha)\psi(\beta(x+\alpha))\overline{dx} = \int_{A_K} \Phi(x)\psi(\beta x)dx = \widehat{\Phi}(\beta).$$

Since $\widehat{\Phi}$ has compact support, there is a parallelotope P_b such that $\widehat{h}_\Phi(\beta) = 0$ if $\beta \notin K \cap P_b$. Regarding $\widehat{h}_\Phi(\beta)$ as Fourier coefficients of h_Φ, we show

Proposition 3. *(Fourier inversion formula) Let Φ be a Schwartz function on A_K, define h_Φ as above. Then $h_\Phi(x) = \sum_{\beta\in K} \widehat{h}_\Phi(\beta)\psi(-\beta x)$ for $x \in A_K/K$. In particular, we have*

$$\sum_{\alpha\in K} \Phi(\alpha) = h_\Phi(0) = \sum_{\beta\in K} \widehat{h}_\Phi(\beta) = \sum_{\beta\in K} \widehat{\Phi}(\beta).$$

Proof. Given $x \in A_K/K$, choose P_b large enough such that $\widehat{h}_\Phi(\beta) = 0$ for $\beta \notin K\cap P_b$ and $h_\Phi(x + P_b^\perp) = h_\Phi(x)$. We start with the right hand side :

$$\sum_{\beta\in K} \widehat{h}_\Phi(\beta)\psi(-\beta x) = \sum_{\beta\in K\cap P_b} \int_{A_K/K} h_\Phi(z)\psi(\beta z)\psi(-\beta x)\overline{dz}$$
$$= \int_{A_K/K} h_\Phi(z) \sum_{\beta\in K\cap P_b} \psi(\beta(z-x))\overline{dz}.$$

Here

$$\sum_{\beta\in K\cap P_b} \psi(\beta(z-x)) = \begin{cases} |K\cap P_b| & \text{if } z-x \in (K\cap P_b)^\perp, \\ 0 & \text{if } z-x \notin (K\cap P_b)^\perp. \end{cases}$$

Clearly, $(K\cap P_b)^\perp \supseteq K^\perp + P_b^\perp$. Thus $K\cap P_b \subseteq (K^\perp + P_b^\perp)^\perp \subseteq K^{\perp\perp}\cap P_b^{\perp\perp} = K\cap P_b$, which implies $K\cap P_b = (K^\perp + P_b^\perp)^\perp$ whence $(K\cap P_b)^\perp = K^\perp + P_b^\perp = K + P_{c^{-1}b}$. Using the same notation as in Chapter 4, §4, we write $|K\cap P_b| = q^{\lambda(b)}$. Hence

$$\sum_{\beta\in K} \widehat{h}_\Phi(\beta)\psi(-\beta x) = q^{\lambda(b)} \int_{K+P_{cb^{-1}}/K} h_\Phi(x)\overline{dz}$$
$$= h_\Phi(x)q^{\lambda(b)} \operatorname{vol}\left((K + P_{cb^{-1}})/K\right) = h_\Phi(x)$$

since

$$\operatorname{vol}\left((K + P_{cb^{-1}})/K\right) = \operatorname{vol}(P_{cb^{-1}}/P_{cb^{-1}} \cap K) = |cb^{-1}| \operatorname{vol}(P_1)/q^{\lambda(cb^{-1})}$$
$$= q^{2g-2}|b^{-1}|q^{1-g}/q^{\lambda(cb^{-1})} = q^{-\lambda(cb^{-1})+g-1+\deg \operatorname{div} b} = q^{-\lambda(b)}$$

by the Riemann–Roch Theorem (Theorem 5, Chapter 4). This proves the inversion formula. □

Remark. As we have seen, the essence of the above proof is the Riemann–Roch theorem. In particular, if Φ is the characteristic function of P_a, i.e., $\Phi = \Phi_a$, then $\widehat{\Phi} = |a||c^{-1}|^{1/2}\Phi_{ca^{-1}}$, by Corollary 1, and the formula $\sum_{\alpha\in K} \Phi_a(\alpha) = \sum_{\beta\in K} \widehat{\Phi}_a(\beta)$ is nothing but the Riemann–Roch Theorem. It is also the Poisson summation formula since $K^{\perp} = K$.

We shall formulate the Poisson summation formula in a more general form ready for future use :

Theorem 3. *(Poisson summation formula) Let Φ be a Schwartz function on A_K. Then for all $x \in I_K$ we have*

$$\sum_{\alpha\in K} \Phi(x\alpha) = |x|^{-1} \sum_{\beta\in K} \widehat{\Phi}(x^{-1}\beta).$$

Proof. Let $g(y) = \Phi(xy)$. Then g is also a Schwartz function on A_K. Its Fourier transform is

$$\begin{aligned}\widehat{g}(z) &= \int_{A_K} g(y)\psi(yz)dy = \int_{A_K} \Phi(xy)\psi(yz)dy \\ &= \int_{A_K} \Phi(y')\psi(x^{-1}y'z)|x|^{-1}dy' = |x|^{-1}\widehat{\Phi}(x^{-1}z)\end{aligned}$$

since $dy' = d(xy) = |x|dy$. Thus the desired formula becomes $\sum_{\alpha\in K} g(\alpha) = \sum_{\beta\in K} \widehat{g}(\beta)$, which was established in Proposition 3. □

---

§3 Analytic continuation and functional equation for $Z(s, \chi, \Phi)$

At each place v of K, choose the Haar measure $d^{\times}x_v$ on $K_v^{\times}$ so that $\mathcal{U}_v$ has volume 1. Denote by $d^{\times}x = \prod_v' d^{\times}x_v$ the measure on I_K.

Lemma 1. *Let χ_v be an unramified character of $K_v^{\times}$ and Φ_v be the characteristic function of $\mathcal{O}_v$. Then*

$$\int_{K_v^{\times}} \Phi_v(x_v)\chi_v(x_v)|x_v|^s d^{\times}x_v = \left(1 - \chi_v(\pi_v)Nv^{-s}\right)^{-1} = L(s, \chi_v)$$

for Re $s > 0$, *where* π_v *is a uniformizer of* K_v.

Proof. Since Φ_v is zero outside $\mathcal{O}_v$, the above integral is actually over $\mathcal{O}_v - \{0\} = \bigcup_{n\geq 0} \mathcal{U}_v\pi_v^n$. On each set $\mathcal{U}_v\pi_v^n$, the function $\Phi_v(x)\chi_v(x)|x|^s$ is constant, equal to $\chi_v(\pi_v)^n|\pi_v|^{ns} = \chi_v(\pi_v)^n Nv^{-ns}$. As $\mathcal{U}_v\pi_v^n$ has volume 1, we get

$$\int_{K_v^\times} \Phi_v(x_v)\chi_v(x_v)|x_v|^s d^\times x_v = \sum_{n=0}^{\infty} \chi_v(\pi_v)^n Nv^{-ns} = \left(1 - \chi_v(\pi_v)Nv^{-s}\right)^{-1}$$

for Re $s > 0$, which is equal to $L(s, \chi_v)$ by definition. □

Given a Schwartz function Φ and a quasi-character χ of the idèle class group $I_K/K^\times$, define the zeta function

$$Z(s, \chi, \Phi) = \int_{I_K} \Phi(x)\chi(x)|x|^s d^\times x.$$

The integral converges absolutely for Re s large. Indeed, there is a finite set S of places of K such that at each place v outside S, χ_v is unramified and $\Phi = \Phi_S\Phi^S$, where Φ_S is a locally constant function with compact support in $\prod_{v\in S} K_v^\times$ and $\Phi^S = \prod_{v\notin S} \Phi_v$ is the product of characteristic function of $\mathcal{O}_v$ over $v \notin S$. Thus

$$\begin{aligned} &Z(s, \chi, \Phi) \\ &= \int_{\prod_{v\in S} K_v^\times} \Phi_S(x)\Big(\prod_{v\in S}\chi_v\Big)(x)|x|^s\Big(\prod_{v\in S} d^\times x_v\Big) \prod_{v\notin S} \int_{K_v^\times} \Phi_v(x_v)\chi_v(x_v)|x_v|_v^s d^\times x_v \\ &= \int_{\prod_{v\in S} K_v^\times} \Phi_S(x)\Big(\prod_{v\in S}\chi_v\Big)(x)|x|^s\Big(\prod_{v\in S} d^\times x_v\Big) \prod_{v\notin S} L(s, \chi_v) \end{aligned}$$

by Lemma 1. The first integral converges absolutely for Re s large and so does the infinite product since it is equal to $L(s, \chi) \cdot \prod_{v\in S} L(s, \chi_v)^{-1}$, where $L(s, \chi)$ converges absolutely for Re s large and $L(s, \chi_v)^{-1}$ is holomorphic in s for all $v \in S$. Observe that $Z(s, \chi|\ |^{s_0}, \Phi) = Z(s + s_0, \chi, \Phi)$. Hence we may assume that either $\chi = \chi_0$ is the trivial character or else $\chi \neq |\ |^t$ for all $t \in \mathbf{C}$. The purpose of this section is to obtain analytic continuation and a functional equation for $Z(s, \chi, \Phi)$.

Denote by r the positive generator of $\deg(I_K)$, and by I_{mr} the set of idèles of degree mr. Then $I_K = \bigcup_{m\in\mathbf{Z}} I_{mr}$, $K^\times I_{mr} = I_{mr}$ and $I_{mr} = I_K^1 x_0^m$, where x_0 is an idèle of degree r. The zeta function $Z(s, \chi, \Phi)$ is the sum of the integrals over I_{mr}. Over I_{mr}, the integral satisfies a functional equation :

Lemma 2.

$$\int_{I_{mr}} \Phi(x)\chi(x)|x|^s d^\times x + \kappa_\chi \Phi(0) q^{-mrs}$$
$$= \int_{I_{-mr}} \widehat{\Phi}(x)\chi^{-1}(x)|x|^{1-s} d^\times x + \kappa_\chi \widehat{\Phi}(0) q^{mr(1-s)},$$

where

$$\kappa_\chi = \kappa_{\chi^{-1}} = \begin{cases} \mathrm{vol}(I_K^1/K^\times) & \textit{if } \chi = \chi_0 \textit{ is the trivial character,} \\ 0 & \textit{if } \chi \neq |\ |^t \textit{ for all } t \in \mathbf{C}. \end{cases}$$

Proof. Denote by $\overline{d^\times x}$ the measure on $I_K/K^\times$ induced from $d^\times x$. We start with

$$(1) \qquad \int_{I_{mr}} \Phi(x)\chi(x)|x|^s d^\times x = q^{-mrs} \int_{I_{mr}/K^\times} \chi(x) \sum_{\alpha \in K^\times} \Phi(x\alpha)\overline{d^\times x}$$
$$= q^{-mrs} \int_{I_{mr}/K^\times} \chi(x) \sum_{\alpha \in K} \Phi(x\alpha)\overline{d^\times x}$$
$$- q^{-mrs} \int_{I_{mr}/K^\times} \chi(x)\Phi(0)\overline{d^\times x}$$
$$= q^{-mrs} \int_{I_{mr}/K^\times} \chi(x)|x|^{-1} \sum_{\alpha \in K} \widehat{\Phi}(x^{-1}\alpha)\overline{d^\times x}$$
$$- q^{-mrs}\Phi(0) \int_{I_{mr}/K^\times} \chi(x)\overline{d^\times x}$$

by Theorem 3. Since

$$\int_{I_{mr}/K^\times} \chi(x)\overline{d^\times x} = \int_{I_K^1/K^\times} \chi(xy_0)\overline{d^\times x} \quad \text{for any idèle } y_0 \text{ of degree } mr$$
$$= \begin{cases} \mathrm{vol}(I_K^1/K^\times)\chi(y_0) & \text{if } \chi \text{ is trivial on } I_K^1/K^\times, \\ 0 & \text{if } \chi \text{ is nontrivial on } I_K^1/K^\times, \end{cases}$$

and χ is trivial on $I_K^1/K^\times$ if and only if $\chi = |\ |^t$ for some $t \in \mathbf{C}$, in which case we have $\chi = \chi_0$ by assumption. Thus the last term in (1) is $\kappa_\chi \Phi(0) q^{-mrs}$. We have shown

$$\int_{I_{mr}} \Phi(x)\chi(x)|x|^s d^\times x + \kappa_\chi \Phi(0) q^{-mrs}$$
$$= q^{mr(1-s)} \int_{I_{mr}/K^\times} \chi(x) \sum_{\alpha \in K^\times} \widehat{\Phi}(x^{-1}\alpha)\overline{d^\times x} + q^{mr(1-s)}\kappa_\chi \widehat{\Phi}(0)$$
$$= q^{mr(1-s)} \int_{I_{mr}} \chi(x)\widehat{\Phi}(x^{-1}) d^\times x + q^{mr(1-s)}\kappa_\chi \widehat{\Phi}(0)$$
$$= \int_{I_{-mr}} \chi^{-1}(x)\widehat{\Phi}(x)|x|^{1-s} d^\times x + \kappa_\chi \widehat{\Phi}(0) q^{mr(1-s)},$$

as desired. □

Now we are ready to derive an analytic continuation of $Z(s,\chi,\Phi)$. By definition,

$$Z(s,\chi,\Phi)=\sum_{m\geq 0}\int_{I_{mr}}\Phi(x)\chi(x)|x|^s d^\times x+\sum_{m<0}\int_{I_{mr}}\Phi(x)\chi(x)|x|^s d^\times x.$$

Since Φ has compact supposrt, $\Phi=0$ on I_{mr} if m is negatively large. So the second sum is a finite sum, and it is a polynomial in q^{rs}, thus is holomorphic for all s. To handle the first sum, we use Lemma 2. For Re $s>1$, we get

$$\begin{aligned}\sum_{m\geq 0}\int_{I_{mr}}\Phi(x)\chi(x)|x|^s d^\times x=&\sum_{m\geq 0}\int_{I_{-mr}}\widehat{\Phi}(x)\chi^{-1}(x)|x|^{1-s}d^\times x\\&+\kappa_\chi\widehat{\Phi}(0)\frac{1}{1-q^{r(1-s)}}\\&-\kappa_\chi\Phi(0)\frac{1}{1-q^{-rs}}.\end{aligned}$$

Again, $\widehat{\Phi}$ has compact support, hence it is zero in I_{-mr} for m positively large. So the right hand side of the above equation gives an continuation of the first sum of $Z(s,\chi,\Phi)$. Put together, we find

$$\begin{aligned}(2)\qquad &Z(s,\chi,\Phi)+\kappa_\chi\Phi(0)\frac{1}{1-q^{-rs}}-\kappa_\chi\widehat{\Phi}(0)\frac{1}{1-q^{r(1-s)}}\\&=\sum_{m\geq 0}\int_{I_{-mr}}\widehat{\Phi}(x)\chi^{-1}(x)|x|^{1-s}d^\times x+\sum_{m<0}\int_{I_{mr}}\Phi(x)\chi(x)|x|^s d^\times x\end{aligned}$$

is a polynomial in q^{rs} and q^{-rs}. In particular, if $\chi\neq|\ |^t$ for all $t\in\mathbf{C}$, then $\kappa_\chi=0$ and $Z(s,\chi,\Phi)$ has a holomorphic continuation. To see functional equation, we use $\hat{\widehat{\Phi}}(x)=\Phi(-x)$ as proved in Proposition 2 and the fact that $\chi(-1)=1$ to express the right hand side of (2) as

$$\begin{aligned}&\sum_{m\leq 0}\int_{I_{mr}}\widehat{\Phi}(x)\chi^{-1}(x)|x|^{1-s}d^\times x+\sum_{m<0}\int_{I_{mr}}\hat{\widehat{\Phi}}(x)\chi(x)|x|^s d^\times x\\&=\sum_{m\leq 0}\int_{I_{mr}}\hat{\widehat{\Phi}}(x)(\chi^{-1})^{-1}(x)|x|^{1-(1-s)}d^\times x\\&\quad+\sum_{m<0}\int_{I_{mr}}\widehat{\Phi}(x)\chi^{-1}(x)|x|^{1-s}d^\times x+\kappa_\chi\Phi(0)-\kappa_\chi\widehat{\Phi}(0).\end{aligned}$$

Here we applied Lemma 2 to the integral over I_0 to change from one integrand to the other. The right hand side of last equality is equal to

$$\begin{aligned}&Z(1-s,\chi^{-1},\widehat{\Phi})+\kappa_\chi\widehat{\Phi}(0)\frac{1}{1-q^{-r(1-s)}}-\kappa_\chi\Phi(0)\frac{1}{1-q^{rs}}+\kappa_\chi\Phi(0)-\kappa_\chi\widehat{\Phi}(0)\\&=Z(1-s,\chi^{-1},\widehat{\Phi})-\kappa_\chi\widehat{\Phi}(0)\frac{1}{1-q^{r(1-s)}}+\kappa_\chi\Phi(0)\frac{1}{1-q^{-rs}}.\end{aligned}$$

Comparing with (2), we obtain

$$Z(s,\chi,\Phi) = Z(1-s,\chi^{-1},\widehat{\Phi}).$$

Finally, we compute $\mathrm{vol}(I_K^1/K^\times)$ from $\mathrm{vol}(\prod_v \mathcal{U}_v) = 1$. By Corollary 5 in Chapter 3, we have $\mathrm{Jac}(K) \cong I_K^1/K^\times \prod_v \mathcal{U}_v$. Since $K^\times \cap \prod_v \mathcal{U}_v = k^\times$ which has cardinality $q-1$, we find $\mathrm{vol}(I_K^1/K^\times) = |\mathrm{Jac}(K)|/(q-1)$. We summarize the above discussion in

Theorem 4. *Let r be the positive generator of $\deg(I_K)$, and let Φ be a Schwartz function on A_K. For a quasi-character $\chi \neq |\ |^t$ for all $t \in \mathbf{C}$, the zeta function $Z(s,\chi,\Phi)$ defined by the integral for $\mathrm{Re}\ s$ large has a holomorphic continuation to the whole s-plane as a polynomial in q^{rs} and q^{-rs}. The zeta function $Z(s,\chi_0,\Phi)$ for the trivial character χ_0 of I_K defined by the integral for $\mathrm{Re}\ s$ large has a meromorphic continuation to the whole s-plane, more precisely, $Z(s,\chi_0,\Phi) + \kappa\Phi(0)\frac{1}{1-q^{-rs}} - \kappa\widehat{\Phi}(0)\frac{1}{1-q^{r(1-s)}}$ is a polynomial in q^{rs} and q^{-rs}. Here $\kappa = |Jac(K)|/(q-1)$. In both cases, $Z(s,\chi,\Phi)$ satisfies the functional equation*

$$Z(s,\chi,\Phi) = Z(1-s,\chi^{-1},\widehat{\Phi}).$$

We shall prove in the next section that the number r is in fact equal to 1.

---

§4 The Zeta function of K (A proof of Theorem 1)

Recall that the zeta function of K is defined as

$$\zeta_K(s) = \prod_v (1 - Nv^{-s})^{-1} = L(s,\chi_0).$$

We saw before that it converges absolutely for $\mathrm{Re}\ s$ large. In fact, it does so for $\mathrm{Re}\ s > 1$.

Proposition 4. *The zeta function of K defined as an infinite product above converges absolutely for $\mathrm{Re}\ s > 1$, and it tends to 1 as $\mathrm{Re}\ s \to \infty$. Further, it has a meromorphic continuation to the whole s-plane. At $s = 1$ it has a simple pole with the residue equal to $\frac{q^{1-g}|\mathrm{Jac}K|}{(q-1)r\log q}$. Here r is the positive generator of $\deg(I_K)$.*

Remark. Observe that the important arithmatic informations of the field K appear in the residue at $s = 1$ of the zeta function of K, and it has a form analogous to the number field case. More precisely, $|\mathrm{Jac}K|$ plays the role of class number, $q-1$ is the number of roots of unity in K, q^{g-1} plays the role of the square root of the absolute value of the discriminant. The regulator does not come in since all global units of K are roots of unity.

Proof. To see the absolute convergence of the infinite product for Re $s > 1$, we first assume that $K = k(T)$ is a rational function field. In this case, each place $v \neq \infty$ corresponds to a monic irreducible polynomial P_v in $k[T]$ such that $\deg v = \deg P_v$. Thus

$$\zeta_K(s)(1-q^{-s}) = \prod_{v\neq\infty}(1-Nv^{-s})^{-1} = \prod_{\substack{P_v\in k[T]\\ \text{monic, irred.}}} \left(1-q^{-(\deg P_v)s}\right)^{-1}$$

$$= \sum_{\substack{m(T)\in k[T]\\ \text{monic}}} q^{-(\deg m(T))s}$$

whenever the series converges absolutely. Given a degree $n \geq 0$, there are q^n monic polynomials in $k[T]$ of degree n. Therefore

$$\zeta_K(s)(1-q^{-s}) = \sum_{n\geq 0} q^n q^{-ns} = \sum_{n\geq 0} q^{n(1-s)}$$

which converges absolutely for Re $s > 1$. And the same is true for $\zeta_K(s)$. The above formula also shows $\zeta_K(s) \to 1$ as Re $s \to \infty$.

Next suppose that K is a separable degree n extension of a rational function field $F = k(T)$. We have

$$\zeta_K(s) = \prod_{\substack{v \text{ place}\\ \text{of } F}} \prod_{\substack{w \text{ place of } K\\ w|v}} (1-Nw^{-s})^{-1}.$$

Let S be a finite set of places of F such that K over F is unramified outside S. Pick any place v of F outside S. We know from Theorems 3 and 4 in Chapter 3 that

$$n = \sum_{w|v}[K_w : F_v] = \sum_{w|v}[k_w : k_v],$$

where k_w (resp k_v) is the residue field of K_w (resp F_v). Let w be any place of K dividing v. Suppose $[K_w : F_v] = m$. Then $Nw = (Nv)^m$. Observe that for $\sigma = \text{Re } s > 1$, $(1-Nw^{-\sigma})^{-1} \leq (1-Nv^{-\sigma})^{-m}$, i.e., $(1-Nv^{-\sigma})^m \leq 1-(Nv^{-\sigma})^m$. This is obvious if $m = 1$; if $m > 1$, divide both sides by $1 - Nv^{-\sigma}$, then the inequality is equivalent to $(1-Nv^{-\sigma})^{m-1} \leq 1+Nv^{-\sigma}+Nv^{-2\sigma}+\cdots+Nv^{-(m-1)\sigma}$, which is true since the right hand side is > 1, while the left hand side is < 1. Hence we have, for $\sigma = \text{Re } s > 1$,

$$1 \leq \zeta_K(\sigma)\prod_{v\in S}\prod_{w|v}(1-Nw^{-\sigma}) = \prod_{\substack{v \text{ place of } F\\ v\notin S}} \prod_{w|v}(1-Nw^{-\sigma})^{-1}$$

$$\leq \prod_{v\notin S}(1-Nv^{-\sigma})^{-n} = \zeta_F(\sigma)^n \prod_{v\in S}(1-Nv^{-\sigma})^n,$$

which implies that ζ_K converges absolutely for Re $s > 1$, and also $\zeta_K(s) \to 1$ as Re $s \to \infty$.

Choose $\Phi = \Phi_1$ to be the characteristic function of the parallelotope $P_1 = \prod_w \mathcal{O}_w$. As computed in the previous section, $Z(s, \chi_0, \Phi_1) = \zeta_K(s)$. The existence of a meromorphic continuation follows from Theorem 4. Further, $\zeta_K(s)$ has a simple pole at $s = 1$ by Theorem 4, with the residue equal to that of $\kappa\widehat{\Phi}_1(0)\frac{1}{1-q^{r(1-s)}}$ as $s = 1$. We have $\kappa = |\mathrm{Jac}K|/(q-1)$, $\widehat{\Phi}_1(0) = |c^{-1}|^{1/2}\Phi_1(0) = q^{1-g}$ by Corollary 1, and $\frac{1}{1-q^{r(1-s)}}$ has residue $1/r \log q$ at $s = 1$. This shows that $\mathrm{Res}_{s=1}\, \zeta_K(s) = \frac{q^{1-g}|\mathrm{Jac}K|}{(q-1)r \log q}$.
□

Now we are ready to show that $\deg(I_K) = \mathbf{Z}$. In fact, we shall prove a stronger result.

Theorem 5. *Let S be a finite set of places of K. Then there is a divisor of K supported outside S with degree 1. In particular,* $\deg(I_K) = \mathbf{Z}$.

Proof. Let l be the positive generator of the image of idèles in I_K supported outside S under the degree map. We want to show that $l = 1$. Denote by L the compositum of K and k_l. Then L is a finite separable extension of K of degree n, say. Let S' be a finite set of places of K containing S such that outside S', the field extension L over K is unramified. Let v be a place of K outside S'. Then $\bigoplus_w L_w = K_v \otimes_K L$, where w runs through places of L dividing v, is the algebra obtained by joining $q^l - 1$ st roots of unity to K_v. In particular, each L_w is obtained from K_v by joining a $q^l - 1$ st roots of unity. Since $\deg v = ml$ for some positive integer m, the residue field of K_v is k_{ml}. This shows in particular that K_v contains all $q^l - 1$ st roots of unity, and hence $L_w \cong K_v$ for all $w \mid v$. We have shown that for each place v of K outside S', there are n places w of L dividing v, with each L_w isomorphic to K_v. So

$$\begin{aligned}
\zeta_L(s) &= \prod_{\substack{v \notin S' \\ v \text{ place of } K}} \prod_{w|v} (1 - Nw^{-s})^{-1} \prod_{v \in S'} \prod_{w|v} (1 - Nw^{-s})^{-1} \\
&= \prod_{v \notin S'} (1 - Nv^{-s})^{-n} \prod_{v \in S'} \prod_{w|v} (1 - Nw^{-s})^{-1} \\
&= \zeta_K(s)^n \prod_{v \in S'} \left((1 - Nv^{-s})^n \prod_{w|v} (1 - Nw^{-s})^{-1}\right).
\end{aligned}$$

For each $v \in S'$, $(1 - Nv^{-s})^n \prod_{w|v} (1 - Nw^{-s})^{-1}$ is holomorphic at $s = 1$ and nonzero there. Since $\zeta_K(s)$ has a simple pole at $s = 1$ by Proposition 4, the above shows that $\zeta_L(s)$ has pole at $s = 1$ of order n. But $\zeta_L(s)$ has only simple pole at $s = 1$ by Proposition 4, we conclude that $n = 1$. This shows that $L = K$, i.e., K contains k_l. As k is the field of constants in K, this implies $k_l = k$, that is, $l = 1$, as desired.
□

Hence the number r in Theorem 4 and Proposition 4 is equal to 1.

Remark. Let K be a global field and let L be a separable degree n field extension of K. A place v of K is said to split completely in L if there are n places of L dividing v. What we have shown above is that for function fields if almost all places of K split completely in L, then $L = K$. In general, Čebotarev density theorem says that the places of K splitting completely in L has density $\frac{1}{n}$. The same is true for number fields.

By Theorem 4, $\zeta_K(s) + \frac{\kappa}{1-q^{-s}} - \frac{\kappa q^{1-g}}{1-q^{1-s}}$ is a polynomial in q^s and q^{-s}. By Proposition 4, $\zeta_K(s)$ tends to 1 as Re $s \to \infty$, so this can happen only if $\zeta_K(s) + \frac{\kappa}{1-q^{-s}} - \kappa\frac{q^{1-g}}{1-q^{1-s}}$ is a polynomial in q^{-s}. Letting $U = q^{-s}$, we can write $\zeta_K(s) = Z_C(q^{-s})$, where $Z_C(U)$ is the zeta function of the underlying nonsingular projective curve C, which then has the form

$$Z_C(U) = \frac{P_1(U)}{(1-U)(1-qU)},$$

with $P_1(U)$ a polynomial in U. We have $P_1(0) = Z_C(0) = \lim_{\mathrm{Re}\, s\to\infty} \zeta_K(s) = 1$, and from

$$\frac{P_1(1)}{q-1} = \mathrm{Res}_{U=1}\, Z_C(U) = \mathrm{Res}_{U=1} \frac{\kappa}{1-U} = \kappa = \frac{|\mathrm{Jac}K|}{q-1}$$

we get $P_1(1) = |\mathrm{Jac}K|$. The functional equation $Z(s, \chi_0, \Phi_1) = Z(1-s, \chi_0, \widehat{\Phi}_1)$ translates as

$$\zeta_K(s) = q^{(1/2-s)(2g-2)}\zeta_K(1-s)$$

since $\widehat{\Phi}_1(x) = |c^{-1}|^{1/2}\Phi_1(c^{-1}x)$ by Corollary 1 and

$$\begin{aligned} Z(1-s, \chi_0, \widehat{\Phi}_1) &= \int_{I_K} |c^{-1}|^{1/2}\Phi_1(c^{-1}x)|x|^{1-s}d^\times x = |c|^{\frac{1}{2}-s}\int_{I_K} \Phi_1(x)|x|^{1-s}d^\times x \\ &= q^{(\frac{1}{2}-s)(2g-2)}L(1-s, \chi_0) = q^{(\frac{1}{2}-s)(2g-2)}\zeta_K(1-s). \end{aligned}$$

Therefore $P_1(U)$ satisfies the functional equation

$$P_1(U) = (q^{\frac{1}{2}}U)^{2g} P_1(\frac{1}{qU}). \tag{3}$$

As $P_1(U)$ is a polynomial in U, this implies that $\deg P_1 \le 2g$. If $\deg P_1 < 2g$, then the right hand side of (3) would equal 0 when evaluated at $U = 0$, contradicting $P_1(0) = 1$. Therefore $\deg P_1 = 2g$. Finally, $P_1(U)$ has integral coefficients since $P_1(U) = Z_C(U)(1-U)(1-qU)$ and $Z_C(U) = \prod_v (1-U^{\deg v})^{-1}$ is a power series in U with integral coefficients. This completes the proof of Theorem 1.

§5 $L(s,\chi)$ for nontrivial χ (A proof of Theorem 2)

In this section we assume that χ is a character of $I_K/K^\times$ not equal to $|\ |^t$ for $t \in \mathbf{C}$. Denote by $f(\chi) = \sum_v n_v v$ the conductor of χ, and by S the support of $f(\chi)$. Then

$$L(s,\chi) = \prod_{v \notin S} (1 - \chi_v(\pi_v) Nv^{-s})^{-1}.$$

By the same proof as in Proposition 4, we see that $L(s,\chi)$ converges absolutely for Re $s > 1$, and $L(s,\chi) \to 1$ as Re $s \to \infty$. Let a be an idèle such that div $a = f(\chi)$, and let Φ be the characteristic function of $1 + P_a$. Then at a place $v \notin S$, χ_v is unramified and Φ restricted to K_v is the characteristic function of $\mathcal{O}_v$; at a place $v \in S$, Φ restricted to K_v is the characteristic function of $1 + \mathcal{P}_v^{n_v}$ on which χ_v is trivial. Hence

$$Z(s,\chi,\Phi) = \prod_{v \notin S} L(s,\chi_v) \prod_{v \in S} \int_{1+\mathcal{P}_v^{n_v}} \chi_v(x_v)|x_v|^s d^\times x_v = L(s,\chi) \prod_{v \in S} \mathrm{vol}(1 + \mathcal{P}_v^{n_v}).$$

We see from Theorem 4 that $L(s,\chi)$ has a holomorphic continuation, which is a polynomial in q^s and q^{-s}. Again, since $L(s,\chi) \to 1$ as Re $s \to \infty$, we get $L(s,\chi) = P(q^{-s},\chi)$, where $P(U,\chi)$ is a polynomial in U. To get functional equation for $L(s,\chi)$, we compute $Z(1-s,\chi^{-1},\widehat{\Phi})$. It follows from Corollary 2 that $\widehat{\Phi} = \psi \cdot \widehat{\Phi}_a = |a||c^{-1}|^{1/2} \psi \Phi_{ca^{-1}}$. Thus, for Re s large,

$$\begin{aligned} Z(1-s,\chi^{-1},\widehat{\Phi}) &= |a||c^{-1}|^{1/2} \int_{I_K} \psi(x)\Phi_{ca^{-1}}(x)\chi^{-1}(x)|x|^{1-s} d^\times x \\ &= |a||c^{-1}|^{1/2} |ca^{-1}|^{1-s} \int_{I_K} \psi(ca^{-1}x)\Phi_1(x)\chi^{-1}(ca^{-1}x)|x|^{1-s} d^\times x. \end{aligned}$$

At a place $v \notin S$, Φ_1 restricted to K_v is the characteristic function of $\mathcal{O}_v$ on which $\psi_v(c_v a_v^{-1} x_v)$ is trivial, and χ_v is unramified. Thus

$$\begin{aligned} &Z(1-s,\chi^{-1},\widehat{\Phi}) \\ =&|a|^s|c|^{\frac{1}{2}-s}\chi^{-1}(ca^{-1})L(1-s,\chi^{-1}) \prod_{v \in S} \int_{\mathcal{O}_v - \{0\}} \psi_v(c_v a_v^{-1} x_v)\chi_v^{-1}(x_v)|x_v|^{1-s} d^\times x_v. \end{aligned}$$

Write $\mathcal{O}_v - \{0\} = \bigcup_{n=0}^{\infty} \mathcal{U}_v \pi_v^n$. For $n \geq 1$,

$$\begin{aligned} &(Nv)^{n(1-s)} \int_{\mathcal{U}_v \pi_v^n} \psi_v(c_v a_v^{-1} x_v)\chi_v^{-1}(x_v)|x_v|^{1-s} d^\times x_v \\ =& \begin{cases} \displaystyle\sum_{y_v \in \mathcal{U}_v/1+\mathcal{P}_v^{n_v-n}} \psi_v(c_v a_v^{-1}\pi_v^n y_v)\chi_v^{-1}(\pi_v^n y_v) \int_{1+\mathcal{P}_v^{n_v-n}} \chi_v^{-1}(x_v) d^\times x_v & \text{if } n < n_v, \\ \displaystyle\int_{\mathcal{U}_v} \chi_v^{-1}(x_v \pi_v^n) d^\times x_v & \text{if } n \geq n_v, \end{cases} \\ =&0 \quad \text{since } \chi_v \text{ is nontrivial on } 1 + \mathcal{P}_v^{n_v-1}, \end{aligned}$$

hence the integral over $\mathcal{O}_v - \{0\}$ is a nonzero constant (a Gauss sum) for each $v \in S$. We can write $Z(1-s, \chi^{-1}, \widehat{\Phi}) = \beta \cdot q^{(\frac{1}{2}-s)(2g-2+\deg f(\chi))} L(1-s, \chi^{-1})$ for a constant β so that the functional equation for $L(s,\chi)$, arising from $Z(s,\chi,\Phi) = Z(1-s,\chi^{-1},\widehat{\Phi})$, reads

$$L(s,\chi) = c(\chi) q^{(\frac{1}{2}-s)(2g-2+\deg f(\chi))} L(1-s, \chi^{-1}),$$

i.e.,

$$P(U,\chi) = c(\chi)(q^{\frac{1}{2}}U)^{(2g-2+\deg f(\chi))} P(\frac{1}{qU}, \chi^{-1})$$

for a nonzero constant $c(\chi)$. Since $P(U,\chi)$ is a polynomial in U, we get that $P(U,\chi^{-1})$ is a polynomial of degree $\leq 2g-2+\deg f(\chi)$. In fact, equality holds since $P(0,\chi) = 1$. This shows that $P(U,\chi^{-1})$ and hence $P(U,\chi)$ is a polynomial of degree $2g-2+\deg f(\chi)$. This completes the proof of Theorem 2.

Remark. It follows from the above computation that

$$c(\chi) = q^{(2-2g-\deg f(\chi))/2} \prod_{v \in S} \sum_{x_v \in \mathcal{U}_v / 1+\mathcal{P}_v^{n_v}} \psi_v(c_v a_v^{-1} x_v) \chi_v^{-1}(c_v a_v^{-1} x_v) \prod_{v \notin S} \chi_v^{-1}(c_v).$$

References

[1] J. Tate : Fourier analysis in number fields and Hecke's zeta–functions. Thesis. Princeton University 1950. Published in Algebraic Number Theory, J. W. S. Cassels and A. Fröhlich edited, Thompson, Washington D.C. (1967), republished by Academic Press, London.

[2] A. Weil : Basic Number Theory, Springer–Verlag, Berlin–Heidelberg–New York (1973).

CHAPTER 6

Character Sum Estimates and Idèle Class Characters

§1 Roots of L–functions

In the previous chapter we studied the analytic behavior of the L–function attached to a character of the idèle class group of a function field. In this section we shall estimate its roots, which will lay the foundation of character sum estimates to be studied in the later sections. For this purpose, some results in class field theory will be assumed without proof. The reader is referred to [2,10] for more details.

As before, K denotes a function field of one variable of genus g_K with the field of constants being a finite field k of q elements. Let $\overline{K}$ be a separable closure of K. Endow the Galois group $Gal(\overline{K}/K)$ with the Krull topology, that is, a neighborhood system of the identity map consists of $Gal(\overline{K}/F)$ as F runs through all finite Galois extensions of K in $\overline{K}$ so that $Gal(\overline{K}/K)$ is a totally disconnected topological group. To each finite-dimensional representation ρ of $Gal(\overline{K}/K)$, Artin defined an L–function $L(s,\rho)$. When ρ is the trivial representation of $Gal(\overline{K}/K)$, the L–function $L(s,\rho)$ is the zeta function $\zeta_K(s)$ of K. A 1–dimensional representation (i.e., a character) ρ of $Gal(\overline{K}/K)$ factors through a finite abelian quotient of $Gal(\overline{K}/K)$, due to its topology. More precisely, there is a finite abelian extension F of K in $\overline{K}$ such that ρ may be regarded as a character of $Gal(F/K) \cong Gal(\overline{K}/K)/Gal(\overline{K}/F)$. By global class field theory, the group $I_K/K^\times N_{F/K}(I_F)$ is isomorphic to $Gal(F/K)$, and there is an idèle class character χ of $I_K/K^\times N_{F/K}(I_F)$ such that $L(s,\chi) = L(s,\rho)$. Conversely, given a character χ of the idèle class group $I_K/K^\times$ of finite order, there is a finite abelian extension F of K such that $I_K/\ker\chi$ is isomorphic to $Gal(F/K)$ and there is a 1–dimensional representation ρ of $Gal(F/K)$ such that $L(s,\chi) = L(s,\rho)$. Thus there is a 1–1 correspondence between characters of $I_K/K^\times$ of finite order and characters of $Gal(\overline{K}/K)$ (or of $Gal(K^{ab}/K)$) such that the corresponding characters have the same L–function. Hence to study the roots of L–functions attached to characters of finite order of the idèle class group $I_K/K^\times$, it suffices to study those attached to characters of $Gal(\overline{K}/K)$. Recall from Proposition 1 in Chapter 5 that any quasi-character η of $I_K/K^\times$ is equal to the product of an idèle class character χ of finite order and $|\ |^{s_0}$ for some $s_0 \in \mathbf{C}$. Thus $L(s,\eta) = L(s,\chi|\ |^{s_0}) = L(s+s_0,\chi)$, and we would know roots of $L(s,\eta)$ from those of $L(s,\chi)$.

Let F be a finite abelian extension of K and let 1_F be the trivial representation of $Gal(\overline{K}/F) = Gal(\overline{F}/F)$. Then the induced representation $\rho = \mathrm{Ind}_{Gal(\overline{K}/F)}^{Gal(\overline{K}/K)} 1_F = \mathrm{Ind}_{\{id\}}^{Gal(F/K)} 1$ is the regular representation of $Gal(F/K)$ of degree $[F:K]$, and it decomposes as a direct sum of the degree 1 representations ρ_i of $Gal(F/K)$, each occurring with multiplicity one, i.e., $\rho = \bigoplus_i \rho_i$. It follows from the properties of Artin L-functions that

$$\zeta_F(s) = L(s,1_F) = L(s,\rho) = \prod_i L(s,\rho_i) = \zeta_K(s) \prod_{\rho_i \neq 1_K} L(s,\rho_i).$$

The field of constants of F is a finite extension of k, denote it by k_n, which has cardinality q^n. By Theorem 1 in Chapter 5, we have

$$\zeta_F(s) = \frac{P_F(q^{-ns})}{(1-q^{-ns})(1-q^{n(1-s)})}, \qquad \zeta_K(s) = \frac{P_K(q^{-s})}{(1-q^{-s})(1-q^{1-s})},$$

where $P_F(u), P_K(u)$ are polynomials in u of degree $2g_F$, $2g_K$, respectively. Further, denote by χ_i the characters of $I_K/K^\times$ such that $L(s,\chi_i) = L(s,\rho_i)$. By Theorems 1 and 2 in Chapter 5, if χ_i is not of the form $|\ |^{s_0}$ for some $s_0 \in \mathbf{C}$, then $L(s,\chi_i) = P(q^{-s},\chi_i)$, where $P(u,\chi_i)$ is a polynomial in u of degree $2g_K - 2 + \deg f(\chi_i)$ with $f(\chi)$ being the conductor of χ; while if $\chi_i = |\ |^{s_0}$ for some $s_0 \in \mathbf{C}$, then $L(s,\chi_i) = \zeta_K(s+s_0)$, in which case we write $P(q^{-s},\chi_i)$ for the numerator of $L(s,\chi_i)$, thus $P(u,\chi_i)$ is a polynomial of degree $2g_K$. Our goal is to prove

Theorem 1. *With notations as above, we have* $P_F(u^n) = \prod_{\chi_i} P(u,\chi_i)$, *where* χ_i *runs through all idèle class characters of* $I_K/K^\times$ *corresponding to* $\rho_i \in \widehat{Gal(F/K)}$, *that is, all characters of* $I_K/K^\times \mathrm{N}_{F/K}(I_F)$.

In view of the relation $\zeta_F(s) = \prod_{\chi_i} L(s,\chi_i)$ and the discussion of the L-functions involved, the key is to study the χ_i's of the form $|\ |^{s_0}$. Recall from Theorem 5 in Chapter 5 that there is an idèle in I_K of degree 1.

Proposition 1. *Let* k_m *be a degree* m *extension of* k *in* $\overline{K}$ *and let* E *be the composite of* K *and* k_m. *Then* E *is an abelian unramified extension of degree* m *over* K. *For any place* v *of* K *and any place* w *of* E *dividing* v, *we have* $\deg w = (\deg v)/d_v$ *and* $[E_w : K_v] = m/d_v$, *where* d_v *is the degree over* k *of the intersection of* k_m *with the residue field of* K_v. *Further,* $\mathrm{N}_{E/K}(I_E)K^\times = I_K^1\langle t\rangle^m$ *for any idèle* t *in* I_K *of degree 1.*

Here the norm map $\mathrm{N}_{E/K}$ from I_E to I_K is the product of local norms. Precisely, given $x = (x_w) \in I_E$, $\mathrm{N}_{E/K}(x) = \left(\prod_{w|v} \mathrm{N}_{E_w/K_v}(x_w)\right) \in I_K$.

Proof. Let ζ be a primitive q^m–1st root in k_m and let $f(x)$ be the irreducible polynomial of ζ over k. Then $\deg f = m$. Clearly, $E = K(\zeta)$ is an abelian extension of K. Further, since the field of constants of K is k, we have $[E : K] = \deg f = m$. Let v be a place of K. Decompose $f(x) = f_1(x)\cdots f_r(x)$ as a product of irreducible polynomials over K_v. Then each $\overline{f_i}(x)$ is irreducible over the residue field κ_v of K_v by Hensel's lemma, and $\deg \overline{f_i} = \deg f_i$. For any place w of E dividing v, E_w is isomorphic to $K_v[x]/(f_i(x))$ for some i. Therefore, $[E_w : F_v] = \deg f_i = [\kappa_w : \kappa_v]$, where κ_w denotes the residue field of E_w. In other words, E_w is unramified over K_v. This being so for all v and w, we conclude that E is unramified over K. Let d_v be the degree of $\kappa_v \cap k_m$ over k. As κ_w is the composite of κ_v and k_m, we get $[E_w : K_v] = [\kappa_w : \kappa_v] = [k_m : k_m \cap \kappa_v] = m/d_v$ and $\deg w = [\kappa_w : k_m] = [\kappa_v : \kappa_v \cap k_m] = (\deg v)/d_v$. Since E_w is unramified over K_v, a uniformizer π_v of K_v is also a uniformizer π_w of E_w.

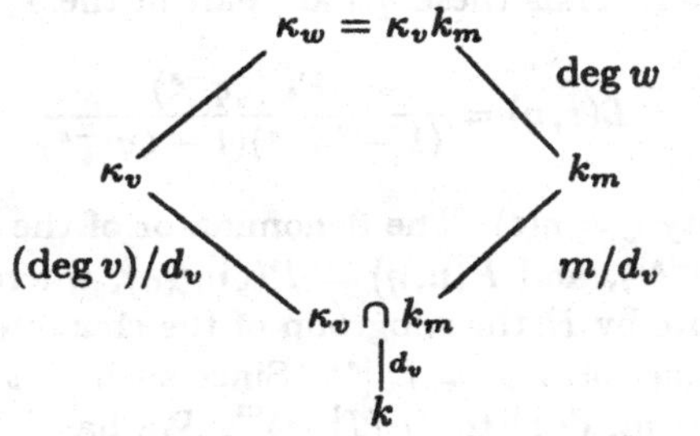

Now let $y = \prod_{w \text{ place of } E} \pi_v^{n_w}$ be an idèle of E such that degree $y = 1$. This means that $n_w \in \mathbb{Z}$, $n_w = 0$ for almost all w and

$$1 = \sum_w n_w \deg w = \sum_{v \text{ place of } K} \Big(\sum_{w|v} n_w\Big)(\deg v)/d_v.$$

Then

$$\begin{aligned} \mathrm{N}_{E/K}(y) &= \prod_{v \text{ place of } K} \prod_{w|v} \mathrm{N}_{E_w/K_v}(\pi_w)^{n_w} \\ &= \prod_v \prod_{w|v} \pi_v^{n_w[E_w:K_v]} \quad (\text{since } \pi_w = \pi_v) \\ &= \prod_v \pi_v^{(\sum_{w|v} n_w)m/d_v} \in I_K \end{aligned}$$

and

$$\deg \mathrm{N}_{E/K}(y) = \sum_v \Big(\sum_{w|v} n_w\Big)\frac{m}{d_v} \deg v = m.$$

Clearly, $\mathrm{N}_{E/K}(I_E^1) \subset I_K^1$. For any idèle t in I_K with degree 1, we have $I_K = I_K^1\langle t\rangle$; and $I_K^1 t^m$ is the set of all idèles in I_K with degree m, in particular, it contains

$\mathrm{N}_{E/K}(y)$. This shows that $\mathrm{N}_{E/K}(I_E)K^\times \subset I_K^1\langle t\rangle^m$. Since $\mathrm{N}_{E/K}(I_E)K^\times$ is an open subgroup of I_K of index m (as a consequence of $I_K/\mathrm{N}_{E/K}(I_E)K^\times \cong Gal(E/K)$) and $I_K^1\langle t\rangle^m$ also has index m in I_K, therefore $\mathrm{N}_{E/K}(I_E)K^\times = I_K^1\langle t\rangle^m$, as wanted. □

Corollary 1. *$\mathrm{N}_{E_w/K_v}(\mathcal{U}_w) = \mathcal{U}_v$ for all places v of K and all palces w of E dividing v.*

Proof of Theorem 1. As observed, the characters χ_i corresponding to ρ_i are the characters in the dual of $I_K/K^\times \mathrm{N}_{F/K}(I_F)$, that is, the characters of I_K whose kernel contain $K^\times \mathrm{N}_{F/K}(I_F)$. The field F contains the subfield E which is the composite of K and k_n. By Proposition 1 above, $I_K^1\langle t\rangle^n = \mathrm{N}_{E/K}(I_E)K^\times \supset \mathrm{N}_{F/K}(I_F)K^\times$, and the characters of $I_K/I_K^1\langle t\rangle^n$ are the characters η of the form $\eta = |\ |^{s_0}$ for some $s_0 \in \mathbf{C}$ such that $\eta^n = 1$. Thus these $\eta's$ are part of the $\chi_i's$ and

$$L(s,\eta) = \frac{P_K(\zeta q^{-s})}{(1-\zeta q^{-s})(1-\zeta q^{1-s})}$$

for an nth root of unity $\zeta = \eta(t)$. The denominator of the product of the $L(s,\eta)'s$ is $(1-q^{-ns})(1-q^{n(1-s)})$, and $P(u,\eta) = P(\zeta u, \chi_0)$, where χ_0 denotes the trivial character of I_K. Denote by H the subgroup of the characters of $I_K/K^\times \mathrm{N}_{F/K}(I_F)$ generated by $\chi_i's$ of the form $\chi_i = |\ |^{s_0}$. Since such $\chi_i's$ are trivial on I_K^1, H is a cyclic group of order m, dual to $I_K/I_K^1\langle t\rangle^m$. We have shown that n divides m. Further, by Proposition 1 again, $I_K^1\langle t\rangle^m = K^\times \mathrm{N}_{E'/K}(I_{E'})$ for the field E' which is the composite of K and k_m. $K^\times \mathrm{N}_{E'/K}(I_{E'})$ contains $K^\times \mathrm{N}_{F/K}(I_F)$, by class field theory, E' is a subfield of F. This shows that k_m is contained in the field of constants k_n of F. Therefore $k_m = k_n$, i.e., $m = n$. We have shown that exactly n characters among χ_i are of the form $|\ |^{s_0}$, hence the remaining characters have associated L-functions equal to polynomials in q^{-s}. This proves the theorem. □

By Theorem 8 in Chapter 4, the Riemann hypothesis for F holds, that is,

$$P_F(U) = \prod_{i=1}^{2g_F}(1-\omega_i U),$$

where $|\omega_i| = q^{n/2}$. Hence

$$P_F(u^n) = \prod_{i=1}^{2g_F}(1-\omega_i u^n) = \prod_{j=1}^{2ng_F}(1-\beta_j u)$$

with $|\beta_j| = q^{1/2}$. This combined with Theorem 1 shows that all roots of $P(u,\chi_i)$ have absolute value $q^{-1/2}$. In view of the analysis at the beginning of this section, we have proven

Theorem 2. *Let χ be an idèle class character of I_K of finite order. Write $P(q^{-s},\chi)$ for the numerator of $L(s,\chi)$. Then the roots of $P(u,\chi)$ have absolute value $q^{-1/2}$.*

Corollary 2. *Let χ be an idèle class character of I_K of finite order which is not of the form $|\ |^{s_0}$. Denote by $f(\chi)$ the conductor of χ. Then $L(s,\chi)=P(q^{-s},\chi)$, where $P(u,\chi)=1+a_1u+a_2u^2+\cdots+a_ru^r$ is a polynomial of degree $r=2g_K-2+\deg f(\chi)$ and*

$$|a_1|\le\big(2g_K-2+\deg f(\chi)\big)\sqrt{q},\quad |a_r|=q^{r/2}.$$

Recall the definition of

$$L(s,\chi)=\prod_v L(s,\chi_v)=\prod_{\substack{v\\ \chi\text{ unr. at }v}}\left(1-\chi_v(\pi_v)(Nv)^{-s}\right)^{-1}.$$

Here $Nv=q^{\deg v}$. Hence a_1 is $\sum\limits_v \chi_v(\pi_v)$ as v runs through the degree 1 places of K where χ is unramified. From Corollary 2 we get

Corollary 3. *With the same notation as in Corollary 2, we have*

$$\left|\sum_{\substack{\deg v=1\\ \chi_v\text{ unr.}}}\chi_v(\pi_v)\right|\le\big(2g_K-2+\deg f(\chi)\big)\sqrt{q}.$$

This is the fundamental inequality on which our character sum estimates in the subsequent sections are based. More precisely, given a character sum, we shall construct an idèle class character χ of finite order such that $\sum\limits_{\substack{\deg v=1\\ \chi_v\text{ unr.}}}\chi_v(\pi_v)$ is the given character sum, then we obtain an estimate using Corollary 3.

Theorem 2 can also be proved without using class field theory. This is done, for instance, in [8] by W. Schmidt.

---

§2 A. Weil's character sum estimates

Let k be a finite field of q elements and $K=k(t)$ be the rational funtion field over k with genus 0. As usual, denote by ∞ the place of K with $P_\infty(t)=\frac{1}{t}$ as its uniformizer. For a nonzero rational function $f(t)$ in K, write $\operatorname{supp} f$ for the collection of places of K (including ∞) at which f has either zero or pole. Clearly $f(x)\in\mathbf{P}^1(k)$ for $x\in\mathbf{P}^1(k)=k\cup\{\infty\}$. Let χ be a multiplicative character of $k^\times$ of order $d>1$. Extend χ to a function of $\mathbf{P}^1(k)$ by letting $\chi(0)=\chi(\infty)=0$ so that $\chi\big(f(x)\big)$ is defined for all $x\in\mathbf{P}^1(k)$. Let ψ be a nontrivial additive character of k and let $g(t)$ be a polynomial of degree n over k. Then $\psi\big(g(x)\big)$ is defined for all $x\in k$.

In 1948 A. Weil proved the following character sum estimates in [9].

Theorem 3. *Let χ, ψ, f, g be as above, with m being the total degree of the places in* supp f *other than* ∞.

(1) Suppose div $f \notin d \cdot \mathcal{D}iv(K)$. Then

$$\left| \sum_{x \in \mathbf{P}^1(k)} \chi(f(x)) \right| \le (m-1)\sqrt{q};$$

(2) Suppose $q > n$ and $(n, q) = 1$. Then

$$\left| \sum_{x \in k} \psi(g(x)) \right| \le (n-1)\sqrt{q};$$

(3) Suppose either div $f \notin d \cdot \mathcal{D}iv(K)$ or $q > n$ and $(n, q) = 1$. Then

$$\left| \sum_{x \in k} \chi(f(x))\psi(g(x)) \right| \le (m+n-1)\sqrt{q}.$$

We shall derive Theorem 3 from the theorem below. At each place v of K other than ∞, we fix a uniformizer π_v to be the monic irreducible polynomial in $k[t]$ whose roots constitute the place v, and choose π_∞ to be $1/t$.

Theorem 4. *Let χ, ψ, f, g be as above. Suppose $g(0) = 0$.*

(1) There exists an idèle class character ω of K such that

(1.a) *ω is unramified outside* supp f*;*

(1.b) *Let v be a place outside* supp f. *If $v \ne \infty$, then*

$$\omega_v(\pi_v) = \chi\left(\prod_{j=1}^{\deg v} f(\beta_{j,v}) \right),$$

where $\beta_{j,v}$ runs through all the roots of π_v, while if $v = \infty$, we have

$$\omega_\infty(\pi_\infty) = \chi(f(\pi_\infty));$$

(1.c) *Let div $f = \sum_v a_v v$. The conductor of ω is $\sum_v v$, where v runs through all places such that d does not divide a_v.*

(2) There exists an idèle class character η of K unramified outside ∞ such that at each place $v \ne \infty$,

$$\eta_v(\pi_v) = \psi\Big(\sum_{1 \le j \le \deg v} g(\beta_{j,v}) \Big),$$

where $\beta_{j,v}$ runs through all roots of π_v. Furthermore, if $\deg g = n$ *is prime to* q, *then* η *has conductor* $(n+1)\infty$.

Assuming Theorem 4, we prove Theorem 3. Given χ and f as in Theorem 3, let ω be as in Theorem 4, (1). Set $S = \{v \in \operatorname{supp} f : d \mid a_v\}$. The condition on div f implies that ω is nontrivial with $\operatorname{cond}\omega = \sum_{v \in \operatorname{supp} f - S} v$. Write m' for the degree of $\operatorname{cond}\omega$. By Corollary 3, we have

$$\left| \sum_{\substack{\deg v=1 \\ \omega_v \text{ unr.}}} \omega_v(\pi_v) \right| \le (m'-2)\sqrt{q},$$

which implies

$$\left| \sum_{\substack{\deg v=1 \\ v \notin \operatorname{supp} f}} \omega_v(\pi_v) \right| \le \Big(m' + \sum_{v \in S} \deg v - 2\Big)\sqrt{q}.$$

At a place $v \notin \operatorname{supp} f$ with $\pi_v = t - a$, we have $\omega_v(\pi_v) = \chi\big(f(a)\big)$, and the same formula holds if $v = \infty \notin \operatorname{supp} f$. Further, at other points $b \in \mathbf{P}^1(k)$, we have $\chi\big(f(b)\big) = 0$ by definition. Thus the above inequality may be written as

$$\left| \sum_{x \in \mathbf{P}^1(k)} \chi\big(f(x)\big) \right| \le \Big(m' + \sum_{v \in S} \deg v - 2\Big)\sqrt{q}.$$

Finally, note that $m' + \sum_{v \in S} \deg v = m$ if $\infty \notin \operatorname{supp} f$, $= m+1$ if $\infty \in \operatorname{supp} f$, hence the upper bound is $\le (m-1)\sqrt{q}$. This proves the first statement of Theorem 3.

For the second statement, we may assume $g(0) = 0$. Let η be the idèle class character of K constructed from ψ and g as in Theorem 4, (2). The condition on n implies $\operatorname{cond}\eta = (n+1)\infty$. From Corollary 3 we get

$$\left| \sum_{\substack{\deg v=1 \\ v \neq \infty}} \eta_v(\pi_v) \right| \le (n-1)\sqrt{q}.$$

Since at a place v of degree 1 and $v \neq \infty$, we have $\pi_v = t - a$ and $\eta_v(\pi_v) = \psi\big(g(a)\big)$, the above inequality can be rewritten as

$$\left| \sum_{x \in k} \psi\big(g(x)\big) \right| \le (n-1)\sqrt{q},$$

which is (2).

For (3), we may also assume $g(0) = 0$. Consider the idèle class character $\xi = \omega\eta$, where ω and η are as above. If $\operatorname{div} f \notin d \cdot \mathcal{D}iv(K)$, then ω is ramified at a place $v \neq \infty$ and hence ξ is ramified with $\operatorname{cond}\xi \leq \sum_{\substack{v \in \operatorname{supp} f \\ v \neq \infty}} v + (n+1)\infty$. If ξ is ramified at ∞, then $\deg \operatorname{cond} \xi \leq m + n + 1$ and Corollary 3 gives

$$\left| \sum_{\substack{\deg v = 1 \\ \xi_v \text{ unr.}}} \xi_v(\pi_v) \right| = \left| \sum_{x \in k} \chi(f(x))\psi(g(x)) \right| \leq (m+n-1)\sqrt{q};$$

while if ξ is unramified at ∞, then $\deg \operatorname{cond} \xi \leq m$, $\omega_\infty = 1$ and $\eta = 1$, Corollary 3 yields

$$\left| \sum_{\substack{\deg v = 1 \\ \xi_v \text{ unr.}}} \xi_v(\pi_v) \right| = \left| \sum_{x \in k} \chi(f(x))\psi(g(x)) + 1 \right| \leq (m-2)\sqrt{q},$$

which clearly implies the third assertion of Theorem 3. Finally, if $\operatorname{div} f \in d \cdot \mathcal{D}iv(K)$, then ω is trivial and

$$\left| \sum_{x \in k} \chi(f(x))\psi(g(x)) \right| = \left| \sum_{\substack{x \in k \\ f(x) \in k^\times}} \xi(g(x)) \right|,$$

hence the desired inequality follows from the second assertion.

Remark. Note that (2) and (3) obviously hold if $n > q$. The character η defined using ψ and g either has $2\infty \leq \operatorname{cond} \eta \leq (n+1)\infty$ or else it is trivial on I_K. As long as η is nontrivial, the estimate (2) holds. Further, when $g(t)$ is a monomial in t, (2) can be shown by an easier way.

Exercise 1. Let $g(t) = ct^n$, $t \in k^\times$. Use Gauss sum to show that

$$\left| \sum_{x \in k} \psi(g(x)) \right| \leq (n-1)\sqrt{q}.$$

Proof of Theorem 4. (1) Write $f(t) = cf_1(t)/f_2(t)$ as a nonzero constant c times a quotient of two monic and coprime polynomials in $k[t]$. Then $f(t) = c \prod_{i=1}^{m} (t - \alpha_i)^{a_i}$, where α_i's are distinct and a_i's are nonzero integers.

Using the given χ and f, we define a character ω of $I_K/K^\times$ as follows. At a place $v \notin \operatorname{supp} f$ and $v \neq \infty$, define ω_v to be trivial on $\mathcal{U}_v$ and

$$\omega_v(\pi_v) = \chi\left(\prod_{j=1}^{\deg v} f(\beta_{j,v}) \right),$$

where $\beta_{j,v}$ runs through all roots of π_v. If $v = \infty \notin \operatorname{supp} f$, then $\deg f_1 = \deg f_2$ so that $f(\infty) = c$ and $\deg f = 0$, in this case define ω_∞ to be 1 on $\mathcal{U}_\infty$ and

$$\omega_\infty(\pi_\infty) = \chi(f(\infty)) = \chi(c).$$

So far, we have defined an unramified character ω on the idèles whose components are trivial at places in $\operatorname{supp} f$, denote this group by $\prod'_{v \notin \operatorname{supp} f} K_v^\times$. Extend ω to a character on $I' = \prod'_{v \notin \operatorname{supp} f} K_v^\times \prod_{v \in \operatorname{supp} f} (1 + \mathcal{P}_v)$ by letting it be 1 on $\prod_{v \in \operatorname{supp} f} (1 + \mathcal{P}_v)$. Since $I_K = K^\times \cdot I'$, ω extends to an idèle class character of $I_K/K^\times$ if and only if, as defined so far, it is trivial on the intersection $K^\times \cap I'$. To check this, let $h(t) \in K^\times \cap I'$. Write $h(t) = h_1(t)/h_2(t)$ as a quotient of two relatively prime polynomials in $k[t]$. Then at $v \in \operatorname{supp} f$, $v \neq \infty$, we have $h_1(t), h_2(t)$ both units in K_v and $h_1(t) \equiv h_2(t) \bmod \pi_v$. This shows that $h_1(t) - h_2(t)$ is divisible by $\prod_{\substack{v \in \operatorname{supp} f \\ v \neq \infty}} \pi_v(t) =: P(t)$ in $k[t]$. Consequently, $h_1(\alpha_i) = h_2(\alpha_i) \neq 0$, i.e., $h(\alpha_i) = 1$, for $i = 1, \cdots, m$. If $\infty \in \operatorname{supp} f$, then $\deg h = \deg h_1 - \deg h_2 = 0$ and h_1, h_2 have the same leading coefficient. By definition,

$$\omega(h) = \prod_v \omega_v(h) = \prod_{\substack{v \notin \operatorname{supp} f \\ v \neq \infty}} \omega_v(h) = \chi(c)^{-\deg h} \prod_{\substack{v \neq \infty \\ \pi_v \text{ divides} \\ h_1 \text{ or } h_2}} \omega_v(h)$$

$$= \chi\left(\prod_{i=1}^m h(\alpha_i)^{a_i}\right) = \chi(1) = 1.$$

Here we used $\omega_v(\pi_v) = \chi(c)^{\deg v} \chi(\prod_{i=1}^m \pi_v(\alpha_i)^{a_i}) \chi(-1)^{\deg v \deg f}$ and $\deg f \deg h = 0$ due to our choice of h. Thus we get an idèle class character ω on $I_K/K^\times$, which has finite order ($\leq d$).

Write $\operatorname{div} f = \sum_{v \in \operatorname{supp} f} a_v v$. Next we show that ω is ramified at the places $v \in \operatorname{supp} f$ such that a_v is not a multiple of d, and is unramified at the other places in $\operatorname{supp} f$. Let $v \in \operatorname{supp} f$, $v \neq \infty$. Suppose $\pi_v(t) = \prod_{i=1}^{\deg v} (t - \alpha_i)$ so that $a_i = a_v$ for $i = 1, \cdots, \deg v$. We know that the residue field κ_v of K_v is $k(\alpha_1)$ and the roots of $\pi_v(t)$ are the conjugates of α_1 under $Gal(\kappa_v/k)$. There is a polynomial $x(t) \in k[t]$ such that $x(\alpha_1)$ generates $\kappa_v^\times$, then $\mathrm{N}_{\kappa_v/k}(x(\alpha_1)) = \prod_{i=1}^{\deg v} x(\alpha_i)$ generates $k^\times$. By Chinese remainder theorem, there are polynomials $h_1(t), h_2(t) \in k[t]$ such that

$$h_1(t) \equiv x(t), \quad h_2(t) \equiv 1 \pmod{\pi_v}$$
$$h_1(t) \equiv h_2(t) \equiv 1 \pmod{\pi_w} \quad \text{for all } w \in \operatorname{supp} f,\ w \neq v, \infty.$$

Adding a multiple of $P(t)$ to h_1 and h_2 if necessary, we may assume that h_1 and h_2 have the same degree and the same leading coefficient. Let $h(t) = h_1(t)/h_2(t) \in K^\times$.

Then $h(t) \equiv x(t) \bmod \pi_v$ and $h(t) \equiv 1 \bmod \pi_w$ for $w \in \operatorname{supp} f$, $w \neq v$. From

$$1 = \omega\big(h(t)\big) = \omega_v\big(h(t)\big) \prod_{w \notin \operatorname{supp} f} \omega_w\big(h(t)\big) = \omega_v\big(h(t)\big)\chi\left(\prod_{i=1}^{m} h(\alpha_i)^{a_i}\right)$$

$$= \omega_v\big(h(t)\big)\chi\left(\prod_{i=1}^{\deg v} h(\alpha_i)\right)^{a_v} = \omega_v\big(h(t)\big)\chi\left(\prod_{i=1}^{\deg v} x(\alpha_i)\right)^{a_v}$$

we obtain $\omega_v\big(h(t)\big) = \chi\Big(\prod_{i=1}^{\deg v} x(\alpha_i)\Big)^{-a_v}$, which is not equal to one if d does not divide a_v since $\prod_{i=1}^{\deg v} x(\alpha_i) = \mathrm{N}_{\kappa_v/k}\big(x(\alpha_1)\big)$ generates $k^\times$ and χ has order d. On the other hand, if $d \mid a_v$, the same argument shows that $\omega_v\big(h(t)\big) = 1$ for all $x(t)$ in $k[t]$ prime to π_v. Hence ω_v is trivial on units in $K_v^\times$, that is, ω_v is unramified. Finally, suppose $\infty \in \operatorname{supp} f$. Choose $h_2(t) = 1+P(t)$ and $h_1(t) = 1+bP(t)$, $b \in k^\times$. Then $h(t) = h_1(t)/h_2(t)$ is $\equiv 1 \big(\bmod P(t)\big)$ and $h(t) \equiv b \big(\bmod \mathcal{P}_\infty\big)$. Then

$$\begin{aligned}1 = \omega(h) &= \omega_\infty\big(h(t)\big) \prod_{w \notin \operatorname{supp} f} \omega_w\big(h(t)\big) \\ &= \omega_\infty\big(h(t)\big) \prod_{i=1}^{m} \chi\big((P(\alpha_i) + b^{-1})/h_1(\alpha_i)\big)^{a_i} \\ &= \omega_\infty(b)\chi(b)^{-\deg f} = \omega_\infty(b)\chi(b)^{a_\infty}\end{aligned}$$

shows that ω_∞ is non-trivial on $\mathcal{U}_\infty$ if and only if d does not divide a_∞. This proves the first statement of Theorem 4.

(2) Given a polynomial $g(t) \in k[t]$ with $g(0) = 0$ and ψ, we shall construct an idèle class character η on $I_K/K^\times$ as follows. First define η_∞ on $K_\infty^\times$. Let η_∞ be trivial at $\pi_\infty = 1/t$ and $k^\times$. For a polynomial

$$h(x) = 1 + a_1 x + \cdots + a_u x^u = (1 - \gamma_1 x)\cdots(1 - \gamma_u x),$$

where $a_i \in k$, $\gamma_j \in \overline{k}$, define

$$\eta_\infty\big(h(\pi_\infty)\big) = \overline{\psi}\big(g(\gamma_1) + \cdots + g(\gamma_u)\big).$$

Since $g(0) = 0$, by adding more terms to h with coefficients 0 if necessary, we may assume $u \geq n$. Suppose $g(t) = b_n t^n + \cdots + b_1 t$, $b_i \in k$ and $b_n \neq 0$. Then $g(\gamma_1)+\cdots+g(\gamma_u) = b_n s_n + b_{n-1}s_{n-1} + \cdots + b_1 s_1$, where $s_i = \sum_{j=1}^{u} \gamma_j^i$ for $i = 1, \cdots, n$. Using Newton's identity (cf. [4], p.135)

$$s_i + a_1 s_{i-1} + a_2 s_{i-2} + \cdots + a_{i-1}s_1 + i a_i = 0$$

repeatedly, we see that $g(\gamma_1)+\cdots+g(\gamma_u)$ is a polynomial in $a_1,\cdots,a_n$ with coefficients in k, in which $-nb_na_n$ is the only term involving a_n, that is, $g(\gamma_1)+\cdots+g(\gamma_u)=-nb_na_n+\tilde{g}(a_1,\cdots,a_{n-1})$ for some polynomial $\tilde{g}$ over k with $n-1$ variables. In particular, this shows that $g(\gamma_1)+\cdots+g(\gamma_u)\in k$ and hence $\eta_\infty(h(\pi_\infty))$ is defined. Moreover, it also shows that $\eta_\infty(h(\pi_\infty))$ depends only on $a_1,\cdots,a_n$. In other words, if two polynomials $h_1,h_2\in k[t]$ with $h_1(0)=h_2(0)=1$ agree up to degree n term, then $\eta_\infty(h_1(\pi_\infty))=\eta_\infty(h_2(\pi_\infty))$. Consequently, any polynomial in π_∞ belonging to $1+\mathcal{P}_\infty^{n+1}$ has value 1 under η_∞. Extend η_∞ to $1+\mathcal{P}_\infty$ by continuity so that it is trivial on $1+\mathcal{P}_\infty^{n+1}$. Further, η_∞ is a character on $1+\mathcal{P}_\infty$ since $\eta_\infty(h_1(\pi_\infty)h_2(\pi_\infty))=\eta_\infty(h_1(\pi_\infty))\eta_\infty(h_2(\pi_\infty))$ for $h_1,h_2\in k[t]$ with $h_1(0)=h_2(0)=1$. Using the decomposition $K_\infty^\times=\langle\pi_\infty\rangle k^\times(1+\mathcal{P}_\infty)$, we combine the above definition of η_∞ to get a character η_∞ of $K_\infty^\times$.

Extend η_∞ to a character η on $K_\infty^\times\prod_{v\neq\infty}\mathcal{U}_v$ by letting η be 1 on $\prod_{v\neq\infty}\mathcal{U}_v$. As $K^\times\cap K_\infty^\times\prod_{v\neq\infty}\mathcal{U}_v=k^\times$ and η_∞ is trivial on $k^\times$, we may further extend η to a character of $I_K/K^\times$ since $I_K=K^\times\cdot K_\infty^\times\prod_{v\neq\infty}\mathcal{U}_v$. It clearly has finite order.

Assume $\gcd(q,n)=1$, we show that η is ramified at ∞. Indeed, we see from the above analysis that $\eta_\infty(h(\pi_\infty))=\overline{\psi}(-nb_na_n)$ for $h(x)=1+a_nx^n$, $a_n\in k$. Since $(q,n)=1$, $b_n\neq 0$, $-nb_na_n$ runs through all elements in k as a_n does so. As ψ is nontrivial, this shows that $\operatorname{cond}\eta=(n+1)\infty$, which has degree $n+1$.

Finally, note that at a place $v\neq\infty$ with the uniformizer $\pi_v=\prod_{j=1}^{\deg v}(t-\beta_{j,v})=t^{\deg v}\prod_{j=1}^{\deg v}(1-\beta_{j,v}\pi_\infty)$, from

$$1=\eta(\pi_v)=\eta_v(\pi_v)\eta_\infty(\pi_v)=\eta_v(\pi_v)\overline{\psi}(\sum_{j=1}^{\deg v}g(\beta_{j,v})),$$

we get that $\eta_v(\pi_v)=\psi(\sum_{j=1}^{\deg v}g(\beta_{j,v}))$, as asserted. This proves (2). □

Exercise 2. Let K be the rational function field over k and let S be a finite set of places of K. Show that $I_K=\left(\prod'_{v\notin S}K_v^\times\prod_{v\in S}1+\mathcal{P}_v^{n_v}\right)K^\times$ for positive integers n_v.

Corollary 4. *Let $f(t)$ be a monic polynomial over k with m distinct roots. If f is not a dth power in $k[t]$, then for a character $\chi\in\widehat{k^\times}$ with order d, we have*

$$\left|\sum_{x\in k}\chi(f(x))\right|\leq(m-1)\sqrt{q}.$$

As a consequence of Theorem 3, we derive estimates for Kloosterman sums.

Corollary 5. *(Weil) Let $b, c \in k$ not both zero. Suppose q is odd. Then for any nontrivial additive character ψ of k we have*

$$\left|\sum_{x\in k^\times} \psi(bx+cx^{-1})\right| \leq 2\sqrt{q}.$$

In particular, when $k = \mathbf{Z}/p\mathbf{Z}$ is a prime field with odd characteristic, we have

$$\left|\sum_{x=1}^{p-1} e^{2\pi i(bx+cx^{-1})/p}\right| \leq 2\sqrt{p}.$$

Proof. If either b or c is zero, then the other is nonzero and $\sum_{x\in k^\times} \psi(bx+cx^{-1}) = -1$ satisfies the inequality. Assume $bc \neq 0$. Let $a = 4bc$. Put $y = bx + cx^{-1}$. Then $y^2 - a = (bx - cx^{-1})^2$ is a square. One checks easily that two elements $x_1, x_2 \in k^\times$ are such that $bx_1 + cx_1^{-1} = bx_2 + cx_2^{-1}$ if and only if either $x_1 = x_2$ or $x_2x_1 = c/b$. Further $x_1^2 = c/b$ is solvable if and only if bc is a square, which is equivalent to $y^2 - a = 0$ is solvable.

Let χ be the quadratic character of $k^\times$, which takes value 1 at squares in $k^\times$ and -1 at nonsquares. Let $f(t) = t^2 - a$ and $g(t) = t$. By Theorem 3, (3), we have

$$\left|\sum_{y\in k} \psi(y)\chi(y^2-a)\right| \leq 2\sqrt{q}.$$

Further, the nontriviality of ψ implies $\sum_{y\in k} \psi(y) = 0$. Corollary 5 will follow from

$$\sum_{x\in k^\times} \psi(bx+cx^{-1}) = \sum_{y\in k} \psi(y)\big(\chi(y^2-a)+1\big).$$

To prove this, we distinguish two cases.

Case I. bc is not a square. Then $y^2 - a$ is never zero, and

$$\sum_{y\in k} \psi(y)\big(\chi(y^2-a)+1\big) = 2\sum_{\substack{y\in k\\ y^2-a\in k^{\times 2}}} \psi(y).$$

Write $y^2 - a = u^2$ for some $u \in k$. Then $(y+u)(y-u) = a = 4bc \neq 0$. We may write $y + u = 2bx$ for some $x \in k^\times$, then $y - u = 2cx^{-1}$. This shows that $y = bx + cx^{-1}$. On the other hand, we have checked that for y of the form $bx + cx^{-1}$, $y^2 - a$ is a nonzero square. Further, the map from x to y is two-to-one. This proves that

$$2\sum_{\substack{y\in k\\ y^2-a\in k^{\times 2}}} \psi(y) = \sum_{x\in k^\times} \psi(bx+cx^{-1}).$$

Case II. $bc = d^2$. Then $y^2 - a = 0$ has two solutions $y = \pm 2d = \pm\left(b \cdot \frac{d}{b} + c \cdot \frac{b}{d}\right)$. We have

$$\sum_{y \in k} \psi(y)\big(\chi(y^2 - a) + 1\big) = 2 \sum_{\substack{y \neq \pm 2d \\ y^2 - a \in k^{\times 2}}} \psi(y) + \psi(2d) + \psi(-2d).$$

The computation in Case I shows for $y \neq \pm 2d$ with $y^2 - a$ a square, the equation $y = bx + cx^{-1}$ is solvable for two x in $k^\times$; and for $y = \pm 2d$, the equation has only one solution $x = \pm d/b$. Again, we get the desired equation. □

Exercise 3. Prove Corollary 5 by constructing an idèle class character χ of $I_K/K^\times$ with $\operatorname{cond}\chi = 2 \cdot 0 + 2 \cdot \infty$ and $\sum_{\substack{\deg v = 1 \\ v \neq 0, \infty}} \chi_v(\pi_v) = \sum_{x \in k^\times} \psi(bx + cx^{-1})$.

We exhibit an application of Theorem 3 to Diophantine equations (cf. [8]). Given polynomials $f_1(x), \cdots, f_r(x)$ over a finite field k with q elements, consider the equatoins

$$y_1^{n_1} = f_1(x), \cdots, y_r^{n_r} = f_r(x).$$

We want to count the number N of simultaneous solutions $(x, y_1, \cdots, y_r) \in k^{r+1}$. Write $d_i = \gcd(n_i, q-1)$. Observe first that $y_i^{n_i} = f_i(x)$ and $y_i^{d_i} = f_i(x)$ have the same number of solutions. To simplify the computations, we assume that each $f_i(x)^j$ is a monic polynomial, which is not a d_ith power of another polynomial for $j \mid d_i$, $j \neq d_i$, and the f_i's are pairwise coprime. As in Chapter 2, §1, denote by $N_i(u)$ the number of solutions to $y^{d_i} = u$, then the function N_i can be expressed as a character sum : $N_i = \sum_{\chi} \chi$, where χ runs through all characters of $k^\times$ with order dividing d_i. Fix a generator χ of $\widehat{k^\times}$, then we may write

$$N_i = \sum_{j=0}^{d_i - 1} \chi^{j e_i}, \quad \text{where } e_i = (q-1)/d_i.$$

Therefore

$$\begin{aligned} N &= \sum_{x \in k} N_1\big(f_1(x)\big) \cdots N_r\big(f_r(x)\big) \\ &= \sum_{x \in k} \sum_{j_1 = 0}^{d_1 - 1} \cdots \sum_{j_r = 0}^{d_r - 1} \chi^{j_1 e_1}\big(f_1(x)\big) \cdots \chi^{j_r e_r}\big(f_r(x)\big) \\ &= q + \sum_{x \in k} \sum_{\substack{0 \le j_i \le d_i - 1 \\ \text{not all } j_i = 0}} \chi\big(f_1^{j_1 e_1} \cdots f_r^{j_r e_r}\big). \end{aligned}$$

Here the number q arises from $j_1 = \cdots = j_r = 0$, which is the main term in N. For each $J = (j_1, \cdots, j_r)$ with $0 \le j_i \le d_i - 1$ and not all $j_i = 0$, the polynomial

$f^J(x) = f_1(x)^{j_1e_1} \cdots f_r(x)^{j_re_r}$ is a monic polynomial, which is not a $q-1$st power of another polynomial. Hence by Corollary 4,

$$\left|\sum_{x\in k}\chi(f^J(x))\right| \le (m-1)\sqrt{q},$$

where m is the number of distinct roots of $f_1(x)\cdots f_r(x)$. This shows that

$$|N-q| \le (d_1\cdots d_r - 1)(m-1)\sqrt{q}.$$

Theorem 5. *(Schmidt [8]) Let $f_1(x), \cdots, f_r(x)$ be monic polynomials in $k[t]$ which are pairwise coprime. Let N be the number of simultaneous solutions to*

$$y_1^{n_1} = f_1(x), \cdots, y_r^{n_r} = f_r(x)$$

in k^{r+1}. Suppose that none of $f_i(x)^j$ is a d_ith power of another polynomial for $j \mid d_i$, $j \ne d_i$, where $d_i = \gcd(n_i, q-1)$. Then

$$|N-q| \le (d_1\cdots d_r - 1)(m-1)\sqrt{q},$$

where m is the number of distinct roots of $f_1(x)\cdots f_r(x)$. In particular, when q is large, we have $N = q + O(\sqrt{q})$.

Take the special case where $k = \mathbf{Z}/p\mathbf{Z}$ is a prime field, p odd, $f_1(x) = x + 1, \cdots, f_r(x) = x + r$, and the equations are

$$y_1^2 = x+1,\ \ y_2^2 = x+2, \cdots,\ \ y_r^2 = x+r.$$

Solutions to the system yield the values $x \in k$ such that $x+1, \cdots, x+r$ are simultaneously squares in k. If one of $x+1, \cdots, x+r$ should be zero, then it gives rise to 2^{r-1} solutions to the system. Denote by N_0 the total number of solutions $(x, y_1, \cdots, y_r)$ to the system such that $f_1(x)\cdots f_r(x) = 0$. Then $N_0 \le r \cdot 2^{r-1}$. Among the remaining solutions, corresponding to one value of x, there are 2 values for each y_i, hence 2^r solutions to the system. Denote by R_r the number of elements x in k such that $x+1, \cdots, x+r$ are quadratic residues. Then $N = 2^rR_r + N_0$. We have shown

Corollary 6. *For p large, the number R_r of elements x in $\mathbf{Z}/p\mathbf{Z}$ such that $x + 1$, $x+2, \cdots$, $x+r$ are quadratic residues* $\bmod p$ *is $R_r = \frac{p}{2^r} + O(\sqrt{p})$.*

Of course, the same conclusion holds if $1, 2, \cdots, r$ are replaced by r distinct numbers.

§3 More character sum estimates

In this section we prove more character sum estimates, which will be used in Chapter 9 when we talk about applications of number theory to combinatorics. The results in this section are extracted from [7].

The field k is the same as in §1, and k_n denotes degree n field extension of k in an algebraic closure $\bar{k}$ of k. Let N_n be the kernel of the norm map $\mathrm{N}_{k_n/k}$. Let t be an element in k_n such that $k_n = k(t)$. Assume $n \geq 2$. Put

$$S_n = \left\{ \frac{t^q + a}{t + a} : a \in k \cup \{\infty\} = \mathbf{P}^1(k) \right\}.$$

Here, when $a = \infty$, ∞/∞ is interpreted as 1. Recall that the Frobenius automorphism $\sigma \in Gal(k_n/k)$ maps x to x^q, hence $t^q + a = \sigma(t + a)$ and S_n is contained in N_n.

Theorem 6. *For each nontrivial character χ of N_n, we have*

$$\left| \sum_{s \in S_n} \chi(s) \right| \leq (n-2)\sqrt{q}.$$

Proof. Let $P(T)$ be the irreducible polynomial of t over k, and let w be the place of $K = k(t)$ consisting of the roots of $P(T)$. Then $\deg w = n$ and $P(T)$ is a uniformizer of K_w. The k-homomorphism from $\mathcal{O}_w$ to k_n sending T to t is surjective with kernel being the ideal generated by $P(T)$, that is, $\mathcal{P}_w$. Hence we get an isomorphism $\mathcal{O}_w/\mathcal{P}_w \cong k_n$, which induces the isomorphism $\mathcal{U}_w/1 + \mathcal{P}_w \cong k_n^\times$. Recall from Theorem 3 in Chapter 1 that $N_n = \{\sigma(x)/x : x \in k_n^\times\}$. This shows that the homomorphism $x \mapsto x/\sigma(x)$ from $k_n^\times$ to N_n induces an isomorphism $k_n^\times/k^\times \cong N_n$. Put together, we get

$$\mathcal{U}_w/k^\times(1 + \mathcal{P}_w) \cong k_n^\times/k^\times \cong N_n.$$

Therefore a character χ of N_n yields a character ω_w on $\mathcal{U}_w$ trivial on $k^\times(1 + \mathcal{P}_w)$, and ω_w has conductor w if χ is nontrivial. Extend ω_w to a character ω on $\langle \pi_\infty \rangle \prod_v \mathcal{U}_v$ by letting ω be trivial on $\langle \pi_\infty \rangle \prod_{v \neq w} \mathcal{U}_v$. As $\langle \pi_\infty \rangle \prod_v \mathcal{U}_v \cap K^\times = k^\times$ and ω is trivial on $k^\times$, we may extend ω to a character of I_K trivial on $K^\times$. Thus ω is an idèle class character of $I_K/K^\times$, which is of finite order, and $\mathrm{cond}\,\omega = w$. By Corollary 3, we have

$$\left| \sum_{\deg v = 1} \omega_v(\pi_v) \right| \leq (n-2)\sqrt{q}.$$

At $v = \infty$, $\omega_\infty(\pi_\infty) = 1 = \chi(1)$. At a place v with $\pi_v = T + a$, we have

$$\omega_v(\pi_v) = \omega_w(T + a)^{-1} = \chi\big((t + a)/\sigma(t + a)\big)^{-1} = \chi\left(\frac{t^q + a}{t + a}\right)$$

from the construction of ω_w. This proves Theorem 6. □

Next let

$$S'_n = \left\{ \frac{1}{t+a} : a \in k \cup \{\infty\} \right\}.$$

When $a = \infty$, $\frac{1}{\infty}$ is interpreted as 0. We prove

Theorem 7. *For each nontrivial additive character ψ of F_n, we have*

$$\left| \sum_{s \in S'_n} \psi(s) \right| \leq (2n-2)\sqrt{q}.$$

Proof. Let w be the same place of $K = k(T)$ as in the previous proof. The k-homomorphism sending T to t yields an isomorphism $\mathcal{O}_w/\mathcal{P}_w \cong k_n$. Since the elements in k_n are the roots of $T^{q^n} - T$ in $\bar{k}$, the polynomial $T^{q^n} - T$ decomposes as a product of monic irreducible polynomials over k of degree dividing n. In particular, $P(T)$, the irreducible polynomial of t, divides $T^{q^n} - T$ exactly once. Hence we may take π_w to be $T^{q^n} - T$. We saw in Chapter 4, §2, that the nonzero elements in k_n lifts to $q^n - 1$st roots of unity in K_w (by Hensel's Lemma). Denote by U the group they form. Then U is contained in the group of units $\mathcal{U}_w$ and, in fact, $\mathcal{U}_w = U(1+\mathcal{P}_w)$. Thus $\mathcal{U}_w/U(1+\mathcal{P}_w^2)$ is isomorphic to $1+\mathcal{P}_w/1+\mathcal{P}_w^2$, which is isomorphic to $\mathcal{O}_w/\mathcal{P}_w$ under the map $1+\pi_w h \mapsto h$. Combined with the map $T \mapsto t$, we may identify $\mathcal{U}_w/U(1+\mathcal{P}_w^2)$ with k_n. Hence a nontrivial character ψ of k_n gives rise to a character η_w of $\mathcal{U}_w$ trivial on $U(1+\mathcal{P}_w^2)$ with conductor $= 2w$. As before, we may extend η_w to an idèle class character η of $I_K/K^\times$ by requiring it to be trivial on $K^\times \prod_{v \neq w} \mathcal{U}_v \langle \pi_\infty \rangle$. This is possible since $k^\times \subset U$ in $K_w^\times$ and η_w is trivial on $k^\times$. Clearly, $\operatorname{cond} \eta = 2w$ and it has degree $2n$. By Corollary 3, we get

$$\left| \sum_{\deg v = 1} \eta_v(\pi_v) \right| \leq (2n-2)\sqrt{q}.$$

We have $\eta_\infty(\pi_\infty) = 1 = \psi(0)$, and the above inequality can be rewritten as

$$\left| 1 + \sum_{a \in k} \eta_w(T+a)^{-1} \right| \leq (2n-2)\sqrt{q}.$$

The last step is to show $\eta_w(T+a)^{-1} = \psi(\frac{1}{t+a})$. First, we need to find $u \in U$ and $h \in \mathcal{O}_w$ such that $(T+a)^{-1} \equiv u(1+\pi_w h) \pmod{\mathcal{P}_w^2}$. Raising this congruence equation to the q^nth power yields $(T+a)^{-q^n} \equiv u^{q^n} \equiv u \pmod{\mathcal{P}_w^2}$. Thus

$$1 + \pi_w h \equiv (T+a)^{q^n}/(T+a) \equiv \frac{T^{q^n}+a}{T+a} \equiv 1 + \pi_w \frac{1}{T+a} \pmod{\mathcal{P}_w^2}.$$

In other words, $(T+a)^{-1}$ in $\mathcal{U}_w/U(1+\mathcal{P}_w^2)$ is identified with $(T+a)^{-1}$ in $\mathcal{O}_w/\mathcal{P}_w$, which is then identified with $(t+a)^{-1}$ in k_n. So we have $\eta_w(T+a)^{-1} = \psi(\frac{1}{t+a})$, as desired. □

Observe that, for $a \in k \cup \{\infty\}$, $\frac{t^q+a}{t+a} = 1 + \frac{t^q-t}{t+a}$, hence $S_n = bS'_n + c$ with $b = t^q - t$ and $c = 1$. In other words, S_n and S'_n are affine transformations of each other. As ψ runs through all nontrivial additive characters of k_n, so does ψ^b. Hence we can reformulate Theorem 7 as

Theorem 7′. *For all nontrivial additive characters ψ of k_n, we have*

$$\left|\sum_{s \in S_n} \psi(s)\right| \leq (2n-2)\sqrt{q}.$$

Consider the case $n = 2$. We observed that $S_2 \subset N_2$. Further, since both sets have the same cardinality $q+1$, they are equal. We record this in

Corollary 7. *For all nontrivial additive character ψ of k_2, we have*

$$\left|\sum_{x \in N_2} \psi(x)\right| \leq 2\sqrt{q}.$$

By studying the action of Frobenius automorphism on certain étale cohomology, Deligne in [3] proved the following estimates for generalized Kloosterman sums.

Theorem 8. *(Deligne) For all nontrivial additive characters ψ of k_n, we have*

$$\left|\sum_{x \in N_n} \psi(x)\right| \leq nq^{\frac{n-1}{2}}.$$

Corollary 7 is a special case of Theorem 7. What we have shown is that for this special case, the estimate follows from the Riemann hypothesis for curves over finite fields.

Theorem 9. *For each nontrivial character (χ, ψ) of $N_n \times F_n$, we have*

$$\left|\sum_{x \in S_n} \chi(s)\psi(s)\right| \leq (2n-2)\sqrt{q}.$$

Proof. If either χ or ψ is trivial, then the inequality follows from Theorems 6 and 7′. Suppose χ and ψ are both nontrivial. Let w be the place corresponding to the irreducible polynomial $P(T)$ of t. Let ω and η be the idèle class characters of $I_K/K^\times$, $K = k(T)$, arising from χ and ψ^b as in the proofs of Theorems 6 and 7. Here $b = t^q - t$ is such that $S = bS' + 1$. Recall that ω has conductor w and η has conductor $2w$. Thus $\omega\eta$ is an idèle class character of $I_K/K^\times$ of finite order, and has conductor $2w$. It follows from the definition of ω and η that at a place v with $\pi_v = t + a$, we have

$$\omega_v\eta_v(\pi_v) = \omega_w\eta_w(T+a)^{-1} = \chi(\frac{t^q+a}{t+a})\psi^b(\frac{1}{t+a}) = \chi(\frac{t^q+a}{t+a})\psi(\frac{t^q+a}{t+a})\psi(-1),$$

and $\omega_\infty\eta_\infty(\pi_\infty) = \chi(1)\psi^b(0) = \chi(1)\psi(1)\psi(-1)$. From Corollary 3 we obtain

$$\left|\sum_{\deg v=1} \omega_v\eta_v(\pi_v)\right| = \left|\sum_{s\in S_n} \chi(s)\psi(s)\psi(-1)\right| \le (2n-2)\sqrt{q},$$

which proves the desired inequality. □

Again, when $n = 2$, we have $S_2 = N_2$ and Theorem 9 for this case becomes

Corollary 8. *For each nontrivial character (χ, ψ) of $N_2 \times F_2$, one has*

$$\left|\sum_{x\in N_2} \chi(x)\psi(x)\right| \le 2\sqrt{q}.$$

This can be regarded as a Gauss sum over N_2, the kernel of the norm map $N_{k_2/k}$. The norm map $N_{k_2/k}$ can be described by a quadratic form $q(x, y)$ over k so that N_2 consists of the solutions to $q(x, y) = 1$, which is an ellipse. This may be viewed as the "compact form" of the hyperbola $xy = 1$ in k^2. Thus the "split" analogue of Corollary 8 should be

$$\left|\sum_{x\in k^\times} \mu(x)\nu(x^{-1})\psi(bx + cx^{-1})\right| = \left|\sum_{x\in k^\times} \mu\nu^{-1}(x)\psi(bx + cx^{-1})\right| \le 2\sqrt{q},$$

where either $\mu\nu^{-1}$ is a nontrivial character of $k^\times$ or b, c are not both zero in k. Note that the characters of k^2 are $\psi(bx + cy)$ as (b, c) runs through all elements in k^2 (cf. Exercise 5 in Chapter 1). This is indeed true as we prove below, the special case $b = c = 1$ was proved by Mordell. The sum is called twisted Kloosterman sum when $\mu\nu^{-1}$ is nontrivial.

Theorem 10. *Let ψ be a nontrivial additive character of k. Suppose either χ is a nontrivial character of $k^\times$ or $(b, c) \in k^2 - \{(0,0)\}$. Then*

$$\left|\sum_{x\in k^\times} \chi(x)\psi(bx + cx^{-1})\right| \le 2\sqrt{q}.$$

Proof. If χ is trivial, the above inequality is Corollary 5, so assume χ nontrivial. Further, if b or c is zero, the twisted Kloosterman sum becomes the usual Gauss sum, hence it has absolute value $\sqrt{q}$. Assume $bc \neq 0$. We shall construct an idèle class character ξ of $K = k(T)$ with conductor $2{\cdot}0+2{\cdot}\infty$ using χ and ψ. Choose T to be the uniformizer at 0. Any element in $\mathcal{U}_0/1+\mathcal{P}_0^2$ can be written as $x(1+yT)$ with $x \in k^\times$, $y \in k$. Define ξ_0 on $\mathcal{U}_0/1+\mathcal{P}_0^2$ by $\xi_0(x(1+yT)) = \chi(x)^{-1}\overline{\psi}(cy)$. Choose $\pi_\infty = 1/T$. Likewise, elements in $\mathcal{U}_\infty/1+\mathcal{P}_\infty^2$ can be written as $x(1+y\pi_\infty)$ with $x \in k^\times$, $y \in k$. Define ξ_∞ on $\mathcal{U}_\infty/1+\mathcal{P}_\infty^2$ by $\xi_\infty(x(1+y\pi_\infty)) = \chi(x)\overline{\psi}(by)$. One checks easily that ξ_0, ξ_∞ so defined are characters of $\mathcal{U}_0$, $\mathcal{U}_\infty$, respectively. Extend ξ_0, ξ_∞ to a character ξ on $\prod_v \mathcal{U}_v\langle\pi_\infty\rangle$ by letting ξ be trivial on $\prod_{v\neq 0,\infty} \mathcal{U}_v\langle\pi_\infty\rangle$. Since ξ is trivial on $\prod_v \mathcal{U}_v\langle\pi_\infty\rangle \cap K^\times = k^\times$, it extends to an idèle class character of $I_K/K^\times$, which has finite order and $\operatorname{cond}\xi = 2\cdot 0 + 2\cdot\infty$. By Corollary 3, we have

$$\left|\sum_{\substack{\deg v=1\\ v\neq 0,\infty}} \xi_v(\pi_v)\right| \leq 2\sqrt{q}.$$

At a place v with $\pi_v = T+a$, $a \in k^\times$, it follows from the definition of ξ that $\xi_v(\pi_v) = \xi_0(T+a)^{-1}\xi_\infty(T+a)^{-1} = \xi_0(a(1+a^{-1}T))^{-1}\xi_\infty(\pi_\infty^{-1}(1+a\pi_\infty))^{-1} = \chi(a)\psi(ca^{-1})\psi(ba)$. This proves the theorem. □

Actually, when Deligne proved the estimates for generalized Kloosterman sums (Theorem 8), he did it for the split case; he has a simple argument deriving the same estimate for nonsplit form from that of the split form. Based on this philosophy and Mordell's result (Theorem 10), he conjectured in [3] that the same estimates should hold for twisted generalized Kloosterman sums. Namely,

Conjecture. *(Deligne) For nontrivial characters (χ, ψ) of $N_n \times F_n$, we have*

$$\left|\sum_{x\in N_n} \chi(x)\psi(x)\right| \leq nq^{\frac{n-1}{2}}.$$

Corollary 8 verifies the above conjecture for the case $n = 2$. It shows that for $n = 2$ the conjecture follows from the Riemann hypothesis for curves over finite fields.

We end this section by proving a character sum estimate first obtained by N. Katz [5] using a geometric method like Deligne. An étale algebra B over k of degree n is a product $F_1 \times \cdots \times F_r$ of finite field extensions F_i of k with total degree n. The field k is diagonally imbedded in B. An element $x = (x_1, \cdots, x_r)$ of B is called *regular* if $F_i = k(x_i)$ for $i = 1, \cdots, r$, and the components of x are pairwise nonconjugate over k, that is, their irreducible polynomials are pairwise coprime.

Theorem 11. *(Katz) Let B be an étale algebra over k of degree n and let x be a regular element of B. Then for every nontrivial character χ of $B^\times$, we have*

$$\left|\sum_{\substack{a\in k\\ x-a\in B^\times}} \chi(x-a)\right| \le (n-1)\sqrt{q}.$$

Proof. Write $B = F_1 \times \cdots \times F_r$, $x = (x_1, \cdots, x_r)$ and $\chi = (\chi_1, \cdots, \chi_r)$, where χ_i is a character of $F_i^\times$. As there are at most q terms in the charcter sum, the inequality trivially holds if $n - 1 \ge \sqrt{q}$. Hence we assume $n - 1 < \sqrt{q}$, which implies $r \le n \le \sqrt{q}$. Denote by $P_i(T)$ the irreducible polynomial of x_i over k. Let $v_1, \cdots, v_r$ be the places of $K = k(T)$ such that $P_i(T)$ is a uniformizer at v_i. Then $v_1, \cdots, v_r$ and ∞ are distinct places of K with the residue fields isomorphic to $F_1, \cdots, F_r, k$, under the map $T \mapsto x_1, \cdots, x_r$, and $T^{-1} \mapsto 0$, respectively. For $1 \le i \le r$, let ω_{v_i} be the character of $\mathcal{U}_{v_i}$, trivial on $1 + \mathcal{P}_{v_i}$ and given by χ_i^{-1} on $\mathcal{U}_{v_i}/1 + \mathcal{P}_{v_i} \cong F_i^\times$. Let ω_∞ be the character of $\mathcal{U}_\infty$, trivial on $1 + \mathcal{P}_\infty$ and given by $\chi_1 \cdots \chi_r$ on $\mathcal{U}_\infty/1 + \mathcal{P}_\infty \cong k^\times$, and define $\omega_\infty(\pi_\infty) = 1$, where $\pi_\infty = \frac{1}{T}$. At $v \ne v_1, \cdots, v_r, \infty$, define ω_v to be trivial on $\mathcal{U}_v$. Then $\omega = \prod_v \omega_v$ is a character of $\prod_v \mathcal{U}_v\langle\pi_\infty\rangle$, which is trivial on $\prod_v \mathcal{U}_v\langle\pi_\infty\rangle \cap K^\times = k^\times$ since for $a \in k^\times$, $\omega(a) = \omega_{v_1}(a) \cdots \omega_{v_r}(a)\omega_\infty(a) = \chi_1^{-1}(a) \cdots \chi_r^{-1}(a)(\chi_1 \cdots \chi_r)(a) = 1$. Thus ω extends to an idèle class character of $I_K/K^\times$ of finite order and with $0 < \operatorname{cond}\omega \le v_1 + \cdots + v_r + \infty$ since at least one of $\chi_1, \cdots, \chi_r$ is nontrivial.

Let Z be the collection of the components of x lying in k, then $a \in k$ is such that $x - a \in B^\times$ if and only if $a \notin Z$. At the places $v = a \in k - Z$, we may choose $\pi_v = T - a$ and find

$$\begin{aligned}\omega_v(\pi_v) &= \omega_{v_1}(T-a)^{-1} \cdots \omega_{v_r}(T-a)^{-1}\omega_\infty(T-a)^{-1}\\ &= \chi_1(x_1 - a) \cdots \chi_r(x_r - a)\omega_\infty\left(\pi_\infty^{-1}(1 - a\pi_\infty)\right)^{-1}\\ &= \chi(x-a).\end{aligned}$$

First assume that ω is ramified at all places in $Z \cup \{\infty\}$. Then, by Corollary 3,

$$\left|\sum_{\substack{\deg v=1\\ \omega_v \text{ unr.}}} \omega_v(\pi_v)\right| = \left|\sum_{a\in k-Z} \chi(x-a)\right| \le (n-1)\sqrt{q}$$

since $\deg\operatorname{cond}\omega \le \deg v_1 + \cdots + \deg v_r + 1 = n + 1$. Next suppose that there are j places in $Z \cup \{\infty\}$ where ω is unramified. Then $\deg\operatorname{cond}\omega \le n + 1 - j$, and by Corollary 3,

$$\left|\sum_{\substack{\deg v=1\\ \omega_v \text{ unr.}}} \omega_v(\pi_v)\right| \le (n-1-j)\sqrt{q},$$

which implies

$$\left|\sum_{a\in k-Z}\chi(x-a)\right|\le(n-j-1)\sqrt{q}+j\le(n-1)\sqrt{q}.$$

This completes the proof of Theorem 11. □

---

§4 Davenport–Hasse identity in general form

Now that we have acquired proficiency on idèle class characters, we are ready to take another point of view of the Davenport–Hasse identity discussed in Chapter 1, §5. First recall the identity briefly. Given a finite field k of q elements, choose a nontrivial multiplicative character χ of $k^\times$ and nontrivial additive character ψ of k; the Gauss sum

$$g(\chi,\psi)=\sum_{x\in k^\times}\chi(x)\psi(x)$$

has absolute value $\sqrt{q}$. The Davenport–Hasse identity relates $g(\chi,\psi)$ to $g(\chi\circ\mathrm{N}_{k_n/k},\psi\circ\mathrm{Tr}_{k_n/k})$, a Gauss sum over the degree n field extension k_n of k (cf. Theorem 6 in Chapter 1) :

$$(-g(\chi,\psi))^n=-g(\chi\circ\mathrm{N}_{k_n/k},\psi\circ\mathrm{Tr}_{k_n/k}).$$

As shown in the proof of Theorem 10, there is an idèle class character ξ of $I_K/K^\times$, where $K=k(T)$, such that $\operatorname{cond}\xi=1\cdot 0+2\cdot\infty$ with ξ_0 on $\mathcal{U}_0/1+\mathcal{P}_0$ given by $\xi_0(x)=\chi(x)^{-1}$ and ξ_∞ on $\mathcal{U}_\infty/1+\mathcal{P}_\infty^2$ given by $\xi_\infty(x(1+y\pi_\infty))=\chi(x)\overline{\psi}(y)$. Here $\pi_\infty=1/T$, $x\in k^\times$, and $y\in k$. Further we have shown that

$$\sum_{\substack{\deg v=1\\ v\neq 0,\infty}}\xi_v(\pi_v)=\sum_{x\in k^\times}\chi(x)\psi(x)=g(\chi,\psi).$$

Since $\operatorname{cond}\xi$ has degree equal to 3, writing $L(s,\xi)=P(q^{-s},\xi)$, we know from Theorem 2 in Chapter 5 that $P(u,\xi)$ is a polynomial in u with degree 1. It then follows from the definition of $L(s,\xi)$ that

$$P(u,\xi)=1+\sum_{\substack{\deg v=1\\ \xi_v\ \text{unr.}}}\xi_v(\pi_v)u=1+g(\chi,\psi)u.$$

Next let E be the composite of K with k_n, i.e., $E=k_n(T)$. The character ξ composed with the norm map $\mathrm{N}_{E/K}$ from I_E to I_K yields an idèle class character $\xi\circ\mathrm{N}_{E/K}$ of I_E. As shown in Proposition 1, E is unramified over K so that for every place v of K and every place w of E dividing v, the local norm N_{E_w/K_v} maps

$\mathcal{U}_w$ onto $\mathcal{U}_v$. There is only one place of E dividing 0 (resp. ∞) in K, which is still denoted by 0 (resp. ∞). Since ξ is unramified outside 0 and ∞, so is $\xi \circ \mathrm{N}_{E/K}$. We study $\xi_0 \circ \mathrm{N}_{E_0/K_0}$ and $\xi_\infty \circ \mathrm{N}_{E_\infty/K_\infty}$. As $\mathrm{N}_{E_0/K_0}(1+\mathcal{P}_{0,E}) = 1+\mathcal{P}_{0,K}$, $\xi_0 \circ \mathrm{N}_{E_0/K_0}$ is trivial on $1+\mathcal{P}_{0,E}$ and on $\mathcal{U}_{0,E}/1+\mathcal{P}_{0,E} \cong k_n^\times$ it is given by

$$\xi_0 \circ \mathrm{N}_{E_0/K_0}(x) = \xi_0(\mathrm{N}_{k_n/k}\, x) = \chi(\mathrm{N}_{k_n/k}\, x)^{-1}.$$

Likewise $\mathrm{N}_{E_\infty/K_\infty}(1+\mathcal{P}^2_{\infty,E}) = 1+\mathcal{P}^2_{\infty,K}$, hence $\xi_\infty \circ \mathrm{N}_{E_\infty/K_\infty}$ is trivial on $1+\mathcal{P}^2_{\infty,E}$ and on $\mathcal{U}_{\infty,E}/1+\mathcal{P}^2_{\infty,E}$ it is given by, with $x \in k_n^\times$ and $y \in k_n$,

$$\begin{aligned}\xi_\infty \circ \mathrm{N}_{E_\infty/K_\infty}\left(x(1+y\pi_\infty)\right) &= \xi_\infty\left((\mathrm{N}_{k_n/k}\, x)\,\mathrm{N}_{E_\infty/K_\infty}(1+y\pi_\infty)\right)\\ &= \xi_\infty\left((\mathrm{N}_{k_n/k}\, x)(1+(\mathrm{Tr}_{k_n/k}\, y)\pi_\infty)\right)\\ &= \chi(\mathrm{N}_{k_n/k}\, x)\overline{\psi}(\mathrm{Tr}_{k_n/k}\, y).\end{aligned}$$

The same argument as above shows that

$$P(u, \xi \circ \mathrm{N}_{E/K}) = 1 + g(\chi \circ \mathrm{N}_{k_n/k}, \psi \circ \mathrm{Tr}_{k_n/k})u.$$

Hence the Davenport–Hasse identity can be reformulated as

$$P(u^n, \xi \circ \mathrm{N}_{E/K}) = \prod_{i=1}^{n} P(\zeta_n^i u, \xi),$$

where ζ_n is a primitive nth root of unity.

Exercise 4. Check that the proof of the Davenport–Hasse identity given in §5, Chapter 1 is indeed the above reformulation, without using idèlic language explicitly.

Once formulated as an identity on $P(u,\omega)$ attached to an idèle class character ω, it is ready to be generalized. Allow K to be any function field with one variable with k being its field of constants. For an idèle class character ω of $I_K/K^\times$, define $P(u,\omega)$ to be the polynomial so that $P(q^{-s},\omega)$ is the numerator of $L(s,\omega)$ as in §1.

Theorem 12. *(Davenport–Hasse identity) Let K be any function field of one variable with k being its field of constants. Let $E = K \cdot k_n$ and let ω be an idèle class character of $I_K/K^\times$. Then*

$$L(s, \omega \circ \mathrm{N}_{E/K}) = \prod_{i=1}^{n} L(s, \eta^i \omega),$$

where η is the character of I_K sending I_K^1 to 1 and any idèle in I_K of degree 1 to ζ_n. Equivalently, we have

$$P(u^n, \omega \circ \mathrm{N}_{E/K}) = \prod_{i=1}^{n} P(\zeta_n^i u, \omega).$$

Proof. By Proposition 1, the extension E over K is everywhere unramified, and, by Corollary 1, for every place v of K and every palce w of E dividing v, local norm N_{E_w/K_v} sends $\mathcal{U}_w$ to $\mathcal{U}_v$, hence $\omega \circ \mathrm{N}_{E/K}$ is ramified at the places w dividing v where ω is ramified, and is unramified elsewhere. Let v be a place of K where ω is unramified. We show that

$$\prod_{w|v} L(s, \omega_v \circ \mathrm{N}_{E_w/K_v}) = \prod_{i=1}^{n} L(s, \eta_v^i \omega_v).$$

Write κ_v for the residue field of K_v. As in Proposition 1, set $d_v = [\kappa_v \cap k_n : k]$. Note that d_v is the g.c.d. of $\deg v$ and n. Then there are d_v places w of E dividing v, $[E_w : K_v] = n/d_v$, and $\deg w = (\deg v)/d_v$. Since E_w is unramified over K_v, we shall choose π_w to be π_v. We start with the left hand side :

$$\begin{aligned}
\prod_{w|v} L(s, \omega_v \circ \mathrm{N}_{E_w/K_v}) &= \prod_{w|v} \left(1 - \omega_v(\pi_v)^{[E_w:K_v]} q^{-n(\deg w)s}\right)^{-1} \\
&= \left(1 - \omega_v(\pi_v)^{n/d_v} q^{-sn \deg v/d_v}\right)^{-d_v} \\
&= \prod_{i=1}^{n/d_v} \left(1 - \omega_v(\pi_v) \zeta_{n/d_v}^i q^{-s \deg v}\right)^{-d_v} \\
&= \prod_{i=1}^{n} (1 - \omega_v(\pi_v)(\zeta_n^i q^{-s})^{\deg v})^{-1} \quad \text{since } d_v = \gcd(n, \deg v) \\
&= \prod_{i=1}^{n} (1 - \omega_v(\pi_v)\eta_v^i(\pi_v) q^{-s \deg v})^{-1} \\
&= \prod_{i=1}^{n} L(s, \eta_v^i \omega_v),
\end{aligned}$$

which is the right hand side. This proves the first statement. In the case that ω is not of the form $|\ |^{s_0}$ with $s_0 \in \mathbf{C}$, the second statement is the same as the first by setting $u = q^{-s}$. Finally, assume that ω is trivial on I_K^1 and ω sends any idèle of degree 1 to a. Then

$$L(s, \omega) = \frac{P(q^{-s}, \omega)}{(1 - aq^{-s})(1 - aq^{1-s})}$$

so that

$$\prod_{i=1}^{n} L(s, \eta^i \omega) = \frac{\prod_{i=1}^{n} P(\zeta_n^i u, \omega)}{(1 - a^n u^n)(1 - a^n q^{n(1-s)})}, \quad \text{where } u = q^{-s}.$$

On the other hand, for any idèle y in I_E of degree 1, $\mathrm{N}_{E/K}(y)$ has degree n by Proposition 1. Thus $\omega \circ \mathrm{N}_{E/K}(y) = a^n$ and

$$L(s, \omega \circ \mathrm{N}_{E/K}) = \frac{P(q^{-ns}, \omega \circ \mathrm{N}_{E/K})}{(1 - a^n q^{-ns})(1 - a^n q^{n(1-s)})}.$$

It is then clear that $\prod_{i=1}^{n} L(s,\eta^i\omega) = L(s,\omega \circ \mathrm{N}_{E/K})$ is equivalent to $\prod_{i=1}^{n} P(\zeta_n^i u,\omega) = P(u^n,\omega \circ \mathrm{N}_{E/K})$. □

Observe that $\eta_1,\cdots,\eta_n$ are the characters of $I_K/K^\times \mathrm{N}_{E/K}(I_E)$. Davenport–Hasse identity can be stated in the following more general form, which we shall not prove.

Theorem 13. *Let K be a function field of one variable over a finite field and let F be a finite abelian extension of K. For any idèle class character ω of $I_K/K^\times$, we have*

$$L(s,\omega \circ \mathrm{N}_{F/K}) = \prod_{\eta} L(s,\eta\omega),$$

where η runs through all characters of $I_K/K^\times \mathrm{N}_{F/K}(I_F)$.

Note that $\eta\omega' s$ are the characters χ of $I_K/K^\times$ such that $\chi \circ \mathrm{N}_{F/K} = \omega \circ \mathrm{N}_{F/K}$, that is, χ and ω "lift" to the same character of $I_F/F^\times$. We may write $I_F = GL_1(A_F)$ so that idèle class characters of $I_F/F^\times$ are "automorphic forms" of $GL_1(A_F)$. The map $\omega \mapsto \omega \circ \mathrm{N}_{F/K}$ is a "lifting" from automorphic forms of $GL_1(A_K)$ to $GL_1(A_F)$. Theorem 12 says that the L–function attached to an automorphic form ξ of $GL_1(A_F)$ which lies in the image of lifting is the product of the L–functions of the automorphic forms ω of $GL_1(A_K)$ which lift to ξ. The group GL_1 may be replaced by a reductive group and the "lifting map" is also called "base change" map. This is a very difficult problem in group representations, it is known only for $GL(2)$ by the work of Langlands [6] and $GL(n)$ by Arthur and Clozel [1].

Write

$$P(u,\omega) = 1 + a_1u + \cdots + a_ru^r = (1-b_1u)\cdots(1-b_ru), \quad \text{where } r = \deg P(u,\omega).$$

Theorem 11 implies that the conductors $f(\omega)$ and $f(\omega \circ \mathrm{N}_{E/K})$ have the same degree. Hence we may write

$$P(u,\omega \circ \mathrm{N}_{E/K}) = 1 + A_1u + \cdots + A_ru^r = (1-B_1u)\cdots(1-B_ru).$$

Theorem 11 clearly implies that, after reordering the roots of $P(u,\omega \circ \mathrm{N}_{E/K})$ if necessary, we have $B_i = b_i^n$ for $i = 1,\cdots,r$. Conversely, if $B_i = b_i^n$ for $i = 1,\cdots,r$, then we have $P(u,\omega \circ \mathrm{N}_{E/K}) = \prod_{i=1}^{n} P(\zeta_n^i u,\omega)$. Hence Theorem 11 is equivalent to

Theorem 12′. *Let E and ω be as in Theorem 12. Then the roots of $P(u,\omega \circ \mathrm{N}_{E/K})$ are the nth power of those of $P(u,\omega)$.*

In Chapter 5 we proved a functional equation relating $L(s,\omega)$ to $L(1-s,\omega^{-1})$. Expressed in terms of $P(u,\omega)$, we may write it as

$$P(u,\omega) = \epsilon(\omega)u^rP(\frac{1}{qu},\omega^{-1}),$$

where $\epsilon(\omega)$ is a constant, and r is the degree of $P(u,\omega)$, which is equal to $2g_K - 2 + \deg f(\omega)$ unless ω is of the form $|\ |^{s_0}$ for some $s_0 \in \mathbf{C}$, in which case $r = 2g_K$. Since $P(u,\omega^{-1})$ has constant term 1, we get

$$\epsilon(\omega) = a_r = (-1)^r b_1 \cdots b_r.$$

Similarly, $\epsilon(\omega \circ \mathrm{N}_{E/K}) = A_r = (-1)^r B_1 \cdots B_r = (-1)^r (b_1 \cdots b_r)^n$. As r and $\deg f(\omega) = \deg f(\omega \circ \mathrm{N}_{E/K})$ have the same parity, this proves

Corollary 9. *Let E and ω be as in Theorem 12. Then*

$$(-1)^{\deg f(\omega \circ \mathrm{N}_{E/K})} \epsilon(\omega \circ \mathrm{N}_{E/K}) = \left((-1)^{\deg f(\omega)} \epsilon(\omega)\right)^n.$$

When $\omega = \xi$ defined at the beginning of this section, this is the original statement of the Davenport–Hasse identity.

---

§5 Zeta functions of certain curves

In Chapter 2 we computed the zeta function of the projective variety defined by the equation $a_0 x_0^n + \cdots + a_r x_r^n = 0$. In this section, we perform a similar computation of the zeta function of the affine curve V defined by $y_1^{n_1} = f_1(x), \cdots, y_r^{n_r} = f_r(x)$, using results in §2 and §4. Here $f_1, \cdots, f_r$ are polynomials over k. To simplify the computations, assume that $f_1, \cdots, f_r$ have only simple roots and they are pairwise coprime. Suppose further that $n_1, \cdots, n_r$ are divisors of $q-1$, the cardinality of $k^\times$. Denote by N_n the number of points of V over k_n. As computed at the end of §2,

$$N_1 = q + \sum_{x \in k} \sum_{\substack{0 \le j_i \le n_i - 1 \\ \text{some } j_i \ne 0}} \chi\left(f_1^{j_1 e_1}(x) \cdots f_r^{j_r e_r}(x)\right),$$

where χ is a fixed generator of the group of characters of $k^\times$ and $e_i = (q-1)/n_i$ for $i = 1, \cdots, r$. Because of the assumption that n_i's divide $q-1$, when computing N_n, we may use $\chi \circ \mathrm{N}_{k_n/k}$ and the same formula to express the error term, in other words,

$$N_n = q^n + \sum_{\substack{0 \le j_i \le n_i - 1 \\ \text{some } j_i \ne 0}} \sum_{x \in k_n} \chi \circ \mathrm{N}_{k_n/k}\left(f_1^{j_1 e_1}(x) \cdots f_r^{j_r e_r}(x)\right).$$

Therefore

$$\sum_{n=1}^{\infty} N_n u^{n-1} = \sum_{n=1}^{\infty} q^n u^{n-1}$$

$$+ \sum_{\substack{0 \le j_i \le n_i - 1 \\ \text{some } j_i \ne 0}} \sum_{n=1}^{\infty} \sum_{x \in k_n} \chi \circ \mathrm{N}_{k_n/k}\left(f_1^{j_1 e_1}(x) \cdots f_r^{j_r e_r}(x)\right) u^{n-1}.$$

Fix an r-tuple $J = (j_1, \cdots, j_r)$ with $0 \le j_i \le n_i - 1$ and some $j_i \ne 0$, we compute the sum over n. The assumption on $f_1, \cdots, f_r$ implies that $f^J(x) = f_1^{j_1 e_1}(x) \cdots f_r^{j_r e_r}(x)$ is a polynomial which is not a q-1st power of another polynomial over k. Denote by ω_J the idèle class character of $I_K/K^\times$, where $K = k(T)$, constructed using χ and f^J as in Theorem 4. Then ω_J ramifies somewhere with $\sum_{\substack{v \in \operatorname{supp} f^J \\ v \ne \infty}} v \le \operatorname{cond} \omega_J \le \sum_{v \in \operatorname{supp} f^J} v$, and

$$P(u, \omega_J) = 1 + a_1 u + \cdots + a_r u^r = (1 - b_1 u) \cdots (1 - b_r u),$$

where

$$\sum_{x \in k} \chi(f^J(x)) = \begin{cases} a_1 & \text{if } \omega_J \text{ ramifies at } \infty, \\ a_1 - 1 & \text{if } \omega_J \text{ is unramified at } \infty. \end{cases}$$

Likewise, if $E_n = K \cdot k_n$ and $P(u, \omega_J \circ \mathrm{N}_{E_n/K}) = 1 + A_1 u + \cdots + A_r u^r$, then we find

$$\sum_{x \in k_n} \chi \circ \mathrm{N}_{k_n/k}(f^J(x)) = \begin{cases} A_1 & \text{if } \omega_J \text{ ramifies at } \infty, \\ A_1 - 1 & \text{if } \omega_J \text{ is unramified at } \infty, \end{cases}$$

by the discussion in the previous section. On the other hand, by Theorem 12′, we have

$$a_1 = -b_1 - \cdots - b_r \quad \text{and} \quad A_1 = -b_1^n - \cdots - b_r^n.$$

Put together, we get, for $|u|$ small,

$$\begin{aligned} &\sum_{n=1}^{\infty} \sum_{x \in k_n} \chi \circ \mathrm{N}_{k_n/k}(f^J(x)) u^{n-1} \\ &= -\frac{b_1}{1 - b_1 u} - \cdots - \frac{b_r}{1 - b_r u} - \begin{cases} 0 & \text{if } \omega_J \text{ ramifies at } \infty, \\ \frac{1}{1-u} & \text{if } \omega_J \text{ is unramified at } \infty, \end{cases} \\ &= \frac{P'(u, \omega_J)}{P(u, \omega_J)} - \begin{cases} 0 & \text{if } \omega_J \text{ ramifies at } \infty, \\ \frac{1}{1-u} & \text{if } \omega_J \text{ is unramified at } \infty, \end{cases} \end{aligned}$$

and

$$\sum_{n=1}^{\infty} N_n u^{n-1} = \frac{q}{1 - qu} + \sum_J \frac{P'(u, \omega_J)}{P(u, \omega_J)} - \frac{m}{1 - u}$$

for some integer $m \ge 0$. If the homogenized equations define a nonsingular curve $\overline{V}$ over k, then, after taking into account the number of points at infinity, we must get

$$\frac{Z'_{\overline{V}}(u)}{Z_{\overline{V}}(u)} = \sum_{n=1}^{\infty} \overline{N}_n u^{n-1} = \frac{q}{1 - qu} + \frac{1}{1 - u} + \sum_J \frac{P'(u, \omega_J)}{P(u, \omega_J)}$$

so that the zeta function of $\overline{V}$ is

$$Z_{\overline{V}}(u) = \frac{\prod_J P(u, \omega_J)}{(1 - u)(1 - qu)}.$$

This is because we know the general shape of $Z_{\overline{V}}(u)$ and only nontrivial $P(u,\omega_J)'s$ have roots with absolute value $q^{-1/2}$.

For example, let V be the plane curve defined by $y^b = \gamma x^a + \delta$, where $\gamma, \delta \in k^\times$ and a, b are prime to the cardinality q of k. Then $f(x) = \gamma x^a + \delta$ has simple roots. Assume further that a and b are coprime, and b divides $q-1$. Let χ be a multiplicative character of $k^\times$ of order b. As discussed above, the number of solutions N_n of the affine curve V as defined over k_n is

$$N_n = q^n + \sum_{x\in k_n} \sum_{j=1}^{b-1} \chi \circ \mathrm{N}_{k_n/k} (f(x))^j.$$

Let ω^j be the idèle class character of $I_K/K^\times$, $K = k(T)$, constructed using χ^j and $f(x)$ as in §2, then $\operatorname{cond}\omega^j$ has degree equal to $a+1$ since $\deg f = a$ is prime to the order of χ^j. Thus $P(u,\omega^j)$ is a polynomial in u of degree $a-1$ and the zeta function of the projective curve $\overline{V}$ is

$$Z_{\overline{V}}(u) = \frac{P(u,\omega)\cdots P(u,\omega^{b-1})}{(1-u)(1-qu)}.$$

On the other hand, the number of solutions N_n of $\gamma x^a - y^b + \delta = 0$ over k_n was also computed in Chapter 2, §1. By the method there, we first rewrite the equation as $\delta^{-1}\gamma x^a - \delta^{-1} y^b + 1 = 0$. Next put $d_0 = \gcd(a, q^n - 1)$, $d_1 = \gcd(b, q^n-1) = b$. Then

$$N_n = q^n + \sum_{\substack{\chi_0^{d_0}=1,\chi_1^b=1,\\ \chi_0,\chi_1,\chi_2\neq 1,\chi_0,\chi_1,\chi_2\in\widehat{k_n^\times}\\ \chi_0\chi_1\chi_2=1}} \chi_0(\delta\gamma^{-1})\chi_1(-\delta)j(\chi_1,\chi_2)$$
$$\sum_{\substack{\chi_0^{d_0}=1,\chi_1^b=1,\\ \chi_0,\chi_1\neq 1,\chi_0,\chi_1\in\widehat{k_n^\times}\\ \chi_0\chi_1=1}} \chi_0(\delta\gamma^{-1})\chi_1(-\delta)j(\chi_1).$$

Since we assumed $\gcd(a,b) = 1$, thus $(d_0, b) = 1$ and the second sum is empty. Recall that $j(\chi_1,\chi_2)$ is the Jacobi sum given by

$$\sum_{\substack{v_1,v_2\in k_n\\ v_1+v_2+1=0}} \chi_1(v_1)\chi_2(v_2) = \frac{1}{q^n} g(\chi_0,\psi)g(\chi_1,\psi)g(\chi_2,\psi).$$

We also computed the zeta function of the projective curve $\overline{V}$ in Chapter 2, §2, which is

$$Z_{\overline{V}}(u) = \frac{\prod_{(\chi_0,\chi_1,\chi_2)} (1 - c(\chi_0,\chi_1,\chi_2)u^{\mu(\chi_0,\chi_1,\chi_2)})}{(1-u)(1-qu)}.$$

Here $c(\chi_0, \chi_1, \chi_2) = \chi_0(\delta\gamma^{-1})\chi_1(-\delta)j(\chi_1, \chi_2)$, χ_0, χ_1, χ_2 are multiplicative characters of a finite extension k_μ, $\mu = \mu(\chi_0, \chi_1, \chi_2)$, of k with $\chi_0^{d_0} = 1, \chi_1^b = 1, \chi_0, \chi_1, \chi_2 \neq 1, \chi_0\chi_1\chi_2 = 1$ and there is no proper subfield k_m of k_μ such that χ_0, χ_1, χ_2 are characters of $k_m^\times$ composed with norm N_{k_μ/k_m}. Further only one triplet is chosen from each Galois orbit.

We compare the two expressions of $Z_{\overline{V}}(u)$ for the special case where V is an elliptic curve defined by $y^2 = \gamma x^3 + \delta = f(x)$ with char $k \neq 2, 3$. Then the numerator of $Z_{\overline{V}}(u)$ is

$$P(u, \omega) = 1 + \sum_{x \in k} \chi(f(x))u + qu^2,$$

where χ is the quadratic character of $k^\times$. To see it from the second expression, we distinguish 2 cases.

Case 1. $q \equiv 1 \pmod 3$ so that $\gcd(3, q-1) = 3$. Let χ_0 be a character of $k^\times$ of order 3. The other order 3 character of $k^\times$ is thus $\overline{\chi_0}$. As above, denote by χ the quadratic character of $k^\times$. Then

$$P(u, \omega) = \left(1 - \chi_0(\gamma^{-1}\delta)\chi(-\delta)j(\chi, \chi\overline{\chi}_0)u\right)\left(1 - \overline{\chi}_0(\delta\gamma^{-1})\chi(-\delta)j(\chi, \chi\chi_0)u\right).$$

Case 2. $q \equiv 2 \pmod 3$. Then $\gcd(3, q-1) = 1$ and $\gcd(3, q^2-1) = 3$. In the second expression of $Z_{\overline{V}}(u)$, we have only one term with χ_0 being a character of $k_2^\times$ of order 3, $\chi_1 = \chi \circ N_{k_2/k}$, $\chi_2 = \chi_0^{-1}\chi_1$, and $\mu(\chi_0, \chi_1, \chi_2) = 2$. We have

$$P(u, \omega) = 1 + \sum_{x \in k} \chi(f(x))u + qu^2 = 1 - c(\chi_0, \chi_1, \chi_2)u^2.$$

This implies

$$\sum_{x \in k} \chi(f(x)) = 0 \quad \text{and} \quad c(\chi_0, \chi_1, \chi_2) = -q.$$

Geometrically, $\overline{V}$ is a finite cover of $\mathbf{P}^1$. If we view the covering map as $(x, y) \mapsto x$, then we take the function field of $\mathbf{P}^1$ to be $K = k(x)$ and that of $\overline{V}$ to be $k(x, y) = F$ with F a Galois extension of K of degree 2. Then $K^\times \mathrm{N}_{F/K}(I_F)$ is an open subgroup of I_K with index 2. The idèle class character ω is the nontrivial character of $I_K/K^\times \mathrm{N}_{F/K}(I_F)$, and

$$\zeta_F(s) = L(s, \omega)\zeta_K(s).$$

If we view the covering map as $(x, y) \mapsto y$, then the function field of $\mathbf{P}^1$ is taken to be $K = k(y)$ and that of $\overline{V}$ to be $k(x, y) = F$. If $q \equiv 2 \pmod 3$, then F is not Galois over K; but if $q \equiv 1 \pmod 3$, then F is Galois over K, with Galois group cyclic of order 3. In this case $K^\times \mathrm{N}_{F/K}(I_F)$ is an open subgroup of I_K with index 3. According to class field theory there are two idèle class characters $\xi, \overline{\xi}$ of $I_K/K^\times \mathrm{N}_{F/K}(I_F)$ of order 3 such that

$$\zeta_F(s) = L(s, \xi)L(s, \overline{\xi})\zeta_K(s).$$

What are ξ and $\overline{\xi}$? From the discussion in Case 1, we know their P-functions are

$$1-\chi_0(\gamma^{-1}\delta)\chi(-\delta)j(\chi,\chi\overline{\chi}_0)u \quad\text{and}\quad 1-\overline{\chi}_0(\gamma^{-1}\delta)\chi(-\delta)j(\chi,\chi\chi_0)u.$$

To see them explicitly, we appeal to the defining equation : $\gamma x^3 = y^2 - \delta$, i.e., $x^3 = h(y)$, where $h(y) = \gamma^{-1}y^2 - \gamma^{-1}\delta$. With $\chi_0 \in \widehat{k^\times}$ of order 3 and h, we can construct an idèle class character ξ of $I_K/K^\times$, which is ramified at ∞ so that cond ξ has degree 3, and hence $P(u,\xi)$ has degree 1. Same argument shows that

$$Z_{\overline{V}}(u) = \frac{P(u,\xi)P(u,\overline{\xi})}{(1-u)(1-qu)} = \frac{P(u,\omega)}{(1-u)(1-qu)}.$$

Thus $P(u,\xi)$ is one of the factors of $P(u,\omega)$, and $\xi,\overline{\xi}$ are what we are looking for.

Exercise 5. Let V be the curve defined by $y^b = \gamma x^a + \delta$, $\gamma,\delta \in k^\times$, a,b coprime integers, not divisible by char k, and $b \mid q-1$ as above. Let χ be a character of $k^\times$ of order b and ω^i be the idèle class character of $I_K/K^\times$, $K = k(T)$, constructed using χ^i and $f(x) = \gamma x^a + \delta$. Show that in the factorization of the numerator of $Z_{\overline{V}}(u)$ in two ways, we have

$$P(u,\omega^i) = \prod_{\chi_0}\left(1 - c(\chi_0,\chi^i,\overline{\chi}_0\overline{\chi}^i)u^{\mu(\chi_0,\chi^i,\overline{\chi}_0\overline{\chi}^i)}\right)$$

where χ_0 runs through the Galois orbits of nontrivial characters on $k_\mu^\times$ with order dividing $\gcd(a, q^\mu - 1)$.

References

[1] J. Arthur and L. Clozel : Simple Algebras, Base Change, and the Advanced Theory of the Trace Formula, Annals of Math. Studies 120, Princeton Univ. Press, Princeton, New Jersey (1989).

[2] J.W.S. Cassels and A. Fröhlich : Algebraic Number Theory, Thompson, Washington (1967), republished by Academic Press, London.

[3] P. Deligne : Cohomologie étale (SGA $4\frac{1}{2}$), Lecture Notes in Math. 569, Springer–Verlag, Berlin, Heidelberg, New York (1977).

[4] N. Jacobson : Basic Algebra I, Freeman, San Francisco (1980).

[5] N. Katz : An estimate for character sums, J. Amer. Math. Soc. 2 (1989), 197-200.

[6] R. Langlands : Base Change for GL_2, Annals of Math. Studies 96, Princeton Univ. Press, Princeton, New Jersey (1980).

[7] W.-C. W. Li : Character sums and abelian Ramanujan graphs, J. of Number Theory 41 (1992), 199-217.

[8] W. Schmidt : Equations over Finite Fields, Lecture Notes in Math. 536, Springer–Verlag, Berlin, Heidelberg, New York (1976).

[9] A. Weil : On some exponential sums, Proc. National Academy of Science 34 (1948), 204-207.

[10] A. Weil : Basic Number Theory, Springer–Verlag, Berlin, Heidelberg, New York (1973).

CHAPTER 7

The Theory of Modular Forms

§1 Classical modular forms

The group $\Gamma = SL_2(\mathbf{Z}) = \left\{\begin{pmatrix} a & b \\ c & d \end{pmatrix} : a, b, c, d \in \mathbf{Z}, ad - bc = 1\right\}$ acts on the Poincaré upper half-plane $\mathcal{H} = \{z \in \mathbf{C} : \text{ Im } z > 0\}$ by fractional linear transformations, that is, $\gamma = \begin{pmatrix} a & b \\ c & d \end{pmatrix}$ maps z to $\frac{az+b}{cz+d}$. The orbit space $\mathcal{H}/\Gamma$ is usually represented by a region $\mathcal{D}$, called a fundamental domain of Γ, typically of the following shape :

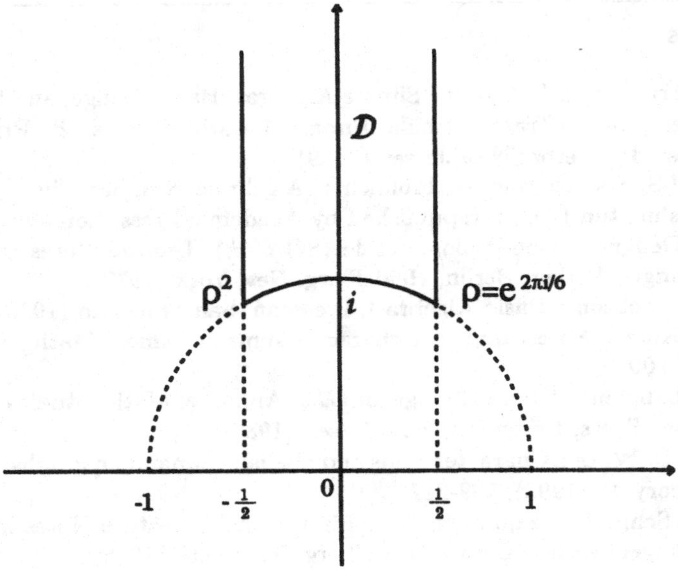

Here the points $-\frac{1}{2}+iy$ and $\frac{1}{2}+iy$ are identified, and on the unit circle, $e^{i\theta}$ and $e^{i(\pi-\theta)}$ are identified for $\frac{\pi}{3} \le \theta \le \frac{\pi}{2}$.

Under the hyperbolic metric $\frac{dx\,dy}{y^2}$ on $\mathcal{H}$, the distance between $-\frac{1}{2}+iy$ and $\frac{1}{2}+iy$ tends to 0 as $y \to \infty$, so $\mathcal{D}$ looks like a punctured sphere; the point $i\infty$ is called a cusp of Γ. The orbit of $i\infty$ under Γ consists of $i\infty$ and the rational points on the real axis, which are also called cusps.

Exercise 1. (i) Express $dx \wedge dy$ in terms of $dz \wedge d\bar{z}$.
(ii) Show that the measure $\frac{dx\,dy}{y^2}$ is Γ-invariant.

Remark. Points in $\mathcal{D}$ parametrize elliptic curves defined over $\mathbf{C}$. Indeed, an elliptic curve over $\mathbf{C}$ is $\mathbf{C}/L$, where L is a lattice, that is, a rank two $\mathbf{Z}$-module in $\mathbf{C}$. Two elliptic curves $\mathbf{C}/L$ and $\mathbf{C}/L'$ are equivalent if there is an analytic group isomorphism from one to the other, say $f(z)$. Thus $f(z+z_1) = f(z)+f(z_1)$ for all $z, z_1 \in \mathbf{C}$. Treating z as a variable and leaving z_1 fixed, we get $f'(z+z_1) = f'(z)$ by differentiating the above equation. Since z_1 is arbitrary, the above shows that $f'(z) = u$ is a constant. As $f(0) = 0$ and f is injective, we conclude that $f(z) = uz$ with $u \in \mathbf{C}^{\times}$. In other words, $\mathbf{C}/L$ and $\mathbf{C}/L'$ are equivalent if and only if $uL = L'$ for some $u \in \mathbf{C}^{\times}$, in which case we say that the two lattices L and L' are equivalent. In an equivalence class of a given lattice, there exists a lattice with basis $\{z, 1\}$, where $z \in \mathcal{H}$. Further, two lattices with basis $\{z, 1\}$ and $\{z', 1\}$ are equivalent if and only if z and z' lie in the same orbit of Γ on $\mathcal{H}$. Hence $\mathcal{H}/\Gamma$ parametrizes equivalence classes of elliptic curves.

Exercise 2. (i) Show that two lattices $L(z) = \mathbf{Z}z + \mathbf{Z}$ and $L(z') = \mathbf{Z}z' + \mathbf{Z}$ are equivalent if and only if there exists $\gamma \in \Gamma$ such that $z' = \gamma z$.
(ii) Describe the isomorphism $\mathbf{C}/L(z) \to \mathbf{C}/L(z')$ when $z' = \gamma z$.

Let H be a subgroup of Γ with finite index. Then a fundamental domain of H is a union of finitely many fundamental domains of Γ, and it has finitely many cusps among the H-orbits of $i\infty$. The compactification $\widehat{\mathcal{H}/H}$ of $\mathcal{H}/H$ is a Riemann surface, i.e., a curve of genus g over $\mathbf{C}$. A holomorphic differential on $\mathcal{H}/H$ can be written as $f(z)dz$, where f is holomorphic on $\mathcal{H}$ such that $f(z)dz = f(\gamma z)d(\gamma z)$ for all $\gamma \in H$. Since for $\gamma = \begin{pmatrix} a & b \\ c & d \end{pmatrix}$ we have $d(\gamma z) = \frac{dz}{(cz+d)^2}$, the condition on f can be restated as $f(z) = (cz+d)^{-2} f(\gamma z)$ for all $\gamma = \begin{pmatrix} \cdot & \cdot \\ c & d \end{pmatrix} \in H$. Similarly, if $f(z)dz^{\otimes k}$ is a holomorphic kth tensor of dz, then f satisfies $f(z) = (cz+d)^{-2k} f(\gamma z)$ for all $\gamma = \begin{pmatrix} \cdot & \cdot \\ c & d \end{pmatrix} \in H$. In general, given a positive integer k, define the "stroke operator $|_k$" by

$$(f\,|_k\,\gamma)(z) = (\det \gamma)^{k/2}(cz+d)^{-k} f(\gamma z)$$

for $\gamma = \begin{pmatrix} a & b \\ c & d \end{pmatrix} \in GL_2(\mathbf{R})$ with positive determinant so that

$$\left(f \mid_k \begin{pmatrix} a & 0 \\ 0 & a \end{pmatrix}\right)(z) = f(z) \quad \text{for } a > 0.$$

We also extend the stroke operator by linearity to linear combinations of matrices of above type.

A modular form for H of weight k is a function f on $\mathcal{H}$ satisfying

(i) f is holomorphic on $\mathcal{H}$;
(ii) $f \mid_k \gamma = f$ for all $\gamma \in H$;
(iii) f is holomorphic at all cusps of H.

Since H has finite index in Γ, H contains some translation $\begin{pmatrix} 1 & m \\ 0 & 1 \end{pmatrix}$ with $m \neq 0$. Let M be the smallest positive integer such that $\begin{pmatrix} 1 & M \\ 0 & 1 \end{pmatrix} \in H$. By (ii), $f(z) = f(z+M)$ so that f has a Fourier expansion

$$f(z) = \sum_{n=-\infty}^{\infty} a_n e^{2\pi i n z/M}.$$

As $e^{2\pi i z/M}$ is a uniformizer at $i\infty$, we say that f is holomorphic at $i\infty$ if the above expansion is a Taylor series in $e^{2\pi i z/M}$, that is, $a_n = 0$ for $n < 0$. We say that f is holomorphic at the cusp $\gamma(i\infty)$, $\gamma \in \Gamma$, if $f \mid_k \gamma^{-1}$ is holomorphic at $i\infty$. The modular forms for H of weight k form a vector space $\mathcal{M}(H,k)$. A modular form f is called a *cusp form* if it vanishes at all cusps. The cusp forms for H of weight k form a subspace $\mathcal{C}(H,k)$. Note that $\mathcal{C}(H,2)$ consists of functions f on $\mathcal{H}$ such that $f(z)dz$ is a holomorphic differential on $\widehat{\mathcal{H}/H}$. Indeed, at the cusp $i\infty$, $q = e^{2\pi i z/M}$ is a uniformizer with $dq = \frac{2\pi i}{M} q dz$ so that $f(z)dz = \frac{M}{2\pi i}\frac{1}{q} f(q)dq$. Thus $f(z)dz$ is holomorphic at $i\infty$ if and only if f vanishes at $i\infty$. The same argument applies to other cusps. This shows that $\dim \mathcal{C}(H,2) =$ the genus of $\widehat{\mathcal{H}/H}$, which is also called the genus of the group H.

The most frequently seen subgroups H of Γ are the following congruence subgroups of Γ :

$$\Gamma(N) = \left\{\gamma \in \Gamma : \gamma \equiv \begin{pmatrix} 1 & 0 \\ 0 & 1 \end{pmatrix} \bmod N\right\} \triangleleft \Gamma,$$

$$\Gamma_1(N) = \left\{\gamma \in \Gamma : \gamma \equiv \begin{pmatrix} 1 & * \\ 0 & 1 \end{pmatrix} \bmod N\right\},$$

$$\text{and} \quad \Gamma_0(N) = \left\{\gamma \in \Gamma : \gamma \equiv \begin{pmatrix} * & * \\ 0 & * \end{pmatrix} \bmod N\right\}.$$

Similar to $\mathcal{H}/\Gamma$, the modular curves $\mathcal{H}/\Gamma(N)$, $\mathcal{H}/\Gamma_1(N)$ and $\mathcal{H}/\Gamma_0(N)$ classify elliptic curves with additional structures as follows. The group of points in an elliptic curve $E = \mathbf{C}/L$ with order dividing N is $E[N] = \frac{1}{N}L/L \cong \mathbf{Z}/N\mathbf{Z} \times \mathbf{Z}/N\mathbf{Z}$. The triple $(E; P, Q)$ denotes an elliptic curve E with two chosen points P, Q of order N such that P and Q generate the group $E[N]$. Two such triples $(E; P, Q)$ and $(E'; P', Q')$ are equivalent if there is an isomorphism from E to E' mapping P to P' and Q to Q'. Similarly, $(E; P)$ denotes an elliptic curve E with one chosen point P of order N, and $(E; P)$ and $(E'; P')$ are equivalent if there is an isomorphism from E to E' mapping P to P'. Finally, $(E; C)$ denotes an elliptic curve E with a cyclic subgroup C of order N; and $(E; C)$ and $(E'; C')$ are equivalent if there is an isomorphism from E to E' mapping C to C'. Write $E(z)$ for the elliptic curve $\mathbf{C}/L(z)$ where $L(z) = \mathbf{Z}z + \mathbf{Z}$ with $z \in \mathcal{H}$, $P(z) = \frac{1}{N} \in E(z)$, $Q(z) = \frac{1}{N}z \in E(z)$ and $C(z)$ for the cyclic group in $E(z)$ generated by $\frac{1}{N}$.

Theorem 1. *(1) An equivalence class of $(E; P, Q)$ is represented by $\big(E(z); P(z), Q(z)\big)$ for some $z \in \mathcal{H}$. Further $\big(E(z); P(z), Q(z)\big)$ and $\big(E(z'); P(z'), Q(z')\big)$ are equivalent if and only if there exists $\gamma \in \Gamma(N)$ such that $z' = \gamma(z)$. Thus $\mathcal{H}/\Gamma(N)$ classifies the equivalence classes of $(E; P, Q)$.*

(2) An equivalence class of $(E; P)$ is represented by $\big(E(z); P(z)\big)$ for some $z \in \mathcal{H}$. Further $\big(E(z); P(z)\big)$ and $\big(E(z'); P(z')\big)$ are equivalent if and only if there exists $\gamma \in \Gamma_1(N)$ such that $z' = \gamma(z)$. Thus $\mathcal{H}/\Gamma_1(N)$ classifies the equivalence classes of $(E; P)$.

(3) An equivalence class of $(E; C)$ is represented by $\big(E(z); C(z)\big)$ for some $z \in \mathcal{H}$. Further $\big(E(z); C(z)\big)$ and $\big(E(z'); C(z')\big)$ are equivalent if and only if there exists $\gamma \in \Gamma_0(N)$ such that $z' = \gamma(z)$. Thus $\mathcal{H}/\Gamma_0(N)$ classifies the equivalence classes of $(E; C)$.

Proof. Given $(E; P, Q)$ with $E = \mathbf{C}/L$, represent P, Q by two complex numbers p, q, respectively. Since P, Q are $\mathbf{Z}$–linearly independent, the quotient q/p is a complex number. Adding to q a suitable lattice point if necessary, we may assume that $\operatorname{Im} q/p > 0$. Since NP, NQ generate the lattice L, multiplication by $\frac{1}{Np}$ gives rise to an isomorphism from $\mathbf{C}/L$ to $\mathbf{C}/L(z)$, where $z = q/p \in \mathcal{H}$, which sends P to $P(z) = \frac{1}{N}$ and Q to $Q(z) = \frac{z}{N}$. This proves the first statement of (1), (2) and (3).

Next let multiplication by $u \in \mathbf{C}^\times$ be an isomorphism from $\big(E(z); P(z), Q(z)\big)$ to $\big(E(z'); P(z'), Q(z')\big)$ for (1), from $\big(E(z); P(z)\big)$ to $\big(E(z'); P(z')\big)$ for (2), and from $\big(E(z); C(z)\big)$ to $\big(E(z'); C(z')\big)$ for (3). In all cases we have that $\{uz, u\}$ is a basis of $L(z')$, in other words, there is a matrix $\gamma = \begin{pmatrix} a & b \\ c & d \end{pmatrix} \in GL_2(\mathbf{Z})$ such that $\begin{pmatrix} uz \\ u \end{pmatrix} = \begin{pmatrix} a & b \\ c & d \end{pmatrix} \begin{pmatrix} z' \\ 1 \end{pmatrix}$. This gives $uz = az' + b$, $u = cz' + d$. Further $z = \frac{uz}{u} = \frac{az'+b}{cz'+d} = \gamma z'$ and $z, z' \in \mathcal{H}$ together imply that $\gamma \in \Gamma = SL_2(\mathbf{Z})$. In cases (i) and (ii), the relation $uP(z) = \frac{u}{N} = \frac{cz'+d}{N} = P(z') = \frac{1}{N}$ in $\mathbf{C}/L(z')$ holds if and only if $c \equiv 0 (\operatorname{mod} N)$ and $d \equiv 1 (\operatorname{mod} N)$. Since $\det \gamma = ad - bc = 1$, the congruence conditions on c and d imply $a \equiv 1 (\operatorname{mod} N)$. This proves (2). For (1), the second relation $uQ(z) = \frac{uz}{N} = \frac{az'+b}{N} = Q(z') = \frac{z'}{N}$ in $\mathbf{C}/L(z')$ holds if and only if $b \equiv 0 (\operatorname{mod} N)$ and $a \equiv 1 (\operatorname{mod} N)$. This proves (1).

As for (3), $uC(z) = \langle\frac{u}{N}\rangle = \langle\frac{cz'+d}{N}\rangle = C(z') = \langle\frac{1}{N}\rangle$ holds if and only if $c \equiv 0 (\bmod N)$ and $\gcd(d, N) = 1$. But the latter is a consequence of $\det\gamma = 1$ and $c \equiv 0 (\bmod N)$. This proves (3). □

Theorem 1 explains why $\mathcal{H}/\Gamma(N)$, $\mathcal{H}/\Gamma_1(N)$, and $\mathcal{H}/\Gamma_0(N)$ are called modular curves.

Back to the congruence subgroups of Γ. Observe that $\Gamma_1(N)$ is a normal subgroup of $\Gamma_0(N)$ and

$$\Gamma_0(N)/\Gamma_1(N) \cong \left\{\begin{pmatrix} a & 0 \\ 0 & d \end{pmatrix} \in SL_2(\mathbf{Z}/N\mathbf{Z})\right\} \cong (\mathbf{Z}/N\mathbf{Z})^\times.$$

Thus $\Gamma_0(N)$ acts on $\mathcal{M}(\Gamma_1(N), k)$ via $|_k$, and this gives rise to a representation of $(\mathbf{Z}/N\mathbf{Z})^\times$ on the space of modular forms for $\Gamma_1(N)$ of weight k, preserving the cusp forms. Hence $\mathcal{M}(\Gamma_1(N), k)$ decomposes as a direct sum over all characters χ of $(\mathbf{Z}/N\mathbf{Z})^\times$ of subspaces $\mathcal{M}(N, k, \chi)$ consisting of functions f on $\mathcal{H}$ satisfying (i), (iii) and

(ii)′ $\qquad (f |_k \gamma)(z) = \chi(d) f(z) \quad \text{for all } \gamma = \begin{pmatrix} \cdot & \cdot \\ c & d \end{pmatrix} \in \Gamma_0(N).$

Such a form is called a modular form of weight k level N and character χ. Likewise, $\mathcal{C}(\Gamma_1(N), k) = \bigoplus_\chi \mathcal{C}(N, k, \chi)$, where χ runs through characters of $(\mathbf{Z}/N\mathbf{Z})^\times$. Note that for $\chi = \chi_0$, the trivial character of $(\mathbf{Z}/N\mathbf{Z})^\times$, the space $\mathcal{M}(N, k, \chi_0)$ is $\mathcal{M}(\Gamma_0(N), k)$.

Put $h = \begin{pmatrix} N & 0 \\ 0 & 1 \end{pmatrix}$. Define

$$\begin{aligned}\widetilde{\Gamma}(N) &= h^{-1}\Gamma(N)h \\ &= \left\{\begin{pmatrix} a & b \\ c & d \end{pmatrix} \in SL_2(\mathbf{Z}) : a \equiv d \equiv 1 \ (\bmod N) \ \text{ and } \ c \equiv 0 \ (\bmod N^2)\right\} \subset \Gamma_0(N^2).\end{aligned}$$

The space $\mathcal{M}(\widetilde{\Gamma}(N), k)$ decomposes as a direct sum of $\mathcal{M}(N^2, k, \chi)$, where χ runs through characters of $(\mathbf{Z}/N^2\mathbf{Z})^\times$ trivial on integers congruent to 1 modulo N. On the other hand, the map $f \mapsto f |_k h$ gives rise to an isomorphism from $\mathcal{M}(\Gamma(N), k)$ to $\mathcal{M}(\widetilde{\Gamma}(N), k)$. Thus the study of modular forms for $\Gamma(N)$, $\Gamma_1(N)$ and $\Gamma_0(N)$ is reduced to studying $\mathcal{M}(N, k, \chi)$, which is a finite-dimensional vector space over $\mathbf{C}$, as can be seen from the Riemann-Roch theorem for $k \geq 2$.

---

§2 Hecke operators

Since $\Gamma_0(N)$ contains $\begin{pmatrix} 1 & 1 \\ 0 & 1 \end{pmatrix}$, every $f(z) \in \mathcal{M}(N, k, \chi)$ has a Fourier expansion $f(z) = \sum_{n=0}^{\infty} a_n e^{2\pi i n z}$, and $a_0 = 0$ if f vanishes at the cusp $i\infty$. To study the arithmetic of the Fourier coefficients a_n of f has been an important subject in the theory of classical modular forms. The first question to ask is to know when the arithmetic function $n \mapsto a_n$ is multiplicative with respect to a prime p. To formulate this question, introduce the Dirichlet series attached to f :

$$D(s,f) = \sum_{n=1}^{\infty} a_n n^{-s}.$$

This series is said to have an Euler product at a place p of $\mathbf{Q}$ if there are complex numbers $c_1, c_p, c_{p^2}, \cdots$ such that

$$D(s,f) = \left(\sum_{p \nmid n} a_n n^{-s}\right)\left(\sum_{r=0}^{\infty} c_{p^r} p^{-rs}\right),$$

that is, $a_{np^r} = a_n c_{p^r}$ for all n prime to p and for all $r \geq 0$. To deal with this question, Hecke defined , for each prime p not dividing the level N, an operator T_p, called the Hecke operator at p nowadays :

$$T_p = p^{k/2-1}\left(\sum_{u=0}^{p-1}\begin{pmatrix} 1 & u \\ 0 & p \end{pmatrix} + \chi(p)\begin{pmatrix} p & 0 \\ 0 & 1 \end{pmatrix}\right),$$

which acts on $\mathcal{M}(N,k,\chi)$ via $|_k T_p$.

Exercise 3. (1) Show that T_p sends $\mathcal{M}(N,k,\chi)$ to itself and preserves the subspace of cusp forms $\mathcal{C}(N,k,\chi)$.

(2) If $f \in \mathcal{M}(N,k,\chi)$ has Fourier expansion $f(z) = \sum_{n=0}^{\infty} a_n e^{2\pi i n z}$, then

$$f \mid_k T_p(z) = \sum_{n \geq 0} \left(a_{np} + \chi(p)p^{k-1} a_{n/p}\right) e^{2\pi i n z}$$

where $a_x = 0$ if x is not an integer.

(3) For distinct primes p, p' not dividing N, T_p commutes with $T_{p'}$.

Theorem 2. *(Hecke) Let $f \in \mathcal{M}(N,k,\chi)$ be such that $D(s,f) \neq 0$. Let p be a prime not dividing N. Then $D(s,f)$ has an Euler product at p if and only if f is an eigenfunction of T_p. If so, the Euler factor has the form*

$$\sum_{r=0}^{\infty} c_{p^r} p^{-rs} = \frac{1}{1 - c_p p^{-s} + \chi(p) p^{k-1-2s}}$$

where $f \mid_k T_p = c_p f$, that is, c_p is the eigenvalue.

To prove this theorem, we shall need

Lemma 1. *Let $f(z) = \sum_{n=0}^{\infty} a_n e^{2\pi i n z}$ be a form in $\mathcal{M}(N,k,\chi)$. Let p be a prime not dividing N. Then*

(1) *If $a_n = 0$ for all n prime to p, then $f = 0$. In particular, if $D(s,f) \neq 0$, then $a_n \neq 0$ for some index n prime to p.*

(2) *If $a_{np} = 0$ for all n, then $f = 0$.*

Proof. (1) By assumption, $f(z) = \sum_{n=0}^{\infty} a_{np} e^{2\pi i n p z}$. Then $f(z) = f(z + \frac{1}{p}) = f \mid_k \begin{pmatrix} p & 1 \\ 0 & p \end{pmatrix}(z)$. On the other hand, there are matrices $\gamma_1, \gamma_2 \in \Gamma_0(N)$ such that $\gamma_1 \begin{pmatrix} p & 1 \\ 0 & p \end{pmatrix} \gamma_2 = \begin{pmatrix} 1 & 0 \\ 0 & p^2 \end{pmatrix}$. In the case that $N = 1$, this is nothing but the elementary divisor theorem. In case $N \neq 1$, this still holds. For instance, we may choose

$$\gamma_1 = \begin{pmatrix} 1 & 0 \\ -pcN & 1 \end{pmatrix} \quad \text{and} \quad \gamma_2 = \begin{pmatrix} a & -1 \\ cN & p \end{pmatrix},$$

where a, c are integers such that $ap + cN = 1$ (which is possible since N and p are coprime). Observe that $f \mid_k \gamma_1 = f$ and $f \mid_k \gamma_2 = \chi(p) f$. Thus

$$\chi(p) f(z) = f \mid_k \gamma_1 \begin{pmatrix} p & 1 \\ 0 & p \end{pmatrix} \gamma_2(z) = f \mid_k \begin{pmatrix} 1 & 0 \\ 0 & p^2 \end{pmatrix}(z) = p^{-k} f(\tfrac{z}{p^2}).$$

In terms of the Fourier expansion of f, this means

$$\sum_{n=0}^{\infty} a_{np} e^{2\pi i n z / p} = \chi(p) p^k \sum_{n=0}^{\infty} a_{np} e^{2\pi i n p z}.$$

If f were nonzero, we would get a contradiction by comparing the leading Fourier coefficients of both sides.

(2) Suppose $a_{np} = 0$ for all n. Then

$$f \mid_k T_p(z) = \chi(p) p^{k-1} \sum_{n \geq 0} a_{n/p} e^{2\pi i n z} = \chi(p) p^{k-1} \sum_{n \geq 0} b_{np} e^{2\pi i n p z} \in \mathcal{M}(N,k,\chi),$$

where $b_{np} = a_n$. By (1), $f \mid_k T_p = 0$, i.e., $a_n = 0$ for all n. Thus $f = 0$. □

Proof of Theorem 2. Write $f(z) = \sum_{n \geq 0} a_n e^{2\pi i n z}$. First assume that $D(s,f)$ has an Euler product at p. Then $a_{mp^r} = a_m c_{p^r}$ for $(m,p) = 1$, $r \geq 0$. Consider $(f \mid_k T_p - c_p f)(z) = \sum_{n \geq 0} (a_{np} + \chi(p) p^{k-1} a_{n/p} - c_p a_n) e^{2\pi i n z} = \sum_{n \geq 0} b_n e^{2\pi i n z}$, which is in $\mathcal{M}(N,k,\chi)$ and whose Fourier coefficients $b_n = 0$ if p does not divide n, thus it is zero by Lemma 1, (1). This shows that $f \mid_k T_p = c_p f$. Conversely, suppose that f is an eigenfunction of T_p with eigenvalue c_p. Then we get

$$a_{np} = c_p a_n - \chi(p) p^{k-1} a_{n/p} \quad \text{for all } n \geq 0.$$

This proves that $a_{np} = a_n c_p$ for n prime to p, and implies inductively that $a_{np^r} = a_n c_{p^r}$ for n prime to p, where c_{p^r} satisfy the recursive relation

$$c_{p^{r+1}} = c_p c_{p^r} - \chi(p)p^{k-1}c_{p^{r-1}},$$

which can be restated as

$$\sum_{r=0}^{\infty} c_{p^r}p^{-rs} = \frac{1}{1 - c_p p^{-s} + \chi(p)p^{k-1-2s}}.$$

This proves the theorem.

Let H be a subgroup of Γ of finite index with a fundamental domain $\mathcal{D}(H)$. Let ω, ω' be two holomorphic 1–forms on the Riemann surface $\widehat{\mathcal{H}/H}$. Then $\omega \wedge \omega'$ is a natural 2–form and one can integrate it over $\mathcal{D}(H)$ to get a definite number. This defines an inner product $\langle \omega, \omega' \rangle$ on differential forms, or equivalently, on cusp forms for H of weight 2. It has a generalization to modular forms for H of weight k, called the Petersson inner product $\langle\,,\rangle$, defined as

$$\begin{aligned}\langle f, g\rangle &= \frac{1}{[SL_2(\mathbf{Z}) : H]}\iint_{\mathcal{D}(H)} f(x+iy)\overline{g(x+iy)}y^k\frac{dxdy}{y^2}\\ &= \frac{1}{[SL_2(\mathbf{Z}) : H]}\frac{i}{2}\iint_{\mathcal{H}/H} f(z)\overline{g(z)}(\text{Im } z)^{k-2}dz\wedge d\bar{z}\end{aligned}$$

since $dz \wedge d\bar{z} = -2idx \wedge dy$. Here we normalize the inner product by dividing out the index of H in $SL_2(\mathbf{Z})$ so that the value is independent of the choice of H for which both f and g are modular forms. The above formula is well-defined when the integral converges. The integral certainly converges if a small neighborhood around each cusp is removed from $\mathcal{D}(H)$, hence the problem is at cusps. If one of f, g, say, f is a cusp form, then $f = O(e^{-cy})$ for some $c > 0$ and g remains bounded in a neighborhood of $i\infty$ so that the integral does converge near $i\infty$, and similarly at other cusps. Consequently the inner product is defined if at least one of f, g is a cusp form. In particular, it is defined on the space of cusp forms $\mathcal{C}(H, k)$.

Petersson showed that the Hecke operators behave nicely with respect to the inner product. More precisely,

Theorem 3. *(Petersson) Let p be a prime not dividing N. Then for $f, g \in \mathcal{C}(N, k, \chi)$, we have $\langle f \mid T_p, g\rangle = \chi(p)\langle f, g \mid T_p\rangle$.*

This shows that each T_p, $p \nmid N$, is skew-Hermitian, and hence is diagonalizable. Further, since different Hecke operators commute, we can diagonalize them simultaneously.

Before proving this theorem, we re-examine the definition of T_p. The matrices occurring in the definition of T_p are actually right coset representatives of a certain double coset which we now explain. Let

$$M_p(N) = \left\{m = \begin{pmatrix} a & b \\ c & d\end{pmatrix} : a, b, c, d \in \mathbf{Z},\ m \equiv \begin{pmatrix} 1 & * \\ 0 & p\end{pmatrix} \pmod{N} \text{ and } \det m = p\right\}.$$

Lemma 2.

$$M_p(N) = \Gamma_1(N)\begin{pmatrix}1 & 0\\ 0 & p\end{pmatrix}\Gamma_1(N) = \bigcup_{u=0}^{p-1}\Gamma_1(N)\begin{pmatrix}1 & u\\ 0 & p\end{pmatrix} \cup \Gamma_1(N)R_p\begin{pmatrix}p & 0\\ 0 & 1\end{pmatrix},$$

where $R_p = \begin{pmatrix}a & -1\\ cN & p\end{pmatrix} \in \Gamma_0(N)$.

Proof. Since $R_p\begin{pmatrix}p & 0\\ 0 & 1\end{pmatrix} = \begin{pmatrix}1 & 0\\ 0 & p\end{pmatrix}\begin{pmatrix}pa & -1\\ cN & 1\end{pmatrix} \in \begin{pmatrix}1 & 0\\ 0 & p\end{pmatrix}\Gamma_1(N)$, clearly we have

$$M_p(N) \supseteq \Gamma_1(N)\begin{pmatrix}1 & 0\\ 0 & p\end{pmatrix}\Gamma_1(N) \supseteq \bigcup_{u=0}^{p-1}\Gamma_1(N)\begin{pmatrix}1 & u\\ 0 & p\end{pmatrix} \cup \Gamma_1(N)R_p\begin{pmatrix}p & 0\\ 0 & 1\end{pmatrix}.$$

Next let $m = \begin{pmatrix}a' & b'\\ c' & d'\end{pmatrix} \in M_p(N)$. If a' and c' are coprime, let $x, y \in \mathbf{Z}$ be such that $xa' + yc' = 1$. Then $\begin{pmatrix}x & y\\ -c' & a'\end{pmatrix} \in \Gamma_1(N)$ and

$$\begin{pmatrix}x & y\\ -c' & a'\end{pmatrix}\begin{pmatrix}a' & b'\\ c' & d'\end{pmatrix} = \begin{pmatrix}1 & *\\ 0 & p\end{pmatrix}.$$

This shows that $m \in \Gamma_1(N)\begin{pmatrix}1 & u\\ 0 & p\end{pmatrix}$, where $u \equiv *(\bmod p)$. If a' and c' are not coprime, then $\gcd(a', c') = p$ since $\gcd(a', c')$ divides $\det m = p$. In this case one checks that $m \in \Gamma_1(N)R_p\begin{pmatrix}p & 0\\ 0 & 1\end{pmatrix}$. This proves the lemma. □

Observe that R_p acts on $\mathcal{C}(N, k, \chi)$ by multiplication by $\chi(p)$. Thus the operator T_p may be interpreted as

$$T_p = p^{k/2-1}\sum_i \alpha_i$$

for *any* right coset representatives of $\Gamma_1(N)$ in $M_p(N)$.

For a matrix $m = \begin{pmatrix}a & b\\ c & d\end{pmatrix}$, denote by m' the matrix $\begin{pmatrix}d & -b\\ -c & a\end{pmatrix}$, that is, $m' = (\det m)m^{-1}$ if m is nonsingular. From

$$R_p\begin{pmatrix}1 & 0\\ 0 & p\end{pmatrix}' = R_p\begin{pmatrix}p & 0\\ 0 & 1\end{pmatrix} = \begin{pmatrix}1 & 0\\ 0 & p\end{pmatrix}\begin{pmatrix}ap & -1\\ cN & 1\end{pmatrix} \in \begin{pmatrix}1 & 0\\ 0 & p\end{pmatrix}\Gamma_1(N)$$

we see that $M_p(N) = R_pM_p(N)'$. Further, as $M_p(N)$ is a single $\Gamma_1(N)$–double coset, every $\Gamma_1(N)$–left coset meets every $\Gamma_1(N)$–right coset, hence there exist $\alpha_1, \cdots, \alpha_{p+1} \in M_p(N)$ which represent both left and right cosets :

$$M_p(N) = \bigcup_i \Gamma_1(N)\alpha_i = \bigcup_i \alpha_i\Gamma_1(N).$$

Therefore,

$$M_p(N) = \bigcup_i \Gamma_1(N)\alpha_i = \bigcup_i R_p(\alpha_i\Gamma_1(N))' = \bigcup_i \Gamma_1(N)R_p\alpha_i'.$$

so that

$$T_p = p^{k/2-1}\sum_i \alpha_i = p^{k/2-1}\sum_i R_p\alpha_i'.$$

Then $\langle f\mid_k T_p, g\rangle = \chi(p)\langle f, g\mid_k T_p\rangle$ is equivalent to $\langle f\mid_k \sum_i \alpha_i, g\rangle = \langle f, g\mid_k \sum_i \alpha_i'\rangle$. For this purpose, it suffices to show $\langle f\mid_k \alpha, g\rangle = \langle f, g\mid_k \alpha'\rangle$ for each $\alpha \in M_p(N)$.

Exercise 4. Let $\alpha \in M_p(N)$. Show that $H = \alpha^{-1}\Gamma(Np)\alpha$ and $H' = (\alpha')^{-1}\Gamma(Np)\alpha'$ are both subgroups of $\Gamma_1(N)$, having the same index in $\Gamma = SL_2(\mathbf{Z})$ as $\Gamma(Np)$. Further, $\alpha^{-1}\mathcal{D}(\Gamma(Np))$ is a fundamental domain of H.

Exercise 5. Denote by $\delta(f,g)(z)$ the form $f(z)\overline{g(z)}(\operatorname{Im} z)^{k-2}dz\wedge d\bar{z}$. Show that for any matrix $\gamma \in M_2(\mathbf{R})$ with $\det\gamma > 0$, we have $\delta(f\mid_k \gamma,\ g\mid_k \gamma)(z) = \delta(f,g)(\gamma z)$.

Given $\alpha \in M_p(N)$, by Exercise 4, $f\mid_k \alpha$, $g\mid_k \alpha'$, f, g are all cusp forms for $\Gamma(Np)$. Further, $f\mid_k \alpha$ and g are also cusp forms for $H = \alpha^{-1}\Gamma(Np)\alpha$. Let $n = [SL_2(\mathbf{Z}) : H] = [SL_2(\mathbf{Z}) : \Gamma(Np)]$. From definition, we have

$$\begin{aligned}
\langle f\mid_k \alpha, g\rangle &= \frac{i}{2n}\iint_{\mathcal{D}(H)} \delta(f\mid_k \alpha, g)(z) \\
&= \frac{i}{2n}\iint_{\alpha^{-1}\mathcal{D}(\Gamma(Np))} \delta(f\mid_k \alpha, g)(z) \quad \text{by Exercise 4} \\
&= \frac{i}{2n}\iint_{\mathcal{D}(\Gamma(Np))} \delta(f\mid_k \alpha, g)(\alpha^{-1}z) \\
&= \frac{i}{2n}\iint_{\mathcal{D}(\Gamma(Np))} \delta(f\mid_k \alpha\mid_k \alpha^{-1}, g\mid_k \alpha^{-1})(z) \quad \text{by Exercise 5} \\
&= \frac{i}{2n}\iint_{\mathcal{D}(\Gamma(Np))} \delta(f, g\mid_k \alpha')(z) \\
&= \langle f, g\mid_k \alpha'\rangle,
\end{aligned}$$

as desired. This completes the proof of Theorem 3. □

Corollary 1. *Let p be a prime not dividing N. Then all eigenvalues of T_p on $\mathcal{C}(N,k,\chi)$ are real if $\chi(p) = 1$, pure imaginary if $\chi(p) = -1$. In general, an eigenvalue λ_p of T_p on $\mathcal{C}(N,k,\chi)$ satisfies $\overline{\lambda_p} = \chi(p)\lambda_p$.*

§3 The structure of $\mathcal{M}(N,k,\chi)$

We study first the structure of the subspace of cusp forms $\mathcal{C}(N,k,\chi)$. For each positive integer M dividing N properly such that χ is a character mod M, the space $\mathcal{C}(M,k,\chi)$ is clearly contained in $\mathcal{C}(N,k,\chi)$. Define, for a positive integer d, the operator $B_d = d^{-k/2}\begin{pmatrix} d & 0 \\ 0 & 1 \end{pmatrix}$ so that its action on a holomorphic function f on $\mathcal{H}$ is given by

$$(f \mid_k B_d)(z) = d^{-k/2} f \mid_k \begin{pmatrix} d & 0 \\ 0 & 1 \end{pmatrix}(z) = f(dz).$$

Exercise 6. (1) Show that B_d maps $\mathcal{M}(N,k,\chi)$ to $\mathcal{M}(Nd,k,\chi)$ and preserves cusp forms.

(2) For a prime p not dividing dN, we have $B_dT_p = T_pB_d$ as maps from $\mathcal{M}(N,k,\chi)$ to $\mathcal{M}(Nd,k,\chi)$.

Thus for each positive divisor d of N/M, the space $\mathcal{C}(M,k,\chi) \mid_k B_d$ is contained in $\mathcal{C}(N,k,\chi)$, and the forms in $\mathcal{C}(M,k,\chi) \mid_k B_d$ are called "push-ups" of forms in $\mathcal{C}(M,k,\chi)$ since they are constructed from forms of level M by a simple change of variable $z \mapsto dz$ which lifts the level to dM. Denote by $\mathcal{C}^-(N,k,\chi)$ the space generated by the forms in $\mathcal{C}(M,k,\chi)$ as M runs through positive proper divisors of N such that χ is a character mod M, as well as their push-ups via B_d with d dividing N/M. The forms in $\mathcal{C}^-(N,k,\chi)$ are called old forms. The Hecke operator T_p for $p\nmid N$ preserves $\mathcal{C}^-(N,k,\chi)$ by Exercise 6, (2). We know from Theorem 3 that $\mathcal{C}^-(N,k,\chi)$ decomposes as a direct sum of common eigenspaces of all T_p, $p\nmid N$. Observe that if $f \in \mathcal{C}(M,k,\chi)$ is a common eigenfunction of all T_p with $p\nmid N$, then so is $f \mid_k B_d$ and it has the same eigenvalue as f for all $d \mid N/M$ (by Exercise 6, (2)); hence each common eigenspace in $\mathcal{C}^-(N,k,\chi)$ has dimension > 1.

Denote by $\mathcal{C}^+(N,k,\chi)$ the orthogonal complement of $\mathcal{C}^-(N,k,\chi)$ in $\mathcal{C}(N,k,\chi)$, which is also invariant under T_p, $p\nmid N$, by Theorem 3. Hence $\mathcal{C}^+(N,k,\chi)$ also decomposes as a direct sum of common eigenspaces of all T_p, $p\nmid N$. Each nonzero form in a common eigenspace is called a *newform* of weight k level N and character χ since these forms are genuinely of level N. Hence the space $\mathcal{C}(N,k,\chi)$ is generated by newforms of weight k character χ and levels M dividing N as well as their push-ups to levels dividing N. Our first characterization of a newform is that the common eigenspace in which it lies is one-dimensional. More precisely.

Theorem 4. *(1) Let $f(z) = \sum_{n=1}^{\infty} a_n e^{2\pi i n z}$ be a newform of weight k level N and character χ. Then $a_1 \neq 0$. So we may assume that f is normalized, that is, $a_1 = 1$. Then for $p\nmid N$, $f \mid_k T_p = a_p f$, that is, a_p is the eigenvalue of T_p on f.*

(2) Let $f(z) = \sum_{n=1}^{\infty} a_n e^{2\pi i n z}$ *and* $g(z) = \sum_{n=1}^{\infty} b_n e^{2\pi i n z}$ *be two normalized newforms of weight* k *character* χ *and levels* N, M*, resp. Here* M *divides* N*. Then* $a_p \neq b_p$ *for infinitely many primes* $p \nmid N$*.*

Proof. (1) For $p \nmid N$, let λ_p be the eigenvalue of T_p on f, i.e., $f \mid T_p = \lambda_p f$. By Theorem 2, $a_p = a_1 \lambda_p$. Thus $a_p = \lambda_p$ if $a_1 = 1$. Assume $a_1 = 0$. Then by Theorem 2 again, $a_{p^r} = a_1 c_{p^r} = 0$ for all $p \nmid N$, $r \geq 0$, and also $a_{p_1^{r_1} \cdots p_l^{r_l}} = a_{p_1^{r_1}} \cdots a_{p_l^{r_l}} = 0$ if $p_1, \cdots, p_l$ are distinct primes not dividing N. We have shown that $a_1 = 0$ would imply $a_n = 0$ if n is prime to N.

(2) Suppose f and g are two normalized newforms of level N, M, respectively, and with the same eigenvalue for almost all T_p, $p \nmid N$. Then $h(z) = f(z) - g(z) = \sum_{n=1}^{\infty} c_n e^{2\pi i n z} \in \mathcal{C}(N, k, \chi)$ which is an eigenfunction for almost all T_p and $c_1 = 0$. The same argument as in (1) shows that there is a positive integer K such that $c_n = 0$ if n is prime to K. To complete the proof of (1) and (2), it suffices to show

Theorem 4′. *Let* $f(z) = \sum_{n=1}^{\infty} a_n e^{2\pi i n z} \in \mathcal{C}(N, k, \chi)$*. If there is a positive square free integer* K *such that* $a_n = 0$ *whenever* $(n, K) = 1$*, then* $f \in \mathcal{C}^-(N, k, \chi)$*.*

Grant Theorem 4′. For Theorem 4, (1), if $a_1 = 0$, then f lies in the intersection of $\mathcal{C}^+(N, k, \chi)$ and $\mathcal{C}^-(N, k, \chi)$, which is $\{0\}$, a contradiction. For (2), if $M = N$, then we get $f - g \in \mathcal{C}^+(N, k, \chi) \cap \mathcal{C}^-(N, k, \chi)$ and hence $f = g$; while if $M \neq N$, then $f - g$ and g both lying in $\mathcal{C}^-(N, k, \chi)$ would imply $f \in \mathcal{C}^-(N, k, \chi)$ and consequently $f = 0$, a contradiction. Thus Theorem 4′ will imply Theorem 4.

We begin by showing the counter part of Lemma 1.

Lemma 3. *Suppose* $f(z) = \sum_{n=1}^{\infty} a_{nq} e^{2\pi i n q z} \in \mathcal{C}(N, k, \chi)$ *with* q *a prime dividing* N*.*

(1) If χ *is a character* mod N/q*, then* $g(z) = \sum_{n=1}^{\infty} a_{nq} e^{2\pi i n z} \in \mathcal{C}(N/q, k, \chi)$*.*

(2) If χ *is not a character* mod N/q*, then* $f = 0$*.*

Proof. $g(z) = \sum_{n=1}^{\infty} a_{nq} e^{2\pi i n z} = q^{k/2} f \mid_k \begin{pmatrix} 1 & 0 \\ 0 & q \end{pmatrix} (z)$. For $\begin{pmatrix} a & b \\ c & d \end{pmatrix} \in \Gamma_0(N)$, we have $\begin{pmatrix} 1 & 0 \\ 0 & q \end{pmatrix}^{-1} \begin{pmatrix} a & b \\ c & d \end{pmatrix} \begin{pmatrix} 1 & 0 \\ 0 & q \end{pmatrix} = \begin{pmatrix} a & bq \\ c/q & d \end{pmatrix}$. Write $\Gamma_0(N/q, q)$ for the conjugate $\begin{pmatrix} 1 & 0 \\ 0 & q \end{pmatrix}^{-1} \Gamma_0(N) \begin{pmatrix} 1 & 0 \\ 0 & q \end{pmatrix}$. Then for $\begin{pmatrix} a & b \\ c & d \end{pmatrix} \in \Gamma_0(N/q, q)$, g satisfies

$$g \mid_k \begin{pmatrix} a & b \\ c & d \end{pmatrix} = \chi(d) g \quad \text{for} \quad \begin{pmatrix} a & b \\ c & d \end{pmatrix} \in \Gamma_0(N/q, q).$$

In particular, g is invariant under the action of

$$\Gamma(N/q,q) = \left\{ \begin{pmatrix} a & b \\ c & d \end{pmatrix} \in \Gamma_0(N/q,q) : a \equiv d \equiv 1 \mod N \right\}$$

and the action of $\begin{pmatrix} 1 & 1 \\ 0 & 1 \end{pmatrix}$.

(1) χ is a character $\bmod N/q$. Then g is invariant under $\Gamma_1(N/q)$. One checks easily that $\Gamma_0(N/q,q)$ is a subgroup of $\Gamma_0(N/q)$ with a set of right coset representatives $R_i \in \Gamma_1(N/q)$. Given $\gamma = \begin{pmatrix} x & y \\ z & w \end{pmatrix} \in \Gamma_0(N/q)$, write $\gamma = \gamma' R_i$ for some i, where $\gamma' = \begin{pmatrix} x' & y' \\ z' & w' \end{pmatrix} \in \Gamma_0(N/q,q)$. We find $w \equiv w' \bmod N/q$. Then $g \mid_k \gamma = g \mid_k \gamma' \mid_k R_i = \chi(w')g = \chi(w)g$ shows $g \in \mathcal{C}(N/q,k,\chi)$.

(2) χ is not a character $\bmod N/q$. Since $\begin{pmatrix} 1 & q \\ N/q & N+1 \end{pmatrix} \in \Gamma(N/q,q)$, g is invariant under

$$\begin{pmatrix} 1 & u \\ 0 & 1 \end{pmatrix} \begin{pmatrix} 1 & q \\ N/q & N+1 \end{pmatrix} \begin{pmatrix} 1 & u' \\ 0 & 1 \end{pmatrix} = \begin{pmatrix} 1+uN/q & u'(1+uN/q)+q+uN+u \\ N/q & u'N/q+N+1 \end{pmatrix}$$
$$= \begin{pmatrix} A & B \\ C & D \end{pmatrix}$$

for all $u, u' \in \mathbf{Z}$. Since χ is not a character $\bmod N/q$, there is some $u' \in \mathbf{Z}$ such that $u'N/q+1 \not\equiv 0 \bmod q$ and $\chi(u'N/q+1) \neq 1$. Further, since $B \equiv u'(1+uN/q)+u \equiv u(1+u'N/q)+u' (\bmod q)$, for our choice of u', there is some $u \in \mathbf{Z}$ such that $B \equiv 0 \bmod q$. With u, u' chosen as above, we have $\begin{pmatrix} A & B \\ C & D \end{pmatrix} \in \Gamma_0(N/q,q)$, and $g = g \mid_k \begin{pmatrix} A & B \\ C & D \end{pmatrix} = \chi(D)g = \chi(u'N/q+1)g$. This is possible only if $g = 0$. Thus $f = g \mid_k B_q = 0$. □

In order to prove Theorem 4′, we introduce two more operators. For convenience, we shall denote by p a prime not dividing N and by q a prime factor of N. The operator playing similar role to T_p at a place q dividing N is

$$U_q = q^{k/2-1} \sum_{u \bmod q} \begin{pmatrix} 1 & u \\ 0 & q \end{pmatrix}.$$

It sends $f(z) = \sum_{n=0}^{\infty} a_n e^{2\pi i n z}$ to

$$f \mid_k U_q(z) = \sum_{n=0}^{\infty} a_{nq} e^{2\pi i n z}.$$

Its properties are summarized in

Exercise 7. (1) U_q sends $\mathcal{C}(N,k,\chi)$ to itself. Further, if $q^2 \mid N$ and χ is a character $\bmod N/q$, then U_q maps $\mathcal{C}(N,k,\chi)$ to $\mathcal{C}(N/q,k,\chi)$.

(2) Show that $T_pU_q = U_qT_p$ and $U_qB_d = B_dU_q$ if q and d are coprime.

For $q \mid N$, write Q for the highest power of q dividing N so that $N = N'Q$ with $(N',Q) = 1$. The Atkin-Lehner W_q^N operator is defined as

$$W_q^N = \begin{pmatrix} Qx & y \\ Nz & Q \end{pmatrix},$$

where x, y, z are integers such that $\det W_q^N = Q$. Write the character χ as $\chi_{N'}\chi_Q$, where χ_M denotes a character mod M. Then W_q^N sends $\mathcal{C}(N,k,\chi_Q\chi_{N'})$ to $\mathcal{C}(N,k,\bar{\chi}_Q\chi_{N'})$, which depends on the choice of the matrix. However, if χ_Q is trivial, that is, if χ is a character $\bmod N/Q$, then we get a well-defined operator on $\mathcal{C}(N,k,\chi)$.

Exercise 8. (1) Suppose χ is a character $\bmod N/Q$, then $W_q^N : \mathcal{C}(N,k,\chi) \to \mathcal{C}(N,k,\chi)$ is independent of the choice of the matrix. Further, $W_q^NT_p = T_pW_q^N$ and $f \mid_k W_q^N \mid_k W_q^N = \chi(Q)f$ for all $f \in \mathcal{C}(N,k,\chi)$. In particular, W_q^N is an automorphism.

(2) If $q \mid N$, $q^2 \nmid N$ and χ is a character $\bmod N/q$, then

$$U_q + q^{k/2-1}W_q^N : \mathcal{C}(N,k,\chi) \to \mathcal{C}(N/q,k,\chi).$$

The purpose of these two operators is to enable us to construct oldforms such that after subtracting them from f, more Fourier coefficients are zero.

Lemma 4. *Let* $f(z) = \sum_{n=1}^{\infty} a_n e^{2\pi i n z} \in \mathcal{C}(N,k,\chi)$ *be such that* $a_n = 0$ *whenever* $(n,K) = 1$. *Then*

(1) $a_n = 0$ *whenever* $(n,N) = 1$.

(2) *If* $q \mid N$ *and* χ *is a character* $\bmod N/q$, *set*

$$\psi = \begin{cases} f \mid_k U_q & \text{if } q^2 \mid N \text{ or } (K,N) = q, \\ (1+q^{-1})^{-1} f \mid_k (U_q + q^{k/2-1}W_q^N) & \text{otherwise.} \end{cases}$$

Then $\psi \in \mathcal{C}(N/q,k,\chi)$ *and* $(f - \psi \mid_k B_q)(z) = \sum_{n=1}^{\infty} b_n e^{2\pi i n z} \in \mathcal{C}(N,k,\chi)$ *satisfies* $b_n = 0$ *whenever* $\big(n,(K,N)\big) = 1$ *or* q.

Proof. For a prime l, introduce the annihilator operator A_l defined by

$$f \mid_k A_l = f - f \mid_k U_l \mid_k B_l,$$

which annihilates the Fourier coefficients of f indexed by multiples of l. Then $f\mid_k A_l \in \mathcal{C}(Nl^2,k,\chi)$. In fact, the level is lower if $l\mid N$ and χ is a character mod N/l. Indeed, in which case we have $f\mid_k A_l \in \mathcal{C}(Nl,k,\chi)$ or $\mathcal{C}(N,k,\chi)$ according to $l\mid N$ or $l^2\mid N$ by Exercises 7 and 6. Suppose $l_1,\cdots,l_r$ are the distinct prime factors of K, and, say, $l_r\nmid N$. Then from hypothesis, $f\mid_k A_{l_1}\mid_k A_{l_2}\mid\cdots\mid_k A_{l_{r-1}} = \Phi\mid B_{l_r}\in\mathcal{C}(Nl_1^2\cdots l_{r-1}^2,k,\chi)$ with $\Phi(z)=\sum_{n=1}^{\infty}c_ne^{2\pi inz}$. Thus $\Phi=0$ by Lemma 1 (1). Therefore $a_n=0$ if $(n,l_1\cdots l_{r-1})=1$. Inductively, we get (1). For (2), observe that in case $\psi=f\mid_k U_q$, we have $\psi\mid_k B_q=\sum_{n=1}^{\infty}a_{nq}e^{2\pi inqz}$, thus $f-\psi\mid_k B_q$ has the desired property. Further, $\psi\in\mathcal{C}(N/q,k,\chi)$ if $q^2\mid N$ by Exercise 7, (1), and if $(K,N)=q$ by Lemma 3, (1). For the remaining case, assume $q^2\nmid N$ so that $ord_qN=1$ and (N,K) has more than one prime factor. Denote by $q_1,\cdots,q_t=q$ the prime factors of (N,K). Here $t\geq 2$. The function $(1+q^{-1})\psi=f\mid_k\left(U_q+q^{k/2-1}W_q^N\right)$, by Exercise 8, (2), lies in $\mathcal{C}(N/q,k,\chi)$. To show that $f-\psi\mid_k B_q$ has the desired property, it suffices to show that $f-\psi\mid_k B_q=\sum_{i=1}^{t-1}f_i\mid B_{q_i}$ for $f_i(z)$ having Fourier expansion of the form $\sum_{n=1}^{\infty}c_ne^{2\pi inz}$. For this purpose, observe that we may write $f=\sum_{i=1}^{t}\Phi_i\mid_k B_{q_i}$, where $\Phi_1=f\mid_k U_{q_1}$, and for $t\geq i\geq 2$, $\Phi_i=f\mid_k A_{q_1}\mid\cdots\mid_k A_{q_{i-1}}\mid_k U_{q_i}$. As already remarked, $\Phi_i\mid_k B_{q_i}\in\mathcal{C}(Nq_1^2\cdots q_i^2,k,\chi)$ for $t>i\geq 1$, and

$$\Phi_t\mid_k B_q=f\mid_k A_{q_1}\mid\cdots\mid_k A_{q_{t-1}}\in\mathcal{C}(Nq_1^2\cdots q_{t-1}^2,k,\chi),$$

which implies $\Phi_t\in\mathcal{C}(Nq_1^2\cdots q_{t-1}^2/q,k,\chi)$ by Lemma 3, (1). Put $M=Nq_1^2\cdots q_{t-1}^2$. Then $\mathrm{ord}_q M=\mathrm{ord}_q N=1$, and we may choose W_q^M to be W_q^N. We compute $\Phi_i\mid_k B_{q_i}\mid_k\left(U_q+q^{k/2-1}W_q^M\right)$. For $1\leq i\leq t-1$, q_i and q are coprime, we have $B_{q_i}U_q=U_qB_{q_i}$ and $B_{q_i}W_q^M=W_q^{M/q_i}B_{q_i}$ so that

$$\Phi_i\mid_k B_{q_i}\mid_k\left(U_q+q^{k/2-1}W_q^M\right)=\Phi_i\mid_k\left(U_q+q^{k/2-1}W_q^{M/q_i}\right)\mid_k B_{q_i}=\Psi_i\mid_k B_{q_i},$$

where $\Psi_i=\Phi_i\mid_k\left(U_q+q^{k/2-1}W_q^{M/q_i}\right)\in\mathcal{C}(M/q,k,\chi)$. From $W_q^M=\begin{pmatrix}qx & y\\ Mz & q\end{pmatrix}$ and $B_qU_q=q^{-1}\sum_{u=0}^{q-1}\begin{pmatrix}1 & u\\ 0 & 1\end{pmatrix}$ we get

$$\Phi_t\mid_k B_q\mid_k\left(U_q+q^{k/2-1}W_q^M\right)=\Phi_t+q^{-1}\Phi_t\mid_k\begin{pmatrix}qx & y\\ Mz/q & 1\end{pmatrix}=(1+q^{-1})\Phi_t$$

since $\Phi_t\in\mathcal{C}(M/q,k,\chi)$. Hence

$$\begin{aligned}f-\psi\mid_k B_q&=\sum_{i=1}^{t}\Phi_i\mid_k B_{q_i}-(1+q^{-1})^{-1}\left(\sum_{i=1}^{t-1}\Psi_i\mid_k B_{q_i}+(1+q^{-1})\Phi_t\right)\mid_k B_q\\&=\sum_{i=1}^{t-1}\left(\Phi_i-(1+q^{-1})^{-1}\Psi_i\mid_k B_q\right)\mid_k B_{q_i}=\sum_{i=1}^{t-1}f_i\mid_k B_{q_i}\end{aligned}$$

with $f_i \in \mathcal{C}(M,k,\chi)$, as desired. □

We are now ready to prove Theorem 4′. With $f(z)$ as given, working modulo $\mathcal{C}^-(N,k,\chi)$ and applying Lemma 4 repeatedly, we may assume that $q_1,\cdots,q_r$ are the prime factors of (N,K) such that χ is not a character mod N/q_i, and $a_n = 0$ whenever $(n, q_1\cdots q_r) = 1$. Let $\Phi_r = f \mid_k A_{q_1} \mid \cdots \mid_k A_{q_{r-1}} \mid_k U_{q_r}$. Then $\Phi_r \mid_k B_{q_r} = f \mid_k A_{q_1} \mid \cdots \mid_k A_{q_{r-1}} \in \mathcal{C}(Nq_1^2\cdots q_{r-1}^2, k, \chi)$. Since χ is not a character mod $Nq_1^2\cdots q_{r-1}^2/q$, we get $\Phi_r = 0$ from Lemma 3, (2). This shows that $a_n = 0$ whenever $(n, q_1\cdots q_{r-1}) = 1$. Proceed by induction, we have $f = 0$. This proves Theorem 4′ and hence Theorem 4. □

Exercise 9. Denote by $\mathcal{N}(M,k,\chi)$ the set of newforms of weight k level M and character χ. (It is empty if no such form exists). Show that $\bigcup_{M\mid N}\bigcup_{d\mid N/M} \mathcal{N}(M,k,\chi) \mid_k B_d$ is a basis of $\mathcal{C}(N,k,\chi)$ over $\mathbf{C}$.

Now let $f(z) = \sum_{n=1}^{\infty} a_n e^{2\pi i n z}$ be a normalized newform of weight k level N and character χ. We know from Theorem 4 (1) that $f \mid_k T_p = a_p f$ for all $p \nmid N$. Thus

$$a_{np} + \chi(p)p^{k-1}a_{n/p} = a_p a_n \quad \text{for all } n \geq 1 \text{ and } p \nmid N,$$

by the definition of T_p. Next let q be a prime factor of N. Then $f \mid_k U_q(z) = \sum_{n=1}^{\infty} a_{nq} e^{2\pi i n z} \in \mathcal{C}(N,k,\chi)$ and it is an eigenfunction of T_p with the same eigenvalue as f for all $p \nmid N$, which in turn implies that $f \mid_k U_q - a_q f$ has the same property for all $p \nmid N$. Since the common eigenspace containing f is 1–dimensional by Theorem 4 (2), and the first Fourier coefficient of $f \mid_k U_q - a_q f$ is 0, we conclude from Theorem 4 (1) that $f \mid_k U_q = a_q f$. Therefore we have

$$a_{nq} = a_q a_n \quad \text{for all } n \geq 1 \text{ and } q \mid N.$$

We can say more about a_q :

Case I. χ is a character mod N/q.

(1) If $q^2 \mid N$, then $f \mid_k U_q = a_q f \in \mathcal{C}(N/q,k,\chi)$. This shows $a_q f = 0$, i.e., $a_q = 0$.

(2) If $q^2 \nmid N$, then by Exercise 8, (2), $f \mid_k (U_q + q^{k/2-1}W_q^N) \in \mathcal{C}(N/q,k,\chi)$. On the other hand, $f \mid_k W_q^N$, like $f \mid_k U_q$, is a common eigenfunction of T_p for all $p \nmid N$ and has the same eigenvalue as f, by Exercise 8 (1). Hence $f \mid_k U_q = a_q f = -q^{k/2-1} f \mid_k W_q^N$. But $f \mid_k W_q^N \mid_k W_q^N = \chi(q) f$, which implies $a_q^2 = \chi(q)q^{k-2}$.

Case II. χ is not a character mod N/q.

Then $|a_q| = q^{(k-1)/2}$ using a theorem of Ogg [16], which we explain below. Introduce the operator

$$H_N = \begin{pmatrix} 0 & -1 \\ N & 0 \end{pmatrix}.$$

Exercise 10. H_N sends $\mathcal{C}(N,k,\chi)$ to $\mathcal{C}(N,k,\overline{\chi})$ with $H_N^2 = \chi(-1)$. Further, $H_N T_p = \overline{\chi}(p) T_p H_N$ for $p \nmid N$. Thus H_N sends $\mathcal{C}^-(N,k,\chi)$ to $\mathcal{C}^-(N,k,\overline{\chi})$ and $\mathcal{C}^+(N,k,\chi)$ to $\mathcal{C}^+(N,k,\overline{\chi})$.

There is another operator K defined by

$$f \mid K(z) = \overline{f(-\overline{z})},$$

that is, if $f(z) = \sum_{n=0}^{\infty} a_n e^{2\pi i n z}$, then

$$f \mid K(z) = \sum_{n=0}^{\infty} \overline{a}_n e^{2\pi i n z},$$

which we sometimes denote by $\overline{f}$. The operator K sends $\mathcal{C}(N,k,\chi)$ to $\mathcal{C}(N,k,\overline{\chi})$ with $K^2 = id$. Further, $KT_p = T_p K$ and $H_N K = (-1)^k K H_N$. If g is a normalized newform of level N weight k and character χ, then $g \mid K = \overline{g}$ is a normalized newform of level N weight k and character $\overline{\chi}$, and $g \mid H_N = \lambda_g \overline{g}$ for some nonzero constant λ_g.

Lemma 5. *(Ogg [16]) Let $q \mid N$. Suppose $N = q^e M$ where $q \nmid M$ and χ is not a character* mod M. *Then $U_q^e H_N U_q^e = q^{k-1} U_q^{e-1} H_N U_q^{e-1}$ on $\mathcal{C}(N,k,\chi)$.*

Proof. $U_q^e = q^{e(\frac{k}{2}-1)} \sum_{u \bmod q^e} \begin{pmatrix} 1 & u \\ 0 & q^e \end{pmatrix}$, and, as an operator,

$$H_N U_q^e = q^{e(\frac{k}{2}-1)} \sum_{u \bmod q^e} \begin{pmatrix} 0 & -1 \\ M & uM \end{pmatrix} = q^{e(\frac{k}{2}-1)} H_M \sum_{u \bmod p^e} \begin{pmatrix} 1 & u \\ 0 & 1 \end{pmatrix}.$$

Since H_N is an isomorphism, we shall compare $(H_N U_q^e)^2$ with $(H_N U_q^{e-1})^2$. As computed above,

$$(H_N U_q^e)^2 = q^{e(k-2)} \sum_{u,v \bmod q^e} H_M \begin{pmatrix} 1 & u \\ 0 & 1 \end{pmatrix} H_M \begin{pmatrix} 1 & v \\ 0 & 1 \end{pmatrix}.$$

Claim that $\sum_{\substack{u,v \bmod q^e \\ (u,q)=1}} H_M \begin{pmatrix} 1 & u \\ 0 & 1 \end{pmatrix} H_M \begin{pmatrix} 1 & v \\ 0 & 1 \end{pmatrix}$ is the zero operator on $\mathcal{C}(N,k,\chi)$.

Indeed, given u prime to q, and any v, we have

$$H_M \begin{pmatrix} 1 & u \\ 0 & 1 \end{pmatrix} H_M \begin{pmatrix} 1 & v \\ 0 & 1 \end{pmatrix} = \begin{pmatrix} -1 & -v \\ uM & uvM-1 \end{pmatrix}$$

$$= \begin{pmatrix} 1+M(v-v') & v-v' \\ M(u(Mv'-1)-Muv+1) & 1+Mu(v'-v) \end{pmatrix} H_M \begin{pmatrix} 1 & 1 \\ 0 & 1 \end{pmatrix} H_M \begin{pmatrix} 1 & v' \\ 0 & 1 \end{pmatrix}$$

for any v'. In particular, there exists a unique $v' \bmod q^e$ such that $u(Mv'-1) - Muv + 1 \equiv 0 \pmod{q^e}$. In this case, $1 + Mu(v'-v) \equiv u \pmod{q^e}$, and $\equiv 1 \pmod{M}$. Since M is prime to q, we may assume that $u \equiv 1 \bmod M$ so that $1 + Mu(v'-v) \equiv u \pmod{N}$. With this v', we have

$$\begin{aligned} &f \mid_k H_M \begin{pmatrix} 1 & u \\ 0 & 1 \end{pmatrix} H_M \begin{pmatrix} 1 & v \\ 0 & 1 \end{pmatrix} \\ =&\chi(u) f \mid_k H_M \begin{pmatrix} 1 & 1 \\ 0 & 1 \end{pmatrix} H_M \begin{pmatrix} 1 & v' \\ 0 & 1 \end{pmatrix} \quad \text{for } f \in \mathcal{C}(N,k,\chi). \end{aligned}$$

As v runs through all integers $\bmod q^e$, so does v'. Hence summing over $v \bmod q^e$ and $u \bmod p^e$ with $(u,q) = 1$ of the above equation yields

$$\begin{aligned} &f \mid_k \sum_{\substack{u,v \bmod q^e \\ (u,q)=1}} H_M \begin{pmatrix} 1 & u \\ 0 & 1 \end{pmatrix} H_M \begin{pmatrix} 1 & v \\ 0 & 1 \end{pmatrix} \\ =&\sum_u \chi(u) f \mid_k \sum_{v'} H_M \begin{pmatrix} 1 & 1 \\ 0 & 1 \end{pmatrix} H_M \begin{pmatrix} 1 & v' \\ 0 & 1 \end{pmatrix} = 0 \end{aligned}$$

since χ is not a character $\bmod M$ and u runs through all elements in $(\mathbf{Z}/N\mathbf{Z})^\times$ congruent to 1 modulo M.

Therefore

$$(H_N U_q^e)^2 = q^{e(k-2)} \sum_{\substack{u \bmod q^{e-1} \\ v \bmod q^e}} H_M \begin{pmatrix} 1 & uq \\ 0 & 1 \end{pmatrix} H_M \begin{pmatrix} 1 & v \\ 0 & 1 \end{pmatrix}$$

on $\mathcal{C}(N,k,\chi)$. But then, as operators, we find

$$H_M \begin{pmatrix} 1 & uq \\ 0 & 1 \end{pmatrix} H_M = H_{qM} \begin{pmatrix} 1 & u \\ 0 & 1 \end{pmatrix} H_{qM}$$

and

$$H_N U_q^{e-1} = q^{(e-1)(\frac{k}{2}-1)} \sum_{u \bmod q^{e-1}} H_{qM} \begin{pmatrix} 1 & u \\ 0 & 1 \end{pmatrix}.$$

Hence

$$\begin{aligned} (H_N U_q^e)^2 &= q^{e(k-2)} \sum_{\substack{u \bmod q^{e-1} \\ v \bmod q^e}} H_{Mq} \begin{pmatrix} 1 & u \\ 0 & 1 \end{pmatrix} H_{Mq} \begin{pmatrix} 1 & v \\ 0 & 1 \end{pmatrix} \\ &= q^{k-2} \cdot q (H_N U_q^{e-1})^2 \\ &= q^{k-1} (H_N U_q^{e-1})^2. \end{aligned}$$

This proves Lemma 5. □

Apply Lemma 5 to f, we get $\lambda_f a_q^e \overline{a}_q^e = q^{k-1}\lambda_f a_q^{e-1}\overline{a}_q^{e-1}$. Hence $|a_q| = q^{\frac{k-1}{2}}$ if $a_q \neq 0$. If $e = 1$, that is, $q \mid N$ and $q^2 \nmid N$, then Lemma 5 implies that U_q is $1-1$, hence $a_q \neq 0$. It remains to show that in case $q^2 \mid N$, and χ is not a character mod N/q, $f \mid_k U_q \neq 0$. Suppose otherwise, that is, $f \mid_k \sum_{u \bmod q} \begin{pmatrix} 1 & u \\ 0 & q \end{pmatrix} = 0$. For any integer v we have

$$\begin{aligned} f \mid_k \begin{pmatrix} 1 & u \\ 0 & q \end{pmatrix} \begin{pmatrix} 1 & 0 \\ vN/q & 1 \end{pmatrix} &= f \mid_k \begin{pmatrix} 1 + vuN/q & -vu^2N/q^2 \\ vN & 1 - vuN/q \end{pmatrix} \begin{pmatrix} 1 & u \\ 0 & q \end{pmatrix} \\ &= \chi(1 - vuN/q) f \mid_k \begin{pmatrix} 1 & u \\ 0 & q \end{pmatrix} \\ &= \zeta_q^{vu} f \mid_k \begin{pmatrix} 1 & u \\ 0 & q \end{pmatrix}, \end{aligned}$$

where $\zeta_q = \chi(1 - N/q)$ is a primitive qth root of unity since χ is not trivial on $1 + \frac{N}{q}\mathbf{Z}$. Thus

$$0 = \sum_{u \bmod q} \zeta_q^{vu} f \mid_k \begin{pmatrix} 1 & u \\ 0 & q \end{pmatrix} \quad \text{for } v = 1, \cdots, q.$$

Viewing this as a system of q equations with matrix (ζ_q^{vu}), which has nonzero determinant $\left(= \pm \prod_{1 \le v < u \le q} (\zeta_q^v - \zeta_q^u) \right)$, we get $f \mid_k \begin{pmatrix} 1 & u \\ 0 & q \end{pmatrix} = 0$ for all u, which implies $f = 0$, a contradiction. □

Remark. We have shown that if $q^2 \mid N$ and χ is not a character of N/q, then U_q on $\mathcal{C}(N, k, \chi)$ is injective, and hence an automorphism.

We summarize the above discussion in

Theorem 5. *Let* $f(z) = \sum_{n=1}^{\infty} a_n e^{2\pi i n z}$ *be a normalized newform of weight* k *level* N *and character* χ. *Let* p *be a prime not dividing* N *and let* q *be a prime factor of* N. *Then*

(1) $f \mid_k T_p = a_p f$ *and* $f \mid_k U_q = a_q f$, *and*

$$\begin{cases} a_{np} + \chi(p)p^{k-1}a_{n/p} = a_p a_n & \text{for all } n \ge 1, \\ a_{nq} = a_q a_n & \text{for all } n \ge 1. \end{cases}$$

Thus the associated Dirichlet series $D(s, f)$ *has Euler product at all primes :*

$$D(s, f) = \sum_{n=1}^{\infty} a_n n^{-s} = \prod_{q \mid N} (1 - a_q q^{-s})^{-1} \prod_{p \nmid N} (1 - a_p p^{-s} + \chi(p)p^{k-1-2s})^{-1}.$$

(2) $|a_q| = q^{(k-1)/2}$ *if* χ *is not a character* $\bmod N/q$.

(3) If χ *is a character* $\bmod N/q$, *then* $a_q = 0$ *if* $q^2 \mid N$, *and* $a_q^2 = \chi(q)q^{k-2}$ *if* $q^2 \nmid N$.

In the latter case, we have $f \mid_k W_q^N = -a_q q^{-k/2+1} f$.

As for the Fourier coefficients a_p with $p \nmid N$, we know from Theorem 4′ that $a_p \neq 0$ infinitely often and from Corollary 1 that $a_p = \chi(p)\overline{a}_p$. Further, it can be shown that $\mathcal{C}(\Gamma_1(N), k)$ has a basis with Fourier coefficients lying in the ring of integers $\mathcal{O}$ of the cyclotomic field $\mathcal{O}(\zeta_N)$ by adjoining Nth roots of unity to $\mathbf{Q}$ such that they also generate the $\mathcal{O}$–module of all forms in $\mathcal{C}(\Gamma_1(N), k)$ with Fourier coefficients in $\mathcal{O}$. In view of the action of T_p in terms of Fourier coefficients, we see that the eigenvalues of T_p are algebraic integers. Concerning their absolute values, Ramanujan first conjectured that for the unique normalized cusp form $\Delta(z) = \sum_{n=1}^{\infty} \tau(n)e^{2\pi i n z}$ for $SL_2(\mathbf{Z})$ of weight 12, $\tau(p)$ has absolute value $|\tau(p)| \leq 2p^{11/2}$. Petersson then conjectured that all eigenvalues λ_p of T_p on $\mathcal{C}(\Gamma_1(N), k)$ satisfy $|\lambda_p| \leq 2p^{(k-1)/2}$. This is the so-called Ramanujan-Petersson conjecture. In 1969 Deligne proved that Weil conjectures imply Ramanujan-Petersson conjecture for forms of weight ≥ 2, and later in 1973 he settled the Weil conjectures, as discussed in Chapter 2. The conjecture for forms of weight 1 was proved by Deligne and Serre [5]. Hence we have

Theorem 6. *(Deligne for* $k \geq 2$, *Deligne-Serre for* $k = 1$, *formerly Ramanujan-Petersson conjecture) Let* $f(z) = \sum_{n=1}^{\infty} a_n e^{2\pi i n z}$ *be a normalized newform of weight* k *level* N *and character* χ. *Then for each prime* $p \nmid N$,

$$1 - a_p p^{-s} + \chi(p)p^{k-1-2s} = (1 - \alpha_p p^{-s})(1 - \beta_p p^{-s}),$$

where $|\alpha_p| = |\beta_p| = p^{(k-1)/2}$.

This clearly implies

$$|a_p| = |\alpha_p + \beta_p| \leq 2p^{(k-1)/2}.$$

The space $\mathcal{C}^+(N, k, \chi)$ was defined using the Petersson inner product, which has two drawbacks. First, it is a transcendental, not algebraic, method; secondly, it is not defined on all modular forms. Now we introduce an algebraic way to characterize the space $\mathcal{C}^+(N, k, \chi)$, which was first used by Serre [22] to characterize the space generated by newforms for $\Gamma_0(q)$ with q a prime.

For a proper divisor M of N, the group $\Gamma_0(N)$ is a subgroup of $\Gamma_0(M)$ and we may choose right coset representatives $R_i \in \Gamma(M)$, i.e., $\Gamma_0(M) = \bigcup_i \Gamma_0(N)R_i$. If χ is a character $\bmod M$, the trace operator from $\mathcal{M}(N, k, \chi)$ to $\mathcal{M}(M, k, \chi)$ defined by

$$\mathrm{Tr}_M^N f = \sum_i f \mid_k R_i$$

is independent of the choice of coset representatives R_i. Further, it preserves cusp forms.

Theorem 7. *Let $f \in \mathcal{C}(N,k,\chi)$. Then $f \in \mathcal{C}^+(N,k,\chi)$ if and only if for all primes $q \mid N$ such that χ is a character* $\bmod N/q$,

$$\mathrm{Tr}^N_{N/q} f = 0 = \mathrm{Tr}^N_{N/q} f \mid_k H_N.$$

Proof. "If" part : By definition, we have to show $\langle f,g\rangle = 0 = \langle f, g \mid_k B_q\rangle$ for all primes q dividing N such that χ is a character $\bmod N/q$, and all $g \in \mathcal{C}(N/q,k,\chi)$. Let q be such a prime and g be such a form. Then for all coset representatives $R_i \in \Gamma(N/q)$ in the definition of $\mathrm{Tr}^N_{N/q}$, we have $R_i^{-1}\Gamma(N)R_i = \Gamma(N)$ so that $R_i^{-1}\mathcal{D}(\Gamma(N)) = \mathcal{D}(\Gamma(N))$. From

$$\begin{aligned}
\langle f,g\rangle &= \frac{1}{[\Gamma:\Gamma(N)]}\iint_{\mathcal{D}(\Gamma(N))} \delta(f,g)(z)\\
&= \frac{1}{[\Gamma:\Gamma(N)]}\iint_{\mathcal{D}(\Gamma(N))} \delta(f,g\mid_k R_i^{-1})(z) \quad \text{since } g \text{ has level } N/q\\
&= \frac{1}{[\Gamma:\Gamma(N)]}\iint_{R_i^{-1}\mathcal{D}(\Gamma(N))} \delta(f,g\mid_k R_i^{-1})(R_i z)\\
&= \frac{1}{[\Gamma:\Gamma(N)]}\iint_{\mathcal{D}(\Gamma(N))} \delta(f\mid_k R_i, g)(z)
\end{aligned}$$

for all i we get

$$\langle \mathrm{Tr}^N_{N/q} f, g\rangle = [\Gamma_0(N/q):\Gamma_0(N)]\,\langle f,g\rangle = 0 \quad \text{since} \quad \mathrm{Tr}^N_{N/q} f = 0.$$

The same argument also shows that $f \mid_k H_N$ is perpendicular to all forms in $\mathcal{C}(N/q,k,\overline{\chi})$ since, by hypothesis, $\mathrm{Tr}^N_{N/q} f \mid_k H_N = 0$. Then

$$\langle f, g\mid_k B_q\rangle = \langle f\mid_k H_N, g\mid_k B_q \mid_k H_N\rangle = q^{-k/2}\langle f\mid_k H_N, g\mid_k H_{N/q}\rangle = 0$$

since $g \mid_k H_{N/q} \in \mathcal{C}(N/q,k,\overline{\chi})$.

"Only if" part : Let q be a prime factor of N such that χ is a character $\bmod N/q$. Then $\mathrm{Tr}^N_{N/q} f \in \mathcal{C}(N/q,k,\chi)$ and $\mathrm{Tr}^N_{N/q} f \mid_k H_N \in \mathcal{C}(N/q,k,\overline{\chi})$. The same computation as "If" part yields

$$\langle \mathrm{Tr}^N_{N/q} f, \mathrm{Tr}^N_{N/q} f\rangle = [\Gamma_0(N/q):\Gamma_0(N)]\,\langle f, \mathrm{Tr}^N_{N/q} f\rangle = 0$$

since $f \in \mathcal{C}^+(N,k,\chi)$ and $\mathrm{Tr}^N_{N/q} f \in \mathcal{C}^-(N,k,\chi)$. This shows $\mathrm{Tr}^N_{N/q} f = 0$. As H_N maps $\mathcal{C}^\pm(N,k,\chi)$ to $\mathcal{C}^\pm(N,k,\overline{\chi})$, the same argument shows $\mathrm{Tr}^N_{N/q} f \mid_k H_N = 0$. □

Now we are ready to discuss the structure of the space $\mathcal{M}(N,k,\chi)$ of modular forms of weight k level N and character χ. Hecke has constructed a space of Eisenstein series $\mathcal{E}(H,k)$ for congruence subgroups $H = \Gamma_0(N), \Gamma_1(N)$ and $\Gamma(N)$ such that

$$\mathcal{M}(H,k) = \mathcal{E}(H,k) \oplus \mathcal{C}(H,k).$$

When the weight k is ≥ 3, $\dim_{\mathbf{C}} \mathcal{E}(H,k)$ equals the number of cusps of H. In fact, $\mathcal{E}(H,k)$ consists of a basis whose restrictions to cusps of H are characteristic functions of the cusps of H. These Eisenstein series are defined as double series, which converge absolutely for $k \geq 3$. When $k = 2$, the double series converge conditionally and Eisenstein series are defined by taking difference of two such double series. Eisenstein series of weight 1 are obtained by analytic continuations. Despite of the fact that there is no "canonical" choice of the space of Eisenstein series for which the above decomposition holds, the choice Hecke made was a good one, for it is invariant under the action of Hecke operators, and $\mathcal{E}(H',k)$ is a subspace of $\mathcal{E}(H,k)$ if H is a subgroup of H'. Like cusp forms, the space $\mathcal{E}(\Gamma_1(N),k)$ decomposes as a direct sum $\bigoplus_{\chi} \mathcal{E}(N,k,\chi)$ as χ runs through all characters mod N. Thus

$$\mathcal{M}(N,k,\chi) = \mathcal{E}(N,k,\chi) \oplus \mathcal{C}(N,k,\chi).$$

The reader is referred to [8, 15, 17] for more details. Further, the operators introduced on cusp forms also act on the space of Eisenstein series. With the trace operator replacing the Petersson inner product, the theory of newforms works equally well on the space of Eisenstein series. Therefore the space $\mathcal{M}(N,k,\chi)$ is generated by normalized newforms of weight k character χ and levels dividing N together with their "push-ups" to forms of level dividing N. Moreover, the normalized new-Eisenstein series of weight k level N and character χ are known : their associated Dirichlet series are products of two Dirichlet L-functions :

$$D(s,f) = L(s,\chi_1)L(s+1-k,\chi_2),$$

where χ_1, χ_2 are characters of $\mathbf{Z}$ of conductors N_1, N_2, respectively, such that $N = N_1N_2$ and $\chi = \chi_1\chi_2$. Strictly speaking, $D(s,f)$ only gives all nonzero Fourier coefficients of f. As we shall see in the next section, the zeroth Fourier coefficient of f is -1 times the residue at $s = 0$ of $D(s,f)$.

Historically, Atkin and Lehner [1] started the theory of newforms, they did it for cusp forms for $\Gamma_0(N)$. Then, using adèlic language, Miyake extended it to $\Gamma_1(N)$ [14]. In her thesis [11], Li explained the reduction from forms for $\Gamma(N)$ to forms of level N^2 and character mod N, and gave several criteria for newforms. The main part of this section is extracted from [11]. The parallel theory of newforms for Eisenstein series was done in Weisinger's thesis [27]. The connection between the theory of newforms and representations of $GL_2(A_{\mathbf{Q}})$ is discussed in the articles by Casselman [2] and Deligne [4], respectively.

§4 Functional equations

One of Hecke's fundamental contribution to the theory of modular forms is to study the analytic property of the L-function attached to a modular form f of weight k :

$$L(s,f)=(2\pi)^{-s}\Gamma(s)D(s,f), \quad \text{where } \Gamma(s)=\int_0^\infty e^{-t}t^{s-1}dt \quad \text{for } \operatorname{Re} s>1.$$

Write f in its Fourier expansion $f(z)=\sum_{n=0}^{\infty}a_n e^{2\pi inz}$. It is easy to show that $a_n=O(n^k)$ for $n\geq 1$, and in fact, we see from Theorem 6 that if f is a cusp form, then $a_n=O(n^{\frac{k-1}{2}+\epsilon})$ for every $\epsilon>0$. At any rate, the associated Dirichlet series $D(s,f)=\sum_{n=1}^{\infty}a_n n^{-s}$ converges absolutely for $\operatorname{Re} s>k+1$ and defines a holomorphic function there. Hecke expressed $L(s,f)$ in terms of the Mellin transform of f :

$$L(s,f)=\int_0^\infty t^{s-1}\big(f(it)-a_0\big)dt=\int_0^\infty t^s\big(f(it)-a_0\big)\frac{dt}{t},$$

then he studied its analytic continuation and functional equation. (Philosophically speaking, $f(it)-a_0$ is viewed as a function on positive reals, which is a multiplicative group $\mathbf{R}^\times_{>0}$. Its quasicharacters are given by $t\mapsto t^s$, $s\in\mathbf{C}$, and $\frac{dt}{t}$ is a Haar measure on $\mathbf{R}^\times_{>0}$. So the Mellin transform of f may be viewed as "Fourier transform" of $f(it)-a_0$ on $\mathbf{R}^\times_{>0}$.) The whole procedure is similar to what we did for the L-function attached to an idèle class character in Chapter 5, hence we sketch it briefly below. Suppose f has level N. Let

$$g(z)=f\mid_k H_N(z)=N^{-k/2}z^{-k}f\Big(\frac{-1}{Nz}\Big)=\sum_{n=0}^{\infty}b_n e^{2\pi inz},$$

which also has the same weight and level as f. In particular,

$$g(iy)=N^{-k/2}(iy)^{-k}f\Big(\frac{i}{Ny}\Big).$$

For $\operatorname{Re} s > \frac{k}{2} + 1$, we write

$$\begin{aligned} L(s,f) &= \int_0^1 t^{s-1}(f(it)-a_0)dt + \int_1^\infty t^{s-1}(f(it)-a_0)dt \\ &= -\frac{a_0}{s} + \int_0^1 t^s f(it)\frac{dt}{t} + \int_1^\infty t^s(f(it)-a_0)\frac{dt}{t} \\ &= -\frac{a_0}{s} + \int_1^\infty t^s(f(it)-a_0)\frac{dt}{t} + \int_1^\infty (Ny)^{-s} f\Big(\frac{i}{Ny}\Big)\frac{dy}{y} \\ &\qquad \text{(letting } t = \frac{1}{Ny} \text{ in 1st integral)} \\ &= -\frac{a_0}{s} + \int_1^\infty t^s(f(it)-a_0)\frac{dt}{t} + i^k N^{\frac{k}{2}-s}\int_1^\infty y^{k-s} g(iy)\frac{dy}{y} \\ &= -\frac{a_0}{s} - i^k N^{\frac{k}{2}-s}\frac{b_0}{k-s} + \int_1^\infty t^s(f(it)-a_0)\frac{dt}{t} \\ &\qquad + i^k N^{\frac{k}{2}-s}\int_1^\infty y^{k-s}(g(iy)-b_0)\frac{dy}{y}. \end{aligned}$$

Since $f(it) - a_0 = O(e^{-ct})$ and $g(it) - b_0 = O(e^{-ct})$ for some positive constant c as t approaches infinity, this shows that the last two integrals define two holomorphic functions in the s-plane, which are bounded in each vertical strip of finite width. Thus we obtained an analytic continuation of $L(s,f)$ to the whole s-plane, holomorphic everywhere except possibly with two simple poles at $s = 0$ and $s = k$. Further,

$$L(s,f) + \frac{a_0}{s} + i^k N^{\frac{k}{2}-s}\frac{b_0}{k-s} = i^k N^{\frac{k}{2}-s}\left(L(k-s,g) + \frac{b_0}{k-s} + i^k N^{s-\frac{k}{2}}\frac{(-1)^k a_0}{s}\right),$$

which implies that

$$L(s,f) = i^k N^{\frac{k}{2}-s} L(k-s,g).$$

We state this in

Theorem 8. *(Hecke) Let* $f(z) = \sum_{n=0}^{\infty} a_n e^{2\pi i n z} \in \mathcal{M}(N,k,\chi)$. *The L-function* $L(s,f)$, *which converges absolutely for* $\operatorname{Re} s \gg 0$, *has an analytic continuation to the whole* s*-plane, holomorphic everywhere except possibly for simple poles at* $s = 0$ *and at* $s = k$ *with the residue at* $s = 0$ *being* $-a_0$. *It is bounded at infinity in each vertical strip of finite width, and satisfies the functional equation*

$$L(s,f) = i^k N^{\frac{k}{2}-s} L(k-s, f\,|_k\, H_N).$$

In particular, if f *is a newform of level* N, *then* $f\,|_k\, H_N = \lambda_f \overline{f}$ *so that the functional equation becomes*

$$L(s,f) = \lambda_f i^k N^{\frac{k}{2}-s} L(k-s,\overline{f}) = \epsilon(s,f) L(k-s,\overline{f}).$$

Further, $L(s,f)$ is holomorphic if f is a cusp form.

Remark. A newform in $\mathcal{M}(N,k,\chi)$ with $a_1 = 1$ is characterized by the fact that $D(s,f)$ has Euler product at all primes and the functional equation having the form $L(s,f) = C_1 N^{\frac{k}{2}-s} L(k-s,\overline{f})$ with a nonzero constant C_1 (cf. Theorem 9 in [11]).

We saw in Theorem 4 that two normalized cuspidal newforms with the same weight and character, and the level of one dividing the other agree if they have the same eigenvalues with respect to almost all Hecke operators T_p. Now we show that this is also true even if the levels are not comparable.

Theorem 9. *(Strong multiplicity one theorem) Let $f(z) = \sum_{n=1}^{\infty} a_n e^{2\pi i n z}$ and $g(z) = \sum_{n=1}^{\infty} a'_n e^{2\pi i n z}$ be two normalized cuspidal newforms of weight k character χ and level N, N', respectively. Suppose $a_p = b_p$ for almost all p. Then $f = g$ (and hence $N = N'$).*

Proof. This can be proved using purely algebraic method, see [11]. Here we shall employ Theorem 8. Suppose $f \neq g$. Denote by K the product of primes p such that $a_p \neq b_p$. Then

$$\frac{L(s,f)}{L(s,g)} = \frac{\prod_{q|K,q|N'} (1 - a'_q q^{-s}) \prod_{p|K,p\nmid N'} (1 - a'_p p^{-s} + \chi(p) p^{k-1-2s})}{\prod_{q|K,q|N} (1 - a_q q^{-s}) \prod_{p|K,p\nmid N} (1 - a_p p^{-s} + \chi(p) p^{k-1-2s})}$$

is a meromorphic function on the s–plane. In view of the functional equations for $L(s,f)$ and $L(s,g)$, we get the functional equation

$$\frac{L(s,f)}{L(s,g)} = C\left(\frac{N}{N'}\right)^{\frac{k}{2}-s} \frac{L(k-s,\overline{f})}{L(k-s,\overline{g})},$$

which holds as quotients of finite products, not merely by analytic continuations. Since the periods in s for $p^{-s} = 1$, $p \mid K$, are all distinct, the above functional equation then implies local functional equations at each prime $p \mid K$. In other words, writing, for $p\nmid N$ or $p\nmid N'$,

$$1 - a_p p^{-s} + \chi(p) p^{k-1-2s} = (1 - \alpha_p p^{-s})(1 - \beta_p p^{-s}),$$

and

$$1 - a'_p p^{-s} + \chi(p) p^{k-1-2s} = (1 - \alpha'_p p^{-s})(1 - \beta'_p p^{-s}),$$

we may split the above functional equation into three types :

(I) $p \mid K$, $p\nmid NN'$:

$$\frac{(1 - \alpha'_p p^{-s})(1 - \beta'_p p^{-s})}{(1 - \alpha_p p^{-s})(1 - \beta_p p^{-s})} = \frac{(1 - \overline{\alpha'_p} p^{-k+s})(1 - \overline{\beta'_p} p^{-k+s})}{(1 - \overline{\alpha_p} p^{-k+s})(1 - \overline{\beta_p} p^{-k+s})} c_p;$$

(II) $p \mid K, p \nmid N, p \mid N'$:

$$\frac{1-a'_p p^{-s}}{(1-\alpha_p p^{-s})(1-\beta_p p^{-s})} = \frac{1-\overline{a'_p}p^{-k+s}}{(1-\overline{\alpha}_p p^{-k+s})(1-\overline{\beta}_p p^{-k+s})} \cdot c_p \cdot (p^{-s})^{e(p)}$$

(and similarly for the symmetric case $p \mid K, p \mid N, p \nmid N'$);

(III) $p \mid K, p \mid N$ and $p \mid N'$:

$$\frac{1-a'_p p^{-s}}{1-a_p p^{-s}} = \frac{1-\overline{a'_p}p^{-k+s}}{1-\overline{a}_p p^{-k+s}} \cdot c_p \cdot (p^{-s})^{e(p)}.$$

Here c_p is a nonzero constant and $e(p)$ is an integer.

Putting $x = p^{-s}$, we show first that $(1-\alpha_p x)(1-\beta_p x)$ and $(x-\overline{\alpha}_p p^{-k})(x-\overline{\beta}_p p^{-k})$ do not have a common zero. This is because $\overline{\alpha}_p p^{-k} = 1/\beta_p$ would imply $p^{2(k-1)} = \overline{\alpha}_p\beta_p\alpha_p\overline{\beta}_p = p^{2k}$, and $\overline{\alpha}_p p^{-k} = 1/\alpha_p$ would imply $|\alpha_p| = p^{k/2}$, which certainly violates Theorem 6, the former Ramanujan-Petersson conjecture. But we don't need such a strong theorem; by a much easier method, Rankin [18] proved that $a_n = O(n^{\frac{k-1}{2}+\frac{1}{5}})$ for n large, and with $|\alpha_p| = p^{k/2}$ for some p would imply that a_{p^n} does not satisfy the above estimate. This observation proves that in case (I), the only possibility is $a_p = \alpha_p + \beta_p = \alpha'_p + \beta'_p = a'_p$, hence Type I cannot exist. Also, Type II cannot exist because both sides have poles and the poles do not agree. In case III, consider first the case $a_p \neq 0$. Then by Theorem 5, $|a_p| = p^{\frac{k}{2}-1}$ or $p^{\frac{k-1}{2}}$, hence $1 - a_p x$ and $x - \overline{a}_p p^{-k}$ do not have a common zero. Thus Type III identity would imply $a_p = a'_p$. Similar argument also shows that $a'_p \neq 0$ would imply $a_p = a'_p$. Hence $a_p = a'_p = 0$, this is again not allowed. This shows that the assumption is wrong and hence $f = g$. □

Exercise 11. (1) Show that Theorem 9 holds if f and g are new Eisenstein series.

(2) Show that a new cusp form and a new Eisenstein series cannot have the same eigenvalues with respect to almost all Hecke operators.

Consider the following question: Given a sequence of complex numbers $a_0, a_1, a_2, \cdots, a_n, \cdots$ satisfying some growth condition $a_n = O(n^c)$, form the function $f(z) = \sum_{n=0}^{\infty} a_n e^{2\pi i n z}$, which is holomorphic on the upper half-plane. It is natural to ask : when is f a modular form of weight k level N and character χ ? This question was first studies by Hecke for the case of full modular group $\Gamma = SL_2(\mathbb{Z})$, that is, $N = 1$ and χ trivial. Hecke obtained a criterion in terms of the analytic behavior of the attached L-function $L(s, f) = (2\pi)^{-s}\Gamma(s)D(s, f)$. If f is a modular form for Γ, then $f = f\mid_k H_1$ so that, by Theorem 8, $L(s, f)$ has the described analytic behavior with the functional equation

$$L(s, f) = i^k L(k-s, f).$$

Conversely, if $L(s,f)$ has the above stated analytic behavior, then working in reverse direction, Hecke showed that f and $f\mid_k H_1$ agree on the half-line $\{it : t>0\}$. As both functions are holomorphic, this implies $f = f\mid_k H_1$ on the upper half-plane. Finally, by noticing that Γ is generated by the translation $\begin{pmatrix}1&1\\0&1\end{pmatrix}$ and inversion H_1, we conclude that f is a modular form for Γ of weight k.

The situation becomes a lot more complicated for the case $N>1$ because the information on $f\mid_k \begin{pmatrix}1&1\\0&1\end{pmatrix} = f$ and $f = f\mid_k H_N$ alone is not enough. This question was answered by Weil in 1967, his answer involved the analytic behavior of the L–function attached to f as well as its many twists f_η, which we now explain.

Given a function $f(z)=\sum_{n=0}^{\infty} a_n e^{2\pi inz}$ and a character η of $\mathbf{Z}$, the twist of f by η is defined as $f_\eta(z)=\sum_{n=0}^{\infty}\eta(n)a_n e^{2\pi inz}$. If f is a modular form, then so is f_η, but with different parameters. In the case that f is a normalized newform of weight k level N and character χ, and $m=\operatorname{cond}\eta$ is prime to N, then it is easy to show that f_η is a normalized newform of weight k level Nm^2 and character $\chi\eta$. In particular, $L(s,f_\eta)$ satisfies the functional equation

$$L(s,f_\eta)=C_\eta(Nm^2)^{\frac{k}{2}-s}L(k-s,\overline{f}_{\overline{\eta}}),$$

where C_η is related to C_1, the constant for $L(s,f)$, as follows :

$$C_\eta = C_1\frac{g(\eta)}{g(\overline{\eta})}\eta(-N)\chi(m)$$

with $g(\eta)$ being the Gauss sum $g(\eta,\psi)$. Here $\psi(x)=e^{2\pi ix/m}$ is the standard choice of the additive character of $\mathbf{Z}/m\mathbf{Z}$. Weil [26] showed that conversely, if we know the analytic behavior of sufficiently many $L(s,f_\eta)$, then we know that f is a modular form. More precisely,

Theorem 10. *(Weil, converse theorem for modular forms) Let* $f(z)=\sum_{n=0}^{\infty}a_n e^{2\pi inz}$ *be a form in* $\mathcal{M}(N,k,\chi)$. *Suppose that* $f\mid_k H_N = C_1(-i)^k\overline{f}$. *Then for all characters* η *with* $\operatorname{cond}\eta = m$ *prime to* N, *the following statement holds :*

$(A)_\eta$ $L(s,f_\eta)$ *has an analytic continuation to the whole s–plane, holomorphic everywhere except possibly for simple poles at* $s=0$ *and* $s=k$, *it is bounded at infinity in each vertical strip of finite width, and satisfies the functional equation*

$$L(s,f_\eta)=C_\eta(Nm^2)^{\frac{k}{2}-s}L(k-s,\overline{f}_{\overline{\eta}}),$$

where

$$C_\eta = C_1\frac{g(\eta)}{g(\overline{\eta})}\eta(-N)\chi(m),\quad \text{and}\quad g(\eta)=g(\eta,\psi)$$

is the Gauss sum as above.

Conversely, suppose $f(z) = \sum_{n=0}^{\infty} a_n e^{2\pi i n z}$, *where* a_n *are complex numbers with* $a_n = O(n^c)$ *for some constant* $c > 0$. *Let* χ *be a character* $\bmod N$. *Assume that* $(A)_\eta$ *holds for* $\eta = 1$ *and for almost all characters* η *with* $\operatorname{cond}\eta = m$ *prime to* N. *Then* f *is in* $\mathcal{M}(N, k, \chi)$ *with*

$$f\,|_k\, H_N = C_1(-i)^k \overline{f}.$$

Further, if $D(s, f) = \sum_{n=1}^{\infty} a_n n^{-s}$ *converges absolutely at* $s = k-\delta$ *for some* $k > \delta > 0$, *then* f *is a cusp form.*

Weil's theorem above has far-reaching consequences in number theory, for it indicates that if an object has an associated L–function which behaves like the L–function attached to a modular form, then this object is related to modular forms. We exhibit three situations.

The first is the idèle class characters of an imaginary quadratic extension K of $\mathbf{Q}$. The field K has only one archimedean place ∞, which is a complex place. A quasi-character of $\mathbf{C}^\times \cong \mathbf{R}^\times_{>0} \times S^1$ maps $z = re^{i\theta} \in \mathbf{C}^\times$ to $r^{s_0} e^{in\theta}$ for some $s_0 \in \mathbf{C}$ and $n \in \mathbf{Z}$. Let χ be an idèle class quasi-character of $I_K/K^\times$. Call χ *algebraic of type* k if χ_∞ sends $z = re^{i\theta}$ to $e^{in\theta}$, where $|n|+1 = k$. Note that an algebraic quasi-character is in fact a character. The L-function attached to an algebraic character χ of $I_K/K^\times$ of type k is

$$L(s, \chi) = \Gamma_{\mathbf{C}}\left(s + \frac{k-1}{2}\right) \prod_{\substack{v \text{ finite place of } K \\ \chi_v \text{ unram.}}} (1 - \chi_v(\pi_v) N v^{-s})^{-1},$$

where $\Gamma_{\mathbf{C}}(s) = (2\pi)^{-s}\Gamma(s)$ and π_v is a uniformizer in the completion K_v of K at the place v. For such a χ, define $f(\chi) = \sum_{n=1}^{\infty} a_n e^{2\pi i n z}$ such that the associated Dirichlet series is

$$D(s, f) = \sum_{n=1}^{\infty} a_n n^{-s} = \prod_{\substack{v \text{ finite} \\ \chi_v \text{ unr.}}} \left(1 - \chi_v(\pi_v) N v^{-s+\frac{k-1}{2}}\right)^{-1}$$

and $L(s, f) = \Gamma_{\mathbf{C}}(s) D(s, f) = L(s + \frac{1-k}{2}, \chi)$. Hecke has shown that $L(s, f)$ has an analytic continuation to the whole s–plane, holomorphic everywhere except possibly for simple poles at $s = 0$ and $s = 1$, is bounded at infinity in each vertical strip of finite width, and satisfies the functional equation $L(s, \chi) = \epsilon(s, \chi) L(1-s, \chi^{-1})$, where $\epsilon(s, \chi) = \text{const.}\ N_{K/\mathbf{Q}}(\mathcal{D}_{K/\mathbf{Q}} \cdot \operatorname{cond}\chi)^{\frac{1}{2}-s}$ with $\mathcal{D}_{K/\mathbf{Q}}$ the different of $K/\mathbf{Q}$, $\operatorname{cond}\chi$ the conductor of χ written multiplicatively (instead of additively as in the previous chapters) and constant arising from Gauss sums. The proof is similar to what we did for function field case in Chapter 5. Further, $L(s, \chi)$ is holomorphic if χ is not principal, that is, not of the form $|\ |^t$ for any $t \in \mathbf{C}$. The analytic behavior of $L(s, \chi)$ implies that $L(s, f)$ satisfies the condition $(A)_1$. Next we consider twists of

f. The Dirichlet characters of $\mathbf{Z}$ are identified with algebraic idèle class characters ξ of $I_{\mathbf{Q}}/\mathbf{Q}^\times$, that is, $\xi_\infty(r) = (\text{sign } r)^\delta$, $\delta = 0$ or 1. If χ is an algebraic character of $I_K/K^\times$ of type k, so is $\chi \cdot \eta \circ N_{K/\mathbf{Q}}$ for any algebraic idèle class character η of $I_{\mathbf{Q}}/\mathbf{Q}^\times$, and $L(s, f_\eta) = L(s + \frac{1-k}{2}, \chi \cdot \eta \circ N_{K/\mathbf{Q}})$. Thus $(A)_\eta$ is also satisfied. We conclude from Theorem 10 that $f(\chi)$ is a modular form of weight k and level $N = N_{K/\mathbf{Q}}(\mathcal{D}_{K/\mathbf{Q}} \operatorname{cond} \chi)$. Further, if χ is also regular, that is, $\chi \neq \xi \circ N_{K/\mathbf{Q}}$ for all idèle class characters ξ of $I_{\mathbf{Q}}/\mathbf{Q}^\times$, then $L(s, \chi)$ and $L(s, \chi \cdot \eta \circ N_{K/\mathbf{Q}})$ are holomorphic for all idèle class characters η of $I_{\mathbf{Q}}/\mathbf{Q}^\times$, and hence $f(\chi)$ is a cusp form. This is because we know the L-series attached to an Eisenstein newform is the product of two Dirichlet L-functions, hence it has poles after twisting by a suitable Dirichlet character.

To describe the character of $f(\chi)$, we first imbed $I_{\mathbf{Q}}$ in I_K. Let v be a place of $\mathbf{Q}$ and w be a place of K dividing v. Then K_w contains $\mathbf{Q}_v$ naturally. If there are two places w and w' dividing v, we imbed $\mathbf{Q}_v$ in $K_w \times K_{w'}$ diagonally. In this way we imbed $I_{\mathbf{Q}}$ in I_K, and χ restricted to $I_{\mathbf{Q}}$ is an algebraic idèle class character χ' of $I_{\mathbf{Q}}/\mathbf{Q}^\times$. Next we observe the Euler factor of $L(s, f)$ at a place $p \nmid N$. If p is inert in K, let w be the only place of K dividing p. Then we may choose $\pi_w = \pi_p$ and we know $Nw = p^2$. The factor is

$$\left(1 - \chi_w(\pi_w) N w^{-s+\frac{k-1}{2}}\right)^{-1} = \left(1 - \chi_w(\pi_p) p^{k-1-2s}\right)^{-1}.$$

If p is split in K, let w, w' be the places of K dividing p. Then $K_w \cong \mathbf{Q}_p \cong K_{w'}$ so that $Nw = Nw' = p$ and we may choose $\pi_w = \pi_p = \pi_{w'}$. The factor is

$$\begin{aligned}&\left(1 - \chi_w(\pi_w) N w^{-s+\frac{k-1}{2}}\right)^{-1} \left(1 - \chi_{w'}(\pi_{w'}) N w'^{-s+\frac{k-1}{2}}\right)^{-1} \\ &= \left(1 - \left(\chi_w(\pi_p) + \chi_{w'}(\pi_p)\right) p^{\frac{k-1}{2}-s} + \chi_w(\pi_p)\chi_{w'}(\pi_p) p^{k-1-2s}\right)^{-1}.\end{aligned}$$

Let $\eta_{K/\mathbf{Q}}$ be the quadratic character of $I_{\mathbf{Q}}/\mathbf{Q}^\times$ which takes value 1 on $\mathbf{Q}^\times N_{K/\mathbf{Q}}(I_K)$ and -1 otherwise. Then the coefficient of p^{k-1-2s} in both cases can be summarized as $(\chi' \eta_{K/\mathbf{Q}})(\pi_p)$. This shows that the character of $f(\chi)$, described adèlically, is $\chi' \eta_{K/\mathbf{Q}}$. We have shown

Theorem 11. *Let χ be an algebraic idèle class character of $I_K/K^\times$ of type k. Then there is a modular form $f(\chi)$ of weight k level $N = N_{K/\mathbf{Q}}(\mathcal{D}_{K/\mathbf{Q}} \operatorname{cond} \chi)$ and character $\chi' \eta_{K/\mathbf{Q}}$, χ' being the restriction of χ to $I_{\mathbf{Q}}$, such that $L(s, f) = L(s + \frac{1-k}{2}, \chi)$. Further, $f(\chi)$ is a cusp form if χ is regular. Moreover, $f(\chi)$ is a newform of level N.*

The last statement is true because $L(s, f)$ is Eulerian at all finite places of $\mathbf{Q}$ and $L(s, f)$ satisfies the functional equation $L(s, f) = \text{const.} \cdot N^{\frac{k}{2}-s} L(k - s, \overline{f})$, as remarked before.

The second application is to irreducible representations of $G = Gal(\overline{\mathbf{Q}}/\mathbf{Q})$. We said in Chapter 6, §1 that to each finite-dimensional (complex) irreducible representation ρ of G, there is attached Artin L-function $L(s, \rho)$. When ρ is 1-dimensional,

$L(s,\rho) = L(s,\chi)$ for an idèle class character χ of $I_{\mathbf{Q}}/\mathbf{Q}^\times$ of finite order, and hence the analytic behavior for each $L(s,\rho)$ is known, by the aforementioned result of Hecke on $L(s,\chi)$. Artin has conjectured that for ρ irreducible of degree ≥ 2, $L(s,\rho)$, which is defined and holomorphic for s on a right half-plane, has a holomorphic continuation to the whole s-plane, is bounded in each vertical strip of finite width, and satisfies the functional equation $L(s,\rho) = \epsilon(s,\rho)L(1-s,\check{\rho})$, where $\check{\rho}$ is the contragredient of ρ. Note that for any character (i.e., degree 1 representation) χ of G, $\rho\otimes\chi$ is again irreducible of the same degree as ρ, hence, according to the conjecture, $L(s,\rho\otimes\chi)$ has the same analytic behavior as $L(s,\rho)$, but with functional equation $L(s,\rho\otimes\chi) = \epsilon(s,\rho\otimes\chi)L(1-s,\check{\rho}\otimes\chi^{-1})$. As any finite-dimensional representation ρ of G factors through a finite quotient, by a theorem of Brauer, $L(s,\rho)$ decomposes as a finite product of $L(s,\chi_i)^{a_i}$, where χ_i is a character of G and $a_i \in \mathbf{Z}$. Therefore Artin's conjecture is established except that the analytic continuation of $L(s,\rho)$ is shown to be meromorphic instead of holomorphic.

Now we concentrate on degree 2 irreducible representations ρ of G. By definition, $L(s,\rho)$ is the product of local L-factors $L_v(s,\rho)$ over all places v of $\mathbf{Q}$. The local L-factor at the archimedean place ∞ depends upon the determinant of the complex conjugation being 1 or -1, and the representation ρ is called even or odd, accordingly. $L_\infty(s,\rho)$ for ρ odd is $\Gamma_{\mathbf{C}}(s) = (2\pi)^{-s}\Gamma(s)$ as before, and for ρ even is $\Gamma_{\mathbf{R}}(s)^2$ or $\Gamma_{\mathbf{R}}(s+1)^2$, where $\Gamma_{\mathbf{R}}(s) = \pi^{-\frac{s}{2}}\Gamma(\frac{s}{2})$, depending on the action of complex conjugation being identity or $-$ identity (cf. [25]). The L-factor at a finite place v is 1 or $(1-a_vNv^{-s})^{-1}$ or $\left(1 - Tr\rho(\mathrm{Frob}_v)Nv^{-s} + (\det\rho)(\mathrm{Frob}_v)Nv^{-2s}\right)^{-1}$ and it is of the last type for almost all places. Note that the determinant $\det\rho$ is a 1-dimensional representation of G, hence is also identified with an algebraic idèle class character of $I_{\mathbf{Q}}/\mathbf{Q}^\times$, or a Dirichlet character of $\mathbf{Z}$. Hence the Artin L-function $L(s,\rho)$ attached to an odd degree 2 irreducible representation ρ of G has the same formality as the L-function attached to a modular form. If the Artin conjecture is true, then there is a cusp form $f(\rho)$ of weight 1 and character $\det\rho$ such that $L\big(s,f(\rho)\big) = L(s,\rho)$. In fact, $f(\rho)$ is a newform of level equal to the Artin conductor of ρ, as seen from the functional equation. Conversely, Deligne and Serre [5] has shown that to each newform f of weight 1 level N character χ, there is an odd degree 2 irreducible complex representation ρ of G such that $L(s,\rho) = L(s,f)$, thus Artin conjecture holds for such ρ. Hence the Artin conjecture for odd degree 2 irreducible representations of G is equivalent to saying that such representations arise from newforms of weight 1. How about even irreducible representations of degree 2 ? Well, they should arise from Maass new forms of weight 1. Maass forms are like modular forms except that they are real analytic instead of holomorphic. Same theory of newforms works through on Maass forms. The reader is referred to [28] for more details on Maass forms.

The third application concerns elliptic curves. Let E be an elliptic curve defined over $\mathbf{Q}$, it has a minimal model, called Néron model, over $\mathbf{Z}$. At each finite place p, $E \bmod p$ is a curve defined over the finite field $\mathbf{Z}/p\mathbf{Z}$, which is again an elliptic curve for almost all p. When $E \bmod p$ is an elliptic curve, we defined its zeta function in

Chapter 2 and studied it in Chapter 5. It has the form

$$\frac{1 - a_p p^{-s} + p^{1-2s}}{(1-p^{-s})(1-p^{1-s})},$$

where $a_p = 1 + p - N_p$ with N_p being the number of $\mathbf{Z}/p\mathbf{Z}$-rational points. In this case define $L_p(s, E)$ to be $(1 - a_p p^{-s} + p^{1-2s})^{-1}$. At a place p where $E \bmod p$ is not an elliptic curve, several possibilities may occur : it may have a node with two rational tangents, or a node with two irrational tangents (in which cases we say that E has a multiplicative reduction at p), or a cusp (in which case we say that E has an additive reduction at p). The local L-factor at p is defined as $(1 - p^{-s})^{-1}, (1 + p^{-s})^{-1}$ and 1, accordingly. There is a positive integer N, called the *conductor* of E, which records the geometric situation of $E \bmod p$, in particular, $p \mid N$ if and only if $E \bmod p$ is not an elliptic curve. The global L-function attached to E is defined as

$$\begin{aligned} L(s,E) &= \Gamma_{\mathbf{C}}(s) \prod_{p \text{ prime}} L_p(s,E) \\ &= (2\pi)^{-s}\Gamma(s)\prod_{q|N}(1 - a_q q^{-s})^{-1} \prod_{p\nmid N} (1 - a_p p^{-s} + p^{1-2s})^{-1}. \end{aligned}$$

The product $\prod_p L_p(s,E) = \zeta(s,E) = \sum_{n=1}^{\infty} a_n n^{-s}$ is also called the Hasse-Weil zeta function attached to E. Its analytic behavior was conjectured by Hasse :

Conjecture. *(Hasse) Let E be an elliptic curve defined over $\mathbf{Q}$. Then $L(s,E)$ as well as its twists $L(s,E,\eta)$ by Dirichlet characters η have holomorphic continuations to the whole s-plane, are bounded in each vertical strip of finite width, and satisfy functional equations*

$$\begin{aligned} L(s,E) &= \epsilon(s,E)L(2-s,E) \\ L(s,E,\eta) &= \epsilon(s,E,\eta)L(2-s,E,\bar{\eta}). \end{aligned}$$

Here $\epsilon(s,E) = C_1 N^{1-s}$ with N being the conductor of E and C_1 a nonzero constant, and for conductor η prime to N, $\epsilon(s,E,\eta)$ is related to $\epsilon(s,E)$ as described in $(A)_\eta$.

In view of Theorem 10, Hasse conjecture amounts to saying that given an elliptic curve E defined over $\mathbf{Q}$, there is a newform $f(E)$ of weight 2 level N and trivial character, i.e., $f(E) \in \mathcal{C}(\Gamma_0(N), 2)$, such that $L(s, f(E)) = L(s,E)$.

Next let g be a nonzero cusp form of weight 2 for $\Gamma_0(N)$ so that $g(z)dz$ is a holomorphic differential on modular curve $X_0(N) = \widehat{\mathcal{H}/\Gamma_0(N)}$, as discussed in §1. This differential defines a homomorphism from $H_1(X_0(N), \mathbf{Z})$ to $\mathbf{C}$ by sending a 1-cycle γ to the integral $\int_\gamma g(z)dz$. Its image is a lattice in $\mathbf{C}$ of rank 2, called the lattice of periods of $g(z)dz$, denoted by L_g. The quotient $\mathbf{C}/L_g$ is an elliptic curve

E_g, and there is a well-defined morphism φ from $X_0(N)$ to E_g induced by the map from $\mathcal{H}$ to $\mathbf{C}$ which sends $\tau \in \mathcal{H}$ to $\int_{i\infty}^{\tau} g(z)dz$. Clearly, $\varphi(i\infty) = 0 \in E_g$.

Let E be an elliptic curve defined over $\mathbf{Q}$ such that its Néron model over $\mathbf{Z}$ is given by the equation $y^2 = f(x) = 4x^3 + g_2x + g_3$. Let ω be the differential $\frac{dx}{y} = \frac{2dy}{f'(x)}$, which generates the space of holomorphic differentials on E. If E is represented as $\mathbf{C}/L$, then ω is dz. Suppose that there is a morphism $\varphi : X_0(N) \to E$ mapping $i\infty$ to 0. It follows from the "congruence relation for modular correspondences" by Eichler-Shimura and Igusa that the pull back differential $\varphi^*\omega$ is in fact equal to $c \cdot h(z)dz$, where $c \in \mathbf{Q}^\times$ and $h(z)$ is a normalized newform of level N in $\mathcal{C}(\Gamma_0(N), 2)$ whose eigenvalue with repect to T_p is $1 + p - N_p$ for all primes $p \nmid N$. In other words, the L–function attached to h is equal to $L(s, E)$ (except possibly for finitely many factors).

Start with an elliptic curve E defined over $\mathbf{Q}$. Granting Hasse conjecture, we obtain a cuspidal newform $f = f(E)$ of weight 2 for $\Gamma_0(N)$ such that $L(s, f) = L(s, E)$. With f, we construct an elliptic curve E_f defined over $\mathbf{Q}$ such that there is a morphism $\varphi : X_0(N) \to E_f$ mapping $i\infty$ to 0. Choosing the canonical differential ω on E_f, we get a cuspidal newform h of weight 2 for $\Gamma_0(N)$ which has the same eigenvalues as f with respect to all T_p, $p \nmid N$. Therefore $h = f$ and $L(s, E) = L(s, E_f)$. It then follows from the Shafarevich conjecture proved by Faltings [7] that E and E_f are isogeneous. (See also the explanation in [23], Chapter 7.) Therefore there is a morphism φ from $X_0(N)$ to E mapping $i\infty$ to 0 such that the pull-back $\varphi^*(\omega_E)$ of Néron differential ω_E on E is a nonzero rational multiple of a newform in $\mathcal{C}(\Gamma_0(N), 2)$. An elliptic curve with this property is called a *Weil curve*. This shows that Hasse conjecture implies

Taniyama-Shimura conjecture. *Every elliptic curve defined over* $\mathbf{Q}$ *is a Weil curve.*

Conversely, if Taniyama-Shimura conjecture holds for E, then $L(s, E) = L(s, f)$ for a cuspidal newform f and hence satisfies the Hasse conjecture. Therefore the Taniyama-Shimura conjecture and the Hasse conjecture are equivalent. See [12, 21, 24] for more details.

K. Ribet [19] has shown that Taniyama-Shimura conjecture implies Fermat's last theorem, namely, for any integer $n \geq 4$, the equation $x^n + y^n = z^n$ has only trivial integral solutions. Recently, A. Wiles [33] and Taylor and Wiles [32] proved the Taniyama-Shimura conjecture for semi-stable elliptic curves over $\mathbf{Q}$, which is enough to imply the truth of Fermat's last theorem. For expository papers on this subject, see the articles [29], [30] and [31].

References

[1] A.O.L. Atkin and J. Lehner : Hecke operators on $\Gamma_0(m)$. Math. Ann. 185 (1970), 134-160.

[2] W. Casselman : On some results of Atkin and Lehner, Math. Ann. 201 (1973), 301-314.

[3] P. Deligne : Formes modulaires et représentations l–adiques. In : Séminaire Bourbaki vol. 1968/69, Lecture Notes in Math. 179, Springer–Verlag, Berlin, Heidelberg, New York (1971).

[4] P. Deligne : Formes modulaires et représentations de $GL(2)$. In : Modular Functions of One Variable II, Lecture Notes in Math. 349, Springer–Verlag, Berlin, Heidelberg, New York (1973), 55-105.

[5] P. Deligne and J.-P. Serre : Formes modulaires de poids 1, Ann. Sci. E.N.S. 7 (1974), 507-530.

[6] P. Deligne and W. Kuyk : Modular Functions of One Variable I, II, Proceedings of the International Summer School on "Modular functions of one variable and arithmetical applications", Univ. of Antwerp, RUCA, July 17 - Aug. 3, 1972, Lecture Notes in Math. 320, 349, Springer–Verlag, Berlin, Heidelberg, New York (1973).

[7] G. Faltings : Endlichkeitssätze für abelsche Varietäten über Zahlkörpern, Invent. Math. 73 (1983), 349-366; Erratum, ibid. 75 (1984), 381.

[8] E. Hecke : Mathematische Werke. Göttingen : Vandenhoeck und Ruprecht (1959).

[9] D. Husemöller : Elliptic Curves, GTM111, Springer–Verlag, Berlin, Heidelberg, New York (1987).

[10] S. Lang : Introduction to Modular Forms, Springer–Verlag, Berlin, Heidelberg, New York (1976).

[11] W.-C. W. Li : Newforms and functional equations, Math. Ann. 212 (1975), 285-315.

[12] B. Mazur : Courbes elliptiques et symboles modulaires. In : Séminaire Bourbaki vol. 1971/72, Lecture Notes in Math. 317, Springer–Verlag, Berlin, Heidelberg, New York (1973).

[13] T. Miyake : Modular Forms, Springer–Verlag, Berlin, Heidelberg, New York (1976).

[14] T. Miyake : On automorphic forms on GL_2 and Hecke operators, Annals of Math. 94 (1971), 174-189.

[15] A.P. Ogg : Modular Forms and Dirichlet Series, Benjamin, New York, Amsterdam (1969).

[16] A.P. Ogg : On the eigenvalues of Hecke operators. Math. Ann. 179 (1969), 101-108.

[17] R. Rankin : Modular Forms and Functions, Cambridge Univ. Press, Cambridge (1977).

[18] R. Rankin : Contributions to the theory of Ramanujan's function $\tau(n)$ and similar arithmetical functions. II. Proc. Camb. Phil. Soc. 35 (1939), 357-372.

[19] K. Ribet : On modular representations of $Gal(\overline{\mathbf{Q}}/\mathbf{Q})$ arising from modular forms, Invent. Math. 100 (1990), 431-476.

[20] J.-P. Serre : A Course in Arithmetic, GTM7, Springer–Verlag, Berlin, Heidelberg, New York (1973).

[21] J.-P. Serre : Abelian l–Adic Representations and Elliptic Curves, Benjamin, New York, Amsterdam (1968).

[22] J.-P. Serre : Formes modulaires et fonctions zêta p–adiques. In : Modular Functions of One Variable III, Lecture Notes in Math. 350, Springer–Verlag, Berlin, Heidelberg, New York (1973), 191-268.

[23] G. Shimura : Introduction to the Arithmetic Theory of Automorphic Functions, Iwanami Shoten and Princeton Univ. Press (1971).

[24] H.P.F. Swinnerton–Dyer and B. J. Birch : Elliptic curves and modular functions. In : Modular Functions of One Variable IV, B. J. Birch and W. Kuyk edited, Lecture Notes in Math. 476, Springer–Verlag, Berlin, Heidelberg, New York (1975), 2-32.

[25] J. Tate : Fourier analysis in number fields and Hecke's zeta–functions. In : Algebraic Number Theory, J.W.S. Cassels and A. Fröhlich edited, Thompson, Washington D.C. (1967), republished by Academic Press, London, 305-347.

[26] A. Weil : Über die Bestimmung Dirichletsher Reihen durch Funktionalgleichungen. Math. Ann. 168 (1967), 149-156.

[27] J. Weisinger : Some results on classical Eisenstein series and modular forms over function fields, Thesis, Harvard University, Cambridge, Mass. (1977).

[28] H. Maass : Lectures on Modular Functions of One Complex Variable. Tata Institute of Fundamental Research, Bombay (1964).

[29] K. Ribet : Galois representations and modular forms, Bulletin of Amer. Math. Soc., to appear.

[30] K. Ribet and B. Hayes Fermat's last theorem and modern arithmetic, American Scientist, Scientific Research Society, March-April (1994), 144-156.

[31] K. Rubin and A. Silverberg : A report on Wiles' Cambridge lectures, Bulletin of Amer. Math. Soc., vol. 31 (1994), 15-38.

[32] R. Taylor and A. Wiles : Ring theoretic properties of certain Hecke algebras, Annals of Math., to appear.

[33] A. Wiles : Modular elliptic curves and Fermat's last theorem, Annals of Math., to appear.

CHAPTER 8

Automorphic Forms and Automorphic Representations

§1 Automorphic forms

In the previous chapter, we studied classical modular forms defined on the Poincaré upper half-plane $\mathcal{H}$. In this section, we shall reformulate them as automorphic forms on $GL_2(A_{\mathbf{Q}})$ and generalize the definition to automorphic forms over global fields.

Denote by $GL_2^0(\mathbf{R})$ the connected component of $GL_2(\mathbf{R})$, which consists 2×2 real matrices with positive determinant, and by $\mathcal{Z}$ the center of GL_2, consisting of the scalar matrices. Thus $\mathcal{Z}(\mathbf{R}) = \left\{ \begin{pmatrix} a & 0 \\ 0 & a \end{pmatrix} \in GL_2(\mathbf{R}) \right\} \cong \mathbf{R}^\times$. The group $GL_2^0(\mathbf{R})$ acts on $\mathcal{H}$ by fractional linear transformations. This action is transitive, with the stabilizer of i being $\mathcal{Z}(\mathbf{R})SO_2(\mathbf{R})$. Every element $g_\infty \in GL_2^0(\mathbf{R})$ can be written in a unique way as

$$g_\infty = z \begin{pmatrix} y & x \\ 0 & 1 \end{pmatrix} r(\theta), \quad \text{where } z > 0,\ x, y \in \mathbf{R},\ y > 0,$$

and

$$r(\theta) = \begin{pmatrix} \cos\theta & \sin\theta \\ -\sin\theta & \cos\theta \end{pmatrix} \in SO_2(\mathbf{R}).$$

Hence sending g_∞ to $g_\infty(i) = x + iy$ gives rise to the identification

$$GL_2^0(\mathbf{R})/\mathcal{Z}(\mathbf{R})SO_2(\mathbf{R}) \approx \mathcal{H}.$$

Let φ be a function on $GL_2^0(\mathbf{R})$ which sends $g_\infty = z \begin{pmatrix} y & x \\ 0 & 1 \end{pmatrix} r(\theta)$ to

$$\varphi(g_\infty) = \varphi\left(\begin{pmatrix} y & x \\ 0 & 1 \end{pmatrix}\right) e^{ik\theta}. \tag{1.1}$$

Define a function f on $\mathcal{H}$ by

$$f(x+iy) = y^{-k/2}\varphi\left(\begin{pmatrix} y & x \\ 0 & 1 \end{pmatrix}\right).$$

Exercise 1. Show that φ is left invariant by $SL_2(\mathbf{Z})$ if and only if $f\,|_k\,\gamma(\tau) := (c\tau+d)^{-k}f(\gamma\tau) = f(\tau)$ for all $\gamma = \begin{pmatrix} a & b \\ c & d \end{pmatrix} \in SL_2(\mathbf{Z})$, $\tau \in \mathcal{H}$.

Therefore a modular form of weight k for the full modular group translates as a function φ on $GL_2^0(\mathbf{R})$ satisfying

$$\varphi(\gamma g_\infty r(\theta) z) = \varphi(g_\infty)e^{ik\theta} \quad \text{for all } \gamma \in SL_2(\mathbf{Z}),\ z > 0,\ r(\theta) \in SO_2(\mathbf{R}),\ g_\infty \in GL_2^0(\mathbf{R}),$$

and with good analytic behavior.

At each finite place p of $\mathbf{Q}$, write $\mathbf{Z}_p$ for the ring of p-adic integers in $\mathbf{Q}_p$. Then $GL_2(\mathbf{Z}_p)$ is an open compact subgroup of $GL_2(\mathbf{Q}_p)$. The adèlic group $GL_2(A_\mathbf{Q})$ is the restricted product of $\{GL_2(\mathbf{Q}_v)\}$ with respect to $\{GL_2(\mathbf{Z}_p)\}$, in which $GL_2(\mathbf{Q})$ is embedded diagonally. Put

$$\Omega = GL_2^0(\mathbf{R})\prod_p GL_2(\mathbf{Z}_p).$$

It is an open subgroup of $GL_2(A_\mathbf{Q})$. It follows from the strong approximation theorem for SL_2 that

$$GL_2(A_\mathbf{Q}) = GL_2(\mathbf{Q})\Omega.$$

Observe that $GL_2(\mathbf{Q}) \cap \Omega = SL_2(\mathbf{Z})$. Thus a function φ on $GL_2^0(\mathbf{R})$ satisfying (1.1) extends uniquely to a function F on $GL_2(A_\mathbf{Q})$ right invariant by $\prod_p GL_2(\mathbf{Z}_p)$ and left invariant by $GL_2(\mathbf{Q})$, i.e., it satisfies

$$F(\gamma g z r(\theta)\beta) = F(g)e^{ik\theta} \quad \text{for all } \gamma \in GL_2(\mathbf{Q}),\ g \in GL_2(A_\mathbf{Q}),\ z > 0,\ r(\theta) \in SO_2(\mathbf{R}), \text{ and } \beta \in \prod_p GL_2(\mathbf{Z}_p).$$

Since the center of $GL_2(A_\mathbf{Q})$ is

$$Z(A_\mathbf{Q}) = \left\{\begin{pmatrix} a & 0 \\ 0 & a \end{pmatrix} \in GL_2(A_\mathbf{Q})\right\} \cong I_\mathbf{Q} = \mathbf{Q}^\times \cdot \mathbf{R}_{>0}\prod_p \mathcal{U}_p,$$

the formula above is valid for all $z \in I_\mathbf{Q}$.

Write $GL_2(A_\mathbf{Q}^f)$ for the subgroup of $GL_2(A_\mathbf{Q})$ consisting of elements whose archimedean components are trivial. Its standard maximal compact subgroup is

$K^f = \prod_p GL_2(\mathbf{Z}_p)$. For each positive integer N, K^f contains the congruence subgroup

$$K_0^f(N) = \left\{ \begin{pmatrix} a & b \\ c & d \end{pmatrix} \in K^f : \mathrm{ord}_p\, c \geq \mathrm{ord}_p\, N \text{ for all } p \right\},$$

which has finite index in K^f. Observe that $GL_2(\mathbf{Q}) \cap GL_2^0(\mathbf{R})K_0^f(N) = \Gamma_0(N)$. Thus, by the same procedure, a modular form of weight k level N and character χ can be lifted to a function F on $GL_2(A_\mathbf{Q})$ satisfying

$$F(\gamma g z r(\theta)\beta) = \chi(z)\chi(d)F(g)e^{ik\theta} \quad \text{for all } \gamma \in GL_2(\mathbf{Q}),\ g \in GL_2(A_\mathbf{Q}),\ z \in \mathcal{Z}(A_\mathbf{Q}),$$
$$r(\theta) \in SO_2(\mathbf{R}), \text{ and } \beta = \begin{pmatrix} a & b \\ c & d \end{pmatrix} \in K_0^f(N),$$

and with good analytic behavior. Here we have identified the Dirichlet character χ with its lifting to an algebraic idèle class character of $I_\mathbf{Q}/\mathbf{Q}^\times$, as explained briefly in §4 of Chapter 6.

Let ω be a quasi–character of the idèle class group $I_\mathbf{Q}/\mathbf{Q}^\times$. A complex–valued function F on $GL_2(A_\mathbf{Q})$ is an automorphic form of character ω if

(1.2) $F(\gamma g z) = \omega(z)F(g)$ for all $\gamma \in GL_2(\mathbf{Q})$, $g \in GL_2(A_\mathbf{Q})$ and $z \in \mathcal{Z}(A_\mathbf{Q}) = I_\mathbf{Q}$.

(1.3) F is right $K(= O_2(\mathbf{R})K^f)$–finite, that is, the space generated by the right translates of F by elements in K is finite–dimensional.

(Note that this condition implies that F is right invariant by $GL_2(\mathbf{Z}_p)$ for almost all p, there is a positive integer N such that the action of right translations by

$$K^f(N) = \left\{ \begin{pmatrix} a & b \\ c & d \end{pmatrix} \in K^f : \mathrm{ord}_p\, b \geq \mathrm{ord}_p\, N,\ \mathrm{ord}_p\, c \geq \mathrm{ord}_p\, N, \right.$$
$$\left. \mathrm{ord}_p\, a - 1 \geq \mathrm{ord}_p\, N,\ \mathrm{ord}_p\, d - 1 \geq \mathrm{ord}_p\, N \ \text{ for all } p \right\}$$

is trivial, and the right translations by $O_2(\mathbf{R})$ on F generate a finite–dimensional representation of $O_2(\mathbf{R})$.)

(1.4) F satisfies some analytic conditions.

Now we elaborate somewhat the analytic conditions. Classical modular forms are holomorphic on the upper half–plane. We interprete this condition on functions defined on $GL_2(A_\mathbf{Q})$. This is a condition at the archimedean place ∞. The Lie algebra $\mathcal{G}$ of $GL_2(\mathbf{R})$ is generated by

$$X = \begin{pmatrix} 0 & 1 \\ 0 & 0 \end{pmatrix}, \quad Y = \begin{pmatrix} 0 & 0 \\ 1 & 0 \end{pmatrix}, \quad U = \begin{pmatrix} 1 & 0 \\ 0 & -1 \end{pmatrix}, \quad Z = \begin{pmatrix} 1 & 0 \\ 0 & 1 \end{pmatrix}.$$

Of these, X, Y, U generate the Lie algebra of $SL_2(\mathbf{R})$; X and U generate the Borel subgroup $B_1 = \left\{ \begin{pmatrix} x & y \\ 0 & z \end{pmatrix} \in SL_2(\mathbf{R}) \right\}$ of $SL_2(\mathbf{R})$; $X - Y$ generates the Lie algebra of $SO_2(\mathbf{R})$; and Z generates the Lie algebra of the center $\mathcal{Z}(\mathbf{R})$. Regard X, Y, U, Z as left invariant differential operators on functions on $GL_2(\mathbf{R})$ by sending a function f to, for instance,

$$(Xf)(g) = \frac{d}{dt} f(g \cdot \exp tX) \,|_{t=0} = \frac{d}{dt} f(g \cdot (1_2 + tX)) \,|_{t=0}, \quad \text{for } g \in GL_2(\mathbf{R}).$$

Here 1_2 denotes the 2×2 identity matrix. The differential operators which are invariant under both left and right translations are those given by the center of the universal enveloping algebra of $\mathcal{G}$. On $SL_2(\mathbf{R})$, this center is generated by the so-called Casimir operator

$$D = \frac{1}{2}U^2 + XY + YX;$$

and on $GL_2(\mathbf{R})$, it is generated by D and Z.

Let φ be a function on $GL_2(\mathbf{R})$ satisfying $\varphi(gzr(\theta)) = \varphi(g)e^{ik\theta}$ for all $g \in GL_2(\mathbf{R})$, $z > 0$ and $r(\theta) \in SO_2(\mathbf{R})$, as we have seen at the beginning of this section. Then we have $(X - Y)\varphi = ik\varphi$. As $U = XY - YX$, this gives

$$(XY + YX)\varphi = (2XY - U)\varphi = 2X(X - ik)\varphi - U(\varphi).$$

Put $f(x + iy) = \varphi\left(y^{-\frac{1}{2}}\begin{pmatrix} y & x \\ 0 & 1 \end{pmatrix}\right)$. Then X and U are in the Lie algebra of B_1, and one finds that $X\varphi = \frac{\partial f}{\partial x} \cdot y$, $U\varphi = \frac{\partial f}{\partial y} \cdot 2y$. Hence $D\varphi$ corresponds to

$$2y^2\left(\frac{\partial^2}{\partial x^2} + \frac{\partial^2}{\partial y^2} - \frac{ik}{y}\frac{\partial}{\partial x}\right)(f) = \Delta f.$$

The operator Δ is an elliptic operator, which may be regarded as a generalized Laplacian–Beltrami operator; it is the usual Laplacian operator if $k = 0$. Viewing a holomorphic function as an eigenfunction of the Laplacian with eigenvalue zero, we are interested in F being an eigenfunction or a finite linear combination of the eigenfunctions of D. As a Maass form is an eigenfunction of the Laplacian with a nonzero eigenvalue, hence our definition of F includes both holomorphic and real analytic forms on the upper half–plane.

Another analytic condition on classical modular forms is the holomorphicity at cusps, which can be expressed as a growth condition. For an adelically defined form F, we require the existence of a nonnegative constant δ such that

$$\left| F\left(\begin{pmatrix} y & 0 \\ 0 & 1 \end{pmatrix}_\infty g\right)\right| = O(y^\delta)$$

for $y \in \mathbf{R}$, $y \to \infty$, uniformly over compact sets with respect to $g \in GL_2(\mathbf{R})K^f$. Note that $g = \begin{pmatrix} 1 & 0 \\ 0 & 1 \end{pmatrix}$ gives the behavior at the cusp $i\infty$, and as g varies in K^f, we get the behavior at other cusps.

Likewise, if F is a global field and ω is a quasi–character of the idèle class group $I_F/F^\times$, we define an *automorphic form* f of $GL_2(A_F)$ of character ω to be a complex–valued function on $GL_2(A_F)$ satisfying

(1.2)′ $f(\gamma g z) = \omega(z) f(g)$ for all $\gamma \in GL_2(F)$, $g \in GL_2(A_F)$ and $z \in I_F = \mathcal{Z}(A_F)$.

(1.3)′ f is right K–finite, where K is the product over v of the standard maximal compact subgroup of $GL_2(F_v)$.

(1.4)′ f satisfies some analytic conditions.

As before, the analytic conditions include growth condition and, if F is a number field, at each archimedean place v, we require the functions on $GL_2(F_v)$ obtained from $f(g)$ by fixing the components of $g \in GL_2(A_F)$ outside v to lie in a finite–dimensional vector space generated by eigenfunctions of the Casimir operator(s). For v real, this is the operator D discussed above; for v complex, there are two Casimir operators D', D'', arising from the center of the universal enveloping algebra of the Lie algebra of $GL_2(\mathbf{C})$.

Let f be an automorphic form for $GL_2(A_F)$. For each $g \in GL_2(A_F)$, the function $x \mapsto f\left(\begin{pmatrix} 1 & x \\ 0 & 1 \end{pmatrix} g\right)$ is continuous on $F\backslash A_F$ since f is left invariant under $GL_2(F)$. Hence we may express it in Fourier expansion. Let ψ be a nontrivial additive character of A_F/F. Recall that other such characters are ψ^ξ for $\xi \in F^\times$. For $\alpha \in F$, let

$$W_\alpha(g) = \int_{F\backslash A_F} f\left(\begin{pmatrix} 1 & x \\ 0 & 1 \end{pmatrix} g\right) \psi^{-\alpha}(x)dx,$$

where dx is the Haar measure on A_F so that A_F/F has measure 1. Then

$$f\left(\begin{pmatrix} 1 & x \\ 0 & 1 \end{pmatrix} g\right) = \sum_{\alpha \in F} W_\alpha(g)\psi(\alpha x) \quad \text{for } x \in A_F.$$

It follows from the definition of W_α and the fact that f is left invariant by the group $\left\{\begin{pmatrix} \beta & 0 \\ 0 & 1 \end{pmatrix} : \beta \in F^\times\right\}$ that for $\alpha \in F^\times$,

$$W_\alpha(g) = W_1\left(\begin{pmatrix} \alpha & 0 \\ 0 & 1 \end{pmatrix} g\right) \text{ and } W_0(g) = W_0\left(\begin{pmatrix} \alpha & 0 \\ 0 & 1 \end{pmatrix} g\right).$$

Writing W for W_1 and keeping in mind that W depends on the choice of ψ, we express the above Fourier expansion as

$$f\left(\begin{pmatrix} 1 & x \\ 0 & 1 \end{pmatrix} g\right) = W_0(g) + \sum_{\alpha \in F^\times} W\left(\begin{pmatrix} \alpha & 0 \\ 0 & 1 \end{pmatrix} g\right) \psi(\alpha x),$$

where W satisfies $W\left(\begin{pmatrix} 1 & x \\ 0 & 1 \end{pmatrix} g\right) = \psi(x)W(g)$, called the Whittaker function attached to f with respect to ψ. We examine briefly its properties. Clearly, W is right K-finite. Modulo a unipotent element on the left, we may write an element $g \in GL_2(A_F)$ as $g = \begin{pmatrix} y & 0 \\ 0 & 1 \end{pmatrix} kz$ with $y \in I_F$, $k \in K$ and $z \in \mathcal{Z}(A_F)$. At a finite place v where f is right invariant by $K_v = GL_2(\mathcal{O}_v)$, we have, for all $u \in \mathcal{O}_v$,

$$\begin{aligned} W\left(\begin{pmatrix} \alpha & 0 \\ 0 & 1 \end{pmatrix} g\right) &= W\left(\begin{pmatrix} \alpha & 0 \\ 0 & 1 \end{pmatrix} g \begin{pmatrix} 1 & u \\ 0 & 1 \end{pmatrix}\right) \\ &= W\left(\begin{pmatrix} \alpha & 0 \\ 0 & 1 \end{pmatrix}\begin{pmatrix} y & 0 \\ 0 & 1 \end{pmatrix}\begin{pmatrix} 1 & u \\ 0 & 1 \end{pmatrix} kz\right) \\ &= W\left(\begin{pmatrix} 1 & \alpha yu \\ 0 & 1 \end{pmatrix}\begin{pmatrix} \alpha & 0 \\ 0 & 1 \end{pmatrix} g\right) \\ &= \psi(\alpha yu) W\left(\begin{pmatrix} \alpha & 0 \\ 0 & 1 \end{pmatrix} g\right). \end{aligned}$$

Recall that the order of ψ_v is the largest integer n such that ψ_v is trivial on $\mathcal{P}_v^{-n}$. Thus $W\left(\begin{pmatrix} \alpha & 0 \\ 0 & 1 \end{pmatrix} g\right) \neq 0$ implies that $\operatorname{ord}_v \alpha y \geq -\operatorname{ord} \psi_v$. In particular, for $F = \mathbf{Q}$, choose ψ to be the standard additive character of $A_{\mathbf{Q}}/\mathbf{Q}$ so that ψ_v has order zero for all finite v. To see the Fourier expansion of f at the cusp "$i\infty$", we choose $g = \begin{pmatrix} y & 0 \\ 0 & 1 \end{pmatrix} \in GL_2(\mathbf{R}) \subset GL_2(A_{\mathbf{Q}})$. Then the above condition $\operatorname{ord}_v \alpha y \geq 0$ for all finite v implies $\alpha \in \mathbf{Z}$, so that the Fourier expansion of f at $i\infty$ is

$$f\left(\begin{pmatrix} 1 & x \\ 0 & 1 \end{pmatrix}\begin{pmatrix} y & 0 \\ 0 & 1 \end{pmatrix}\right) = W_0\left(\begin{pmatrix} y & 0 \\ 0 & 1 \end{pmatrix}\right) + \sum_{n \in \mathbf{Z}\setminus\{0\}} W\left(\begin{pmatrix} n & 0 \\ 0 & 1 \end{pmatrix}\begin{pmatrix} y & 0 \\ 0 & 1 \end{pmatrix}\right)\psi(nx).$$

An automorphic form f is called a *cusp form* if its constant Fourier coefficient W_0 is zero. Denote by $\mathcal{A}(GL_2(A_F), \omega)$ the space of automorphic forms for $GL_2(A_F)$ with character ω and by $\mathcal{A}^0(GL_2(A_F), \omega)$ the subspace of cusp forms in $\mathcal{A}(GL_2(A_F), \omega)$.

At a finite place v, choose a uniformizer ϖ_v and write H_v for the K_v-double coset representated by $\begin{pmatrix} \varpi_v & 0 \\ 0 & 1 \end{pmatrix}$. By the elementary divisor theorem, H_v consists of the matrices in $GL_2(F_v)$ with integral entries and of determinant in $\varpi_v \mathcal{U}_v$. Hence it is independent of the choice of the uniformizer ϖ_v, and it can be written as

$$H_v = \bigcup_{u \in \mathcal{O}_v/\mathcal{P}_v} \begin{pmatrix} \varpi_v & u \\ 0 & 1 \end{pmatrix} K_v \cup \begin{pmatrix} 1 & 0 \\ 0 & \varpi_v \end{pmatrix} K_v,$$

a disjoint union of $1 + Nv$ right K_v-cosets. The Hecke operator T_v acts on the space $\mathcal{A}(GL_2(A_F), \omega)$ by convolution with $\frac{1}{Nv}$ times the characteristic function of H_v. If $f \in \mathcal{A}(GL_2(A_F), \omega)$ is right invariant by K_v, then so is $T_v f$ and we have

$$\begin{aligned} (T_v f)(g) &= \frac{1}{Nv}\left[\sum_{u \in \mathcal{O}_v/\mathcal{P}_v} f\left(g\begin{pmatrix} \varpi_v & u \\ 0 & 1 \end{pmatrix}\right) + f\left(g\begin{pmatrix} 1 & 0 \\ 0 & \varpi_v \end{pmatrix}\right)\right] \\ &= \frac{1}{Nv}\left[\sum_u f\left(g\begin{pmatrix} \varpi_v & u \\ 0 & 1 \end{pmatrix}\right) + \omega_v(\varpi_v) f\left(g\begin{pmatrix} \varpi_v^{-1} & 0 \\ 0 & 1 \end{pmatrix}\right)\right]. \end{aligned}$$

For more details of the material discussed in this section, the reader is referred to [6], [10] and [20]. Before discussing automorphic representations of $GL_2(A_F)$, we need to know local representations, which are the topic of the next two sections.

§2 Representations of $GL_2(F)$ for F a nonarchimedean local field

Throughout this section, F denotes a nonarchimedean local field with q elements in its residue field. As usual, write $\mathcal{O}$ for its ring of integers, $\mathcal{P}$ the unique maximal ideal of $\mathcal{O}$, ϖ a uniformizer of $\mathcal{O}$ so that $\mathcal{P} = \varpi\mathcal{O}$, and $\mathcal{U}$ for the group of units $\mathcal{O}^\times$. The valuation $|\ |$ on $F^\times$ is trivial on $\mathcal{U}$ with $|\varpi| = q^{-1}$. We shall be concerned with complex representations of $GL_2(F)$, that is, actions of $GL_2(F)$ on a finite or infinite–dimensional vector space V over $\mathbf{C}$ endowed with discrete topology. A representation of $GL_2(F)$ on V is called *smooth* if

(2.1) for every $v \in V$, the stabilizer of v in $GL_2(F)$ is an open subgroup. In other words, the representation yields a continuous map from $GL_2(F) \times V$ to V.

It is said to be *admissible* if it is smooth and

(2.2) for every open subgroup H of $GL_2(\mathcal{O})$, the space V^H of vectors in V fixed by H is finite–dimensional.

The representation is *irreducible* if the only $GL_2(F)$ invariant subspaces of V are $\{0\}$ and V. We shall study admissible irreducible representations of $GL_2(F)$.

Let π be an admissible irreducible representation of $GL_2(F)$. By Schur's lemma, its restriction to the center $\mathcal{Z}(F)$ of $GL_2(F)$ is given by a quasi–character ω of $F^\times$, i.e., $\pi\left(\begin{pmatrix} a & 0 \\ 0 & a\end{pmatrix}\right) = \omega(a)Id$ for all $a \in F^\times$. Call ω the *central character* of π. Further, one can show that either π is one–dimensional, given by $\pi(g) = \chi\circ\det(g)$ for some quasi–character χ of $F^\times$, or else π is infinite–dimensional. So we concentrate on infinite–dimensional admissible irreducible representations for the remainder of this section, and simply call them representations if no ambiguity arises.

To such a representation π, there are usually attached two models, called Kirillov model $\mathbf{K}_\pi$ and Whittaker model $\mathbf{W}_\pi$, respectively. Both spaces consist of certain kinds of functions, simple in $\mathbf{K}_\pi$, complicated in $\mathbf{W}_\pi$, with the action of $GL_2(F)$ complicated on $\mathbf{K}_\pi$ and simple on $\mathbf{W}_\pi$. Each model has its own merit, and meets our special needs. We start with $\mathbf{K}_\pi$. A locally constant ($\mathbf{C}$–valued) function on $F^\times$ with compact support is called a Schwartz function on $F^\times$. Denote by $\mathcal{S}(F^\times)$ the space of Schwartz functions on $F^\times$. The space $\mathbf{K}_\pi$ is generated by $\mathcal{S}(F^\times)$ and its image under the action $\pi(w)$ of the Weyl element $w = \begin{pmatrix} 0 & 1 \\ -1 & 0\end{pmatrix}$. In particular, functions in $\mathbf{K}_\pi$ are locally constant on $F^\times$ with support contained in a compact set in F. Fix a nontrivial additive character ψ of F of order 0. The action of the Borel subgroup $B(F) = \left\{\begin{pmatrix} a & b \\ 0 & d\end{pmatrix} \in GL_2(F)\right\}$ on $\mathbf{K}_\pi = \mathbf{K}_\pi(\psi)$ can be easily described :

(2.3) For all $v \in \mathbf{K}_\pi$, $\left(\pi\left(\begin{pmatrix}1 & x\\ 0 & 1\end{pmatrix}\begin{pmatrix}z & 0\\ 0 & z\end{pmatrix}\begin{pmatrix}a & 0\\ 0 & 1\end{pmatrix}\right)v\right)(t) = \omega(z)\psi(xt)v(at)$,
where $x \in F$, $z, a, t \in F^\times$.

Note that the above action only depends on ψ and ω. To specify π, we only need to describe the action of $\pi(w)$ since $GL_2(F)$ is generated by $B(F)$ and w. In order to do this, we introduce some notation. We know $F^\times = \bigcup_{n\in\mathbf{Z}} \mathcal{U}\varpi^n$, and on each open subset $\mathcal{U}\varpi^n$, a locally constant function is a linear combination of characters on $\mathcal{U}$. Let U be an indeterminate. Identify characters of $\mathcal{U}$ with those of $F^\times$ trivial at ϖ. For a character χ of $\mathcal{U}$, denote by χU^n the function on $F^\times$ such that

$$\chi U^n(t) = \begin{cases} \chi(t) & \text{if } \operatorname{ord} t = n,\\ 0 & \text{otherwise.}\end{cases}$$

Thus functions in $\mathcal{S}(F^\times)$ are finite linear combinations of χU^n, while functions inK_π are formal linear combinations $\sum_{n,\chi} c_{n,\chi}\chi U^n$, with finitely many nonzero $c_{n,\chi}$ for each fixed n, and n bounded from below but not necessarily from above. The actual situation depends on the representation π. The representation π is determined by the action of $\pi(w)$ on $\mathcal{S}(F^\times)$, which in turn is determined by $\pi(w)\chi U^0$ for all $\chi \in \widehat{\mathcal{U}}$. Write ω_0 for the restriction of ω to $\mathcal{U}$. It follows from the relation

$$\begin{pmatrix}a & 0\\ 0 & 1\end{pmatrix} w = w\begin{pmatrix}1 & 0\\ 0 & a\end{pmatrix} = w\begin{pmatrix}a^{-1} & 0\\ 0 & 1\end{pmatrix}\begin{pmatrix}a & 0\\ 0 & a\end{pmatrix}$$

that with $v = \pi(w)\chi U^0$ one has $v(tu) = (\omega_0\chi^{-1})(u)v(t)$ for all $u \in \mathcal{U}$, $t \in F^\times$. In other words, on each open set $\mathcal{U}\varpi^m$, v is a constant multiple of $\omega_0\chi^{-1}U^m$; denote this constant by $\gamma_m(\omega_0^{-1}\chi, \psi)$. Let $\gamma(\omega_0^{-1}\chi, \psi, U)$ be the formal Laurent series in U defined by these constants :

$$\gamma(\omega_0^{-1}\chi, \psi, U) = \sum_{m\in\mathbf{Z}} \gamma_m(\omega_0^{-1}\chi, \psi)U^m$$

so that

(2.4) $\pi(w)\chi U^0 = \omega_0\chi^{-1} \cdot \gamma(\omega_0^{-1}\chi, \psi, U)$.

This way of analyzing π is in terms of generators of $GL_2(F)$, hence π must preserve relations. Most relations are preserved by definition, only two key relations remain :

(2.5) $\pi(w)^2 = \omega(-1)$,

(2.6) $\pi(w)\pi\left(\begin{pmatrix}1 & 1\\ 0 & 1\end{pmatrix}\right)\pi(w) = \omega(-1)\pi\left(\begin{pmatrix}1 & -1\\ 0 & 1\end{pmatrix}\right)\pi(w)\pi\left(\begin{pmatrix}1 & -1\\ 0 & 1\end{pmatrix}\right)$,

arising from $w^2 = \begin{pmatrix}-1 & 0\\ 0 & -1\end{pmatrix}$ and $\left(\begin{pmatrix}1 & 1\\ 0 & 1\end{pmatrix}w\right)^3 = \begin{pmatrix}1 & 0\\ 0 & 1\end{pmatrix}$. Here we replace (2.6) by

(2.6)′ $\pi(w)\left[\pi\left(\begin{pmatrix}1 & 1\\ 0 & 1\end{pmatrix}\right)\pi(w) - \pi(w)\right] + \omega(-1)$

$$= \omega(-1)\left[\pi\left(\begin{pmatrix}1 & -1\\ 0 & 1\end{pmatrix}\right)\pi(w) - \pi(w)\right]\pi\left(\begin{pmatrix}1 & -1\\ 0 & 1\end{pmatrix}\right)$$
$$+\,\omega(-1)\pi(w)\pi\left(\begin{pmatrix}1 & -1\\ 0 & 1\end{pmatrix}\right).$$

Observe that (2.5) and (2.6) are equivalent to (2.5) and (2.6)′. The advantage of (2.6)′ is that the operator in each bracket preserves the space $\mathcal{S}(F^\times)$ on which the action of w is explicitly given. Certainly (2.5) and (2.6)′ yield identities on γ. To facilitate our expression of the identities, put, for a quasi-character α of $F^\times$ with $\alpha_0 = \alpha\,|_{\mathcal{U}}$,

$$\gamma(\alpha,\psi,U) = \sum_m \gamma_m(\alpha_0,\psi)\alpha(\varpi)^m U^m = \gamma(\alpha_0,\psi,\alpha(\varpi)U)$$

and define Gauss sum

$$\Gamma(\alpha,\psi,U) = \sum_n \Gamma_n(\alpha,\psi)U^n \quad \text{with} \quad \Gamma_n(\alpha,\psi) = \int_{\mathcal{U}\varpi^n} \alpha(x)\psi(x)d^\times x,$$

where $d^\times x$ is the Haar measure on $F^\times$ such that $\mathcal{U}$ has measure 1. Note that $\Gamma(\alpha,\psi,U) = \Gamma(\alpha_0,\psi,\alpha(\varpi)U)$.

Exercise 2. Show that $\Gamma(\alpha,\psi,U)$ is a monomial in U for α ramified; and for α unramified, $\Gamma(\alpha,\psi,U)$ is a rational function in U with a simple pole.

The identity (2.5) yields

(Complement formula CF) $\qquad \gamma(\chi,\psi,U)\gamma(\omega^{-1}\chi^{-1},\psi,U^{-1}) = \omega(-1),$

while (2.6)′ yields

(Multiplicative formula MF) Given any quasi-characters α,β of $F^\times$ there exists a positive integer $M(\alpha,\beta,\gamma)$ such that for all integers $M \ge M(\alpha,\beta,\gamma)$ the following identity holds as a formal power series in A and B :

$$\sum_{\chi\in\widehat{\mathcal{U}},\text{cond }\chi\le M} \text{const}_U\, \Gamma(\alpha\chi^{-1},\psi,AU^{-1})\Gamma(\beta\chi^{-1},\psi,BU^{-1})\gamma(\chi,\psi,U)$$
$$= (\alpha\beta\omega)(-1)\Gamma(\alpha^{-1}\beta^{-1}\omega^{-1},\psi,A^{-1}B^{-1})\gamma(\alpha,\psi,A)\gamma(\beta,\psi,B)$$
$$+\begin{cases} 0 & \text{if } \alpha\beta\omega \text{ is ramified,}\\ \sum_{n\ge -M}(\alpha\beta\omega)(\varpi)^n A^n B^n & \text{if } \alpha\beta\omega \text{ is unramified.}\end{cases}$$

Here $\text{const}_U(\cdots)$ means the coefficient of U^0 in $(\cdots)$.

Exercise 3. Prove the complement formula (CF) and the multiplicative formula (MF).

We summarize the above discussion in

Theorem 1. *(The existence of Kirillov model [10], [13]) Let π be an infinite-dimensional admissible irreducible representation of $GL_2(F)$ with central character ω. Then given any nontrivial additive character ψ of F, π can be realized on a space $\mathbf{K}_\pi(\psi)$ of $\mathbf{C}$-valued locally constant functions on $F^\times$ supported within compact subsets of F, containing the space $\mathcal{S}(F^\times)$ of Schwartz functions on $F^\times$, on which $B(F)$ acts via (2.3) and the Weyl element w acts on $\mathcal{S}(F^\times)$ via (2.4). The γ-function occurring in (2.4) satisfies the identity (MF). Further, $\mathbf{K}_\pi(\psi)$ is spanned by $\mathcal{S}(F^\times)$ and $\pi(w)\mathcal{S}(F^\times)$.*

Denote by γ_π the γ-function attached to π. The identity (MF) contains a lot of information about γ, here we exhibit several.

Theorem 2. *([8]) For each quasi-character α of $F^\times$, let $\gamma(\alpha, \psi, U)$ be a Laurent series in U such that $\gamma(\alpha,\psi,U) = \gamma(\alpha_0, \psi, \alpha(\varpi)U)$, where $\alpha_0 = \alpha\,|_{\mathcal{U}}$. Suppose that there is a quasi-character ω of $F^\times$ such that (MF) holds. Then*

(i) *γ satisfies (CF);*

(ii) *Either there are quasi-characters μ, ν of $F^\times$ with $\mu\nu = \omega$ such that for all quasi-characters α of $F^\times$,*

$$\gamma(\alpha, \psi, U) = (q-1)^2 q^{-2} \Gamma(\alpha\mu, \psi, q^{-\frac{1}{2}}U)\Gamma(\alpha\nu, \psi, q^{-\frac{1}{2}}U),$$

or else $\gamma(\alpha,\psi,U)$ is a monomial in U with degree $\leq \min(-2, -1-\operatorname{cond}\alpha^2\omega)$ for all α;

(iii) *(deep twist property of γ) Denote by r the lowest power of U occurring in $\gamma(1,\psi,U)$. Then for all quasi-characters α of $F^\times$ with $\operatorname{cond}\alpha \geq -r$, we have*

$$\gamma(\alpha, \psi, U) = (q-1)^2 q^{-2} \Gamma(\alpha, \psi, q^{-\frac{1}{2}}U)\Gamma(\alpha\omega, \psi, q^{-\frac{1}{2}}U).$$

As a consequence of the third statement above, we get

Corollary 1. *([8]) The quasi-character ω in (MF) satisfied by γ is unique.*

Denote by $\mathcal{A}(F^\times)$ the group of quasi-characters of $F^\times$ on which we endow the analytic structure such that each connected component is isomorphic to $\mathbf{C}^\times$, consisting of the quasi-characters agreeing with each other on $\mathcal{U}$, and the connected components are parametrized by the characters of $\mathcal{U}$. Write $\gamma(\alpha,\psi)$ for $\gamma(\alpha,\psi,1)$. We may regard γ as a function on $\mathcal{A}(F^\times)$. Theorem 2, (ii), says that γ is a rational function on $\mathcal{A}(F^\times)$, that is, a rational function on each connected component of $\mathcal{A}(F^\times)$, with at most two poles, and in fact it is equal to a monomial on all except possibly two components. Further, it is not holomorphic if and only if it is a product of two Γ-functions, in which case it has two poles if $\mu\nu^{-1} \neq |\ |^{\pm 1}$ and one pole if $\mu\nu^{-1} = |\ |^{\pm 1}$. Here μ, ν are as in Theorem 2, (ii).

Next theorem says that the invariant γ attached to a representation of $GL_2(F)$ actually determines the representation and is characterized by the identity (MF).

Theorem 3. *(Characterization of γ_π [13], [8]) A rational function $\gamma(\alpha,\psi,U) = \gamma(\alpha_0,\psi,\alpha(\varpi)U)$ on $\mathcal{A}(F^\times)$ is equal to γ_π for some infinite-dimensional admissible irreducible representation π of $GL_2(F)$ with central character ω if and only if γ satisfies (MF) with the same ω. Further γ_π determines the representation π.*

We have seen that γ_π satisfies (MF). Conversely, given a rational γ satisfying (MF), we want to construct a representation π via its Kirillov model $\mathbf{K}_\pi$. The action of Borel subgroup is given by (2.3), the key is to give the action of w on the space of Schwartz functions $\mathcal{S}(F^\times)$. Certainly, we define the action of w by (2.4). Observe that if γ is a monomial, then $\pi(w)$ sends $\mathcal{S}(F^\times)$ to itself so that $\mathbf{K}_\pi = \mathcal{S}(F^\times)$. In the case that γ is the product of two Γ-functions, it follows from (2.4) that $\mathcal{S}(F^\times)$ has index 2 or 1 in $\mathbf{K}_\pi = \mathcal{S}(F^\times) + \pi(w)\mathcal{S}(F^\times)$, depending on γ having 2 or 1 poles. Further the actions defined so far extend to an action of $GL_2(F)$ since the relations (2.5) and (2.6)$'$ are preserved due to (CF) and (MF). This sketches the proof of Theorem 3.

The above proof leads to the following classification of infinite-dimensional admissible irreducible representations π of $GL_2(F)$. We call π a *principal series*, a *special representation*, or a *supercuspidal representation* if the codimension of $\mathcal{S}(F^\times)$ in bk_π is 2, 1, or 0, respectively. Or equivalently, γ_π has 2, 1 or 0 poles on $\mathcal{A}(F^\times)$. In the first two cases, $\gamma_\pi(\alpha,\psi) = (q-1)^2q^{-2}\Gamma(\alpha\mu,\psi,q^{-\frac{1}{2}})\Gamma(\alpha\nu,\psi,q^{-\frac{1}{2}})$ for some quasi-characters μ,ν with $\mu\nu^{-1} \neq |\ \ |^{\pm 1}$ or $\mu\nu^{-1} = |\ \ |^{\pm 1}$, accordingly. Such a representation π can also be obtained from a representation $\rho(\mu,\nu)$ of $GL_2(F)$ induced from the character τ on $B(F)$ defined by

$$\tau\left(\begin{pmatrix} a & b \\ 0 & d \end{pmatrix}\right) = \mu(a)\nu(d)|a/d|^{1/2}.$$

Jacquet and Langlands in [10] showed that when $\mu\nu^{-1} \neq |\ \ |^{\pm 1}$, $\rho(\mu,\nu)$ is irreducible, called $\pi(\mu,\nu)$; while if $\mu\nu^{-1} = |\ \ |^{\pm 1}$, $\rho(\mu,\nu)$ has a unique constituent $\sigma(\mu,\nu)$ which is infinite-dimensional, the other constituent being 1-dimensional. They then proved that the γ-factor attached to $\pi(\mu,\nu)$ or $\sigma(\mu,\nu)$ is γ_π above. It can be shown directly ([8]) that given any pair of quasi-characters μ,ν of $F^\times$, the function γ on $\mathcal{A}(F^\times)$ defined by

$$\gamma(\alpha,\psi) = (q-1)q^{-2}\Gamma(\alpha\mu,\psi,q^{-\frac{1}{2}})\Gamma(\alpha\nu,\psi,q^{-\frac{1}{2}})$$

satisfies (MF) with $\omega = \mu\nu$, hence there is a principal series or special representation π induced from μ,ν. The supercuspidal representations for $GL_2(F)$, and $GL_N(F)$ in general, may also be constructed from the inducing procedure, not from a parabolic subgroup, but from compact-mod-center subgroups, done by Carayol [3] and Kutzko [11, 12].

Of course, the Kirillov model $\mathbf{K}_\pi(\varphi)$ of π exists for all nontrivial additive character φ of F. Writing $\varphi = \psi^t$ with ψ of order zero, we have $\operatorname{ord}\psi^t = \operatorname{ord} t$ and

$$\gamma_\pi(\chi,\psi^t,U) = q^{-\operatorname{ord} t}(\omega^{-1}\chi^{-2})(t)\gamma_\pi(\chi,\psi,U)U^{-2\operatorname{ord} t}.$$

Next we introduce the Whittaker model of an infinite-dimensional admissible irreducible representation π of $GL_2(F)$. Let $bk_\pi = \mathbf{K}_\pi(\psi)$ be its Kirillov model with respect to the nontrivial additive character ψ of F. For each function $v \in bk_\pi$, define a Whittaker function W_v on $GL_2(F)$ by

$$W_v(g) = \big(\pi(g)v\big)(1) \quad \text{for all} \quad g \in GL_2(F).$$

Let $\mathbf{W}_\pi = \mathbf{W}_\pi(\psi) = \{W_v : v \in \mathbf{K}_\pi(\psi)\}$, and define an action π' of $GL_2(F)$ on $\mathbf{W}_\pi$ by right translations, that is, for $g' \in GL_2(F)$

$$(\pi'(g')W_v)(g) = W_v(gg') = \big(\pi(gg')v\big)(1) = \big(\pi(g)(\pi(g')v)\big)(1) = W_{\pi(g')v}(g).$$

This shows that π' is equivalent to π. The space $\big(\pi, \mathbf{W}_\pi(\psi)\big)$ is called the Whittaker model of π. Note that a Whittaker function W_v satisfies

$$\begin{aligned} W_v\left(\begin{pmatrix} 1 & x \\ 0 & 1 \end{pmatrix} g\right) &= \left(\pi\begin{pmatrix} 1 & x \\ 0 & 1 \end{pmatrix}\pi(g)v\right)(1) = \psi(x)\big(\pi(g)v\big)(1) \\ &= \psi(x)W_v(g), \quad x \in F, \quad g \in GL_2(F) \\ W_v\left(\begin{pmatrix} z & 0 \\ 0 & z \end{pmatrix} g\right) &= \omega(z)W_v(g), \quad z \in F^\times, \quad g \in GL_2(F) \\ \text{and} \qquad W_v\left(\begin{pmatrix} a & 0 \\ 0 & 1 \end{pmatrix}\right) &= v(a). \end{aligned}$$

Recall that the local L–function attached to a quasi–character μ of $F^\times$ is defined as

$$L(\mu, U) = \begin{cases} (1 - \mu(\varpi)U)^{-1} & \text{if } \mu \text{ is unramified,} \\ 1 & \text{if } \mu \text{ is ramified.} \end{cases}$$

For a representation π of $GL_2(F)$, its associated L–function is defined as

$$\begin{aligned} &L(\pi, U) \\ &= L(\mu, U)L(\nu, U) \quad && \text{if } \pi = \pi(\mu, \nu) \text{ is a principal series,} \\ &= L(\mu, U) && \text{if } \sigma = \pi(\mu, \nu) \text{ is a special representation with } \mu\nu^{-1} = |\ |, \\ &= 1 && \text{if } \pi \text{ is a supercuspidal representation.} \end{aligned}$$

Given a Whittaker function $W \in \mathbf{W}_\pi(\psi)$ and an element $g \in GL_2(F)$, define two formal power series :

$$\Phi(g, U, W) = \int_{F^\times} W\left(\begin{pmatrix} a & 0 \\ 0 & 1 \end{pmatrix} g\right) |a|^{-\frac{1}{2}} U^{\operatorname{ord} a} d^\times a,$$

and

$$\widetilde{\Phi}(g, U, W) = \int_{F^\times} W\left(\begin{pmatrix} a & 0 \\ 0 & 1 \end{pmatrix} g\right) \omega(a)^{-1} |a|^{-\frac{1}{2}} U^{\operatorname{ord} a} d^\times a,$$

which converge for U of small absolute value. As the action of $GL_2(F)$ on Whittaker functions are right translations, it suffices to know $\Phi(g, U, W)$ and $\widetilde{\Phi}(g, U, W)$ for all $W \in \mathbf{W}_\pi(\psi)$ and for one element $g \in GL_2(F)$. For example, if we take $W = W_v$ with $v = \chi U^m \in \mathcal{S}(F^\times)$, then

$$\Phi(1_2, U, W_v) = \begin{cases} 0 & \text{if } \chi \text{ is ramified,} \\ q^{\frac{1}{2}m} U^m & \text{if } \chi \text{ is unramified,} \end{cases}$$

and

$$\widetilde{\Phi}(w, U, W_v) = \begin{cases} 0 & \text{if } \chi \text{ is ramified,} \\ q^{\frac{1}{2}m} U^{-m} \gamma_\pi(\omega^{-1}, \psi, q^{\frac{1}{2}} U) & \text{if } \chi \text{ is unramified.} \end{cases}$$

Further, if π is a principal series $\pi(\mu, \nu)$ or a special representation $\sigma(\mu, \nu)$, then with $v = \pi(w)\nu_0 U^0$ we find

$$\Phi(1_2, U, W_v) = \begin{cases} 0 & \text{if } \mu \text{ is ramified,} \\ \gamma_\pi(\chi_0, \psi, q^{\frac{1}{2}} U) & \text{if } \mu \text{ is unramified, (where } \chi_0 \text{ is the trivial character)} \end{cases}$$

and

$$\widetilde{\Phi}(w, U, W_v) = \begin{cases} 0 & \text{if } \mu \text{ is ramified,} \\ \omega(-1) & \text{if } \mu \text{ is unramified.} \end{cases}$$

Observe that $\Phi(g, U, W)/L(\pi, U)$ is a finite linear combination in powers of U for all $g \in GL_2(F)$ and all $W \in \mathbf{W}_\pi$, hence it gives an analytic continuation of $\Phi(g, U, W)$. Moreover, $L(\pi, U)$ is the "least common multiple" of the denominators of $\Phi(g, U, W)'s$ when expressed as a rational function in U.

The contragredient $\widetilde{\pi}$ of π is defined as follows. Let V be the space of π and $\widehat{V}$ the space of linear functionals on V. Hence there is a pairing

$$\langle\, ,\, \rangle : V \times \widehat{V} \longrightarrow \mathbf{C} \quad \text{via}$$
$$\langle v, \widehat{v} \rangle = \widehat{v}(v).$$

Define the action π^* of $GL_2(F)$ on $\widehat{V}$ by

$$\langle v, \pi^*(g)\widehat{v} \rangle = \langle \pi(g^{-1})v, \widehat{v} \rangle$$

(so that if V were finite-dimensional, then $\pi^*(g) = {}^t\left(\pi(g^{-1})\right) = {}^t\,\pi(g)^{-1}$ and this would be the contragredient.) For V infinite-dimensional the representation π^* of $GL_2(F)$ on $\widehat{V}$ is usually not admissible. To get an admissible representation, consider the subspace $\widetilde{V}$ of smooth duals of V, they are generated by

$$\int_S \pi^*(g)\widehat{v}\,dg, \quad \widehat{v} \in \widehat{V},$$

as S runs through all open compact subsets of $GL_2(F)$. Then $\widetilde{V}$ is $\pi^*(GL_2(F))$-invariant, and the action is admissible irreducible if π is. Call this representation $\widetilde{\pi}$

the contragredient of π. If ω is the central character of π, then $\tilde{\pi}$ is isomorphic to $\pi \otimes \omega^{-1}$, and

$$\gamma_{\tilde{\pi}}(\chi, \psi, U) = \gamma_{\pi}(\chi\omega^{-1}, \psi, U), \qquad \mathbf{W}_{\tilde{\pi}} = \mathbf{W}_{\pi} \otimes (\omega^{-1} \circ \det).$$

In particular, if $\pi = \pi(\mu, \nu)$ or $\sigma(\mu, \nu)$, then $\tilde{\pi} = \pi(\nu^{-1}, \mu^{-1})$ or $\sigma(\nu^{-1}, \mu^{-1})$.

From the definition above, we get

$$L(\tilde{\pi}, U) = \begin{cases} L(\mu^{-1}, U)L(\nu^{-1}, U) & \text{if } \pi = \pi(\mu, \nu), \\ L(\nu^{-1}, U) & \text{if } \pi = \sigma(\mu, \nu) \text{ with } \mu\nu^{-1} = |\ |, \\ 1 & \text{if } \pi \text{ is supercuspidal.} \end{cases}$$

Put

$$\varepsilon(\pi, \psi, U) = \gamma_{\pi}(\omega^{-1}, \psi, q^{-\frac{1}{2}}U^{-1}) \frac{L(\pi, U)}{L(\tilde{\pi}, q^{-1}U^{-1})}.$$

Note that $\varepsilon(\pi, \psi, U)$ is a monomial in U.

The integral $\tilde{\Phi}(g, U, W)$ can be viewed as an integral $\Phi(g, U, W')$ for some $W' \in \mathbf{W}_{\pi} \otimes (\omega^{-1} \circ \det)$, that is, the Whittaker model of $\tilde{\pi}$. Hence the same remark for $\tilde{\Phi}(g, U, W)/L(\tilde{\pi}, U)$ holds. Using the complement formula

$$\gamma_{\pi}(\chi_0, \psi, q^{\frac{1}{2}}U)\gamma_{\pi}(\omega^{-1}, \psi, q^{-\frac{1}{2}}U^{-1}) = \omega(-1),$$

one checks from the explicit computations above that the analytically continued functions $\Phi(g, U, W)/L(\pi, U)$ and $\tilde{\Phi}(wg, q^{-1}U^{-1}, W)/L(\tilde{\pi}, q^{-1}, U^{-1})$ have ratio equal to $\varepsilon(\pi, \psi, U)^{-1}$, which is independent of W. We have shown

Theorem 4. *(Local functional equations [10]) Let π be an admissible irreducible infinite-dimensional representation of $GL_2(F)$ with central character ω. Let $L(\pi, U)$, $L(\tilde{\pi}, U)$ and $\varepsilon(\pi, \psi, U)$ be as above. For each Whittaker function $W \in \mathbf{W}_{\pi}(\psi)$ of π and $g \in GL_2(F)$, define the two integrals $\Phi(g, U, W)$ and $\tilde{\Phi}(g, U, W)$ as before. Then $L(\pi, U)^{-1}$ (resp. $L(\tilde{\pi}, U)^{-1}$) is the least degree polynomial in U with constant term 1 such that for all g and W, $\Phi(g, U, W)/L(\pi, U)$ (resp. $\tilde{\Phi}(g, U, W)/L(\tilde{\pi}, U)$) has an analytic continuation to all U in $\mathbf{C}^{\times}$ as a finite linear combination in powers of U. Further, the continued functions satisfy the functional equation*

$$\frac{\tilde{\Phi}(wg, q^{-1}U^{-1}, W)}{L(\tilde{\pi}, q^{-1}U^{-1})} = \varepsilon(\pi, \psi, U) \frac{\Phi(g, U, W)}{L(\pi, U)}$$

for all g and W

We end this section by discussing the "new vectors" in the space V of a representation π of $GL_2(F)$. This is worked out by Deligne [4], and may be regarded as the local theory of new forms. We shall see this in the Kirillov model $\mathbf{K}_\pi(\psi)$ of π. Write $n(\psi)$ for the order of ψ. First we look at functions v in $\mathbf{K}_\pi(\psi)$ such that

$$\pi\left(\begin{pmatrix} a & b \\ 0 & d \end{pmatrix}\right) v = \omega(d)v \quad \text{for all } a, d \in \mathcal{U} \text{ and } b \in \mathcal{O}.$$

In view of the action of $B(F)$ on v as described in (2.3), this amounts to saying that for all $x \in \operatorname{Supp} v$, $\psi(\mathcal{O}x) = 1$, and $v(ux) = v(x)$ for all $u \in \mathcal{U}$. Hence we find that v is supported within $\varpi^{-n(\psi)}\mathcal{O} = \mathcal{P}^{-n(\psi)}$ and $v(x)$ depends only on the order of x. There are many such vectors, for instance, $\chi_0 U^m \in \mathcal{S}(F^\times)$ with $m \geq -n(\psi)$ and χ_0 the trivial character of $\mathcal{U}$. Denote by V_0 the space of such functions in $\mathbf{K}_\pi(\psi)$. Next, as π is smooth, each v in V_0 is stabilized by an open subgroup of $GL_2(F)$, hence in particular by $\begin{pmatrix} 1 & 0 \\ \mathcal{P}^m & 1 \end{pmatrix}$ for some m large. Let n be the smallest integer such that there exists a nonzero function $v_0 \in V_0$ stabilized by $\begin{pmatrix} 1 & 0 \\ \mathcal{P}^n & 1 \end{pmatrix}$. Denote by G'_n the subgroup generated by $\begin{pmatrix} 1 & 0 \\ \mathcal{P}^n & 1 \end{pmatrix}$ and $\begin{pmatrix} 1 & \mathcal{O} \\ 0 & 1 \end{pmatrix}$. Then

$$G'_n = \begin{cases} \left\{\begin{pmatrix} a & b \\ c & d \end{pmatrix} \in SL_2(\mathcal{O}) : a-1,\ d-1 \text{ and } c \in \mathcal{P}^n\right\} & \text{if } n > 0, \\ SL_2(\mathcal{O}) & \text{if } n = 0, \\ SL_2(F) & \text{if } n < 0. \end{cases}$$

v_0 is invariant by G'_n. If $n < 0$, then the space spanned by $GL_2(F)v_0$ is the space spanned by $\pi\left(\begin{pmatrix} \varpi^m & 0 \\ 0 & 1 \end{pmatrix}\right) v_0$, $m \in \mathbf{Z}$, which is a nontrivial proper subspace of $\mathbf{K}_\pi(\psi)$, contradicting the irreducibility of π. Thus $n \geq 0$ and $n \geq \operatorname{cond}\omega$ since the action of $\left\{\begin{pmatrix} d^{-1} & 0 \\ 0 & d \end{pmatrix} : d \in \mathcal{U}\right\}$ on v_0 is multiplication by $\omega(d)$ on one hand, and trivial on the other hand. Define, for $k \geq 0$, the congruence subgroups

$$G_k = \left\{\begin{pmatrix} a & b \\ c & d \end{pmatrix} \in GL_2(\mathcal{O}) : c \in \mathcal{P}^k\right\},$$

and let

$$X_k = \left\{v \in \mathbf{K}_\pi(\psi) : \pi\left(\begin{pmatrix} a & b \\ c & d \end{pmatrix}\right) v = \omega(d)v \quad \text{for all } \begin{pmatrix} a & b \\ c & d \end{pmatrix} \in G_k\right\}.$$

Thus $v_0 \in X_n$ and $X_n \subseteq X_{n+1} \subseteq \cdots$ is a filtration of V_0.

Theorem 5. *(Deligne [4]) The space X_k is $k-n+1$ dimensional with a basis given by $\pi\left(\begin{pmatrix} \varpi^{-i} & 0 \\ 0 & 1 \end{pmatrix}\right) v_0$, $0 \leq i \leq k-n$.*

Proof. By the minimality of n, $X_k = \{0\}$ if $k < n$. Further, it is easy to check that the space $\pi\left(\begin{pmatrix} \varpi^{-i} & 0 \\ 0 & 1 \end{pmatrix}\right) X_k$ is contained in the space X_{k+i} for $i \geq 0$, and the vectors $\pi\left(\begin{pmatrix}\varpi^{-i} & 0\\ 0 & 1\end{pmatrix}\right)v_0$, $i \geq 0$, are linearly independent. It remains to show that the dimension of the quotient space $X_k/\pi\left(\begin{pmatrix}\varpi^{-1} & 0\\ 0 & 1\end{pmatrix}\right)X_{k-1}$ is ≤ 1. Indeed, let $v_1, v_2 \in X_k - \{0\}$. Then there are two constants a, b, not both zero, such that $av_1 + bv_2$ is supported in $\mathcal{P}^{-n(\psi)+1}$, and hence $\pi\left(\begin{pmatrix}\varpi & 0\\ 0 & 1\end{pmatrix}\right)(av_1 + bv_2) \in V_0$ and is invariant under $\begin{pmatrix}1 & 0\\ \mathcal{P}^{k-1} & 1\end{pmatrix}$. In case $k = n$, this shows $av_1 + bv_2 = 0$ by the minimality of n, and in case $k > n$, this shows $av_1 + bv_2 \in \pi\left(\begin{pmatrix}\varpi^{-1} & 0\\ 0 & 1\end{pmatrix}\right)X_{k-1}$, whence theorem follows.

By Theorem 5, the space X_n is 1–dimensional, and the nonzero vectors there are called "new vectors" of π. We have seen from the above proof that a new vector is nonzero on $\mathcal{U}\varpi^{-n(\psi)}$. The integer n is called the (exponent of the) conductor of π. Denote by v_0 the normalized new vector with $v_0(\mathcal{U}\varpi^{-n(\psi)}) = 1$. Say π is unramified or of class one if $n = 0$.

As seen before, the $GL_2(\mathcal{O})$ double coset $H = GL_2(\mathcal{O})\begin{pmatrix}\varpi & 0\\ 0 & 1\end{pmatrix}GL_2(\mathcal{O}) = \bigcup_{u\in\mathcal{O}/\mathcal{P}}\begin{pmatrix}\varpi & u\\ 0 & 1\end{pmatrix}GL_2(\mathcal{O}) \cup \begin{pmatrix}1 & 0\\ 0 & \varpi\end{pmatrix}GL_2(\mathcal{O})$. The action of the Hecke operator T on the representation space of π is convolution with q^{-1} times the characteristic function of H. In other words,

$$Tv = q^{-1}\int_{GL_2(\mathcal{O})}\sum_{u\in\mathcal{O}/\mathcal{P}}\pi\left(\begin{pmatrix}\varpi & u\\ 0 & 1\end{pmatrix}\right)\pi(k)v + \pi\left(\begin{pmatrix}1 & 0\\ 0 & \varpi\end{pmatrix}\right)\pi(k)v\,dk.$$

If v_0 is invariant by $GL_2(\mathcal{O})$, i.e., $n = 0$, then Tv_0 is also invariant by $GL_2(\mathcal{O})$, and hence $Tv_0 = \lambda v_0$ for some eigenvalue λ. In this case we have v_0 supported in $\mathcal{O}\varpi^{-n(\psi)}$ and

$$\begin{aligned} Tv_0(x) &= q^{-1}\left[\sum_{u\in\mathcal{O}/\mathcal{P}}\psi(ux)v_0(\varpi x) + \omega(\varpi)v_0(\varpi^{-1}x)\right] \\ &= v_0(\varpi x) + q^{-1}\omega(\varpi)v_0(\varpi^{-1}x) \qquad \text{for } x \in \mathcal{O}\varpi^{-n(\psi)}. \end{aligned}$$

Since $v_0(x)$ depends only on $\operatorname{ord} x$, the above relation yields

$$\lambda v_0(\varpi^m) = v_0(\varpi^{m+1}) + q^{-1}\omega(\varpi)v_0(\varpi^{m-1}) \qquad \text{for } m \geq -n(\psi),$$

or equivalently,

$$\begin{aligned} \Phi(1_2, U, W_{v_0}) &= \int_{F^\times} v_0(a)|a|^{-\frac{1}{2}}U^{\operatorname{ord} a}d^\times a \\ &= \left(1 - \lambda q^{\frac{1}{2}}U + \omega(\varpi)U^2\right)^{-1}\left(q^{\frac{1}{2}}U\right)^{-n(\psi)}. \end{aligned}$$

In view of Theorem 4, we get immediately that if π is of class one then it is a principal series $\pi(\mu,\nu)$ induced from two unramified quasi-characters μ,ν of $F^\times$ with $\mu\nu=\omega$ and

$$L(\pi,U)^{-1}=L(\mu,U)^{-1}L(\nu,U)^{-1}=1-\lambda q^{\frac{1}{2}}U+\mu\nu(\varpi)U^2,$$

where λ is the eigenvalue of the Hecke operator T on a new vector of π. Conversely, suppose $\pi=\pi(\mu,\nu)$ is a principal series induced from two unramified quasi-characters μ,ν with $\omega=\mu\nu$. We want to show that π has conductor 0. For this purpose, choose an additive character ψ of order 0. Then we know $\gamma_\pi(\chi,\psi,U)$ explicitly by Theorem 2, (ii); in particular, $\gamma_\pi(\chi_0,\psi,U)$ is a rational function in U with denominator equal to

$$\left(1-\mu(\varpi)q^{-\frac{1}{2}}U\right)\left(1-\nu(\varpi)q^{-\frac{1}{2}}U\right)=1+\alpha U+\omega(\varpi)q^{-1}U^2,$$

where $\alpha=-\mu(\varpi)q^{-\frac{1}{2}}-\nu(\varpi)q^{-\frac{1}{2}}$, and numerator equal to

$$\left(1+\alpha U+\omega(\varpi)q^{-1}U^2\right)\gamma(\chi_0,\psi,U)=q^{-1}U^{-2}\left[\omega(\varpi)^{-1}+\omega(\varpi)^{-1}q\alpha U+qU^2\right].$$

Consider the function v in $\mathbf{K}_\pi(\psi)$ given by

$$v=c_{-2}\chi_0U^{-2}+c_{-1}\chi_0U^{-1}-\omega(\varpi)qc_{-2}\pi(w)\chi_0U^0+(c_{-1}-\alpha qc_{-2})\pi(w)\chi_0U^{-1},$$

where c_{-1},c_{-2} are constants to be determined. It follows from the explicit expression of $\gamma(\chi_0,\psi,U)$ above and the action $\pi(w)\chi_0U^m=\omega(\varpi)^{-m}\gamma(\chi_0,\psi,U)U^{-m}$ from (2.4) that $\pi(w)v=v$. Further, one checks that v is supported in $\mathcal{P}^{-1}$. Choose a nonzero c_{-2}, then there is a unique choice of c_{-1} such that v is supported in $\mathcal{O}$. As the vectors χ_0U^{-2}, χ_0U^{-1}, $\pi(w)\chi_0U^0$, and $\pi(w)\chi_0U^{-1}$ are linearly independent, we get a nonzero vector v in V_0 and invariant by $SL_2(\mathcal{O})$, hence it is a new vector of π. This shows that π has conductor 0, as claimed. We have shown that $\pi=\pi(\mu,\nu)$ is a principal series induced from two unramified quasi-characters μ,ν of $F^\times$ if and only if its conductor is zero, and in which case $\Phi(1_2,U,W_{v_0})=(q^{\frac{1}{2}}U)^{-n(\psi)}L(\pi,U)$, where W_{v_0} is the normalized new vector of π in its Whittaker model $\mathbf{W}_\pi(\psi)$.

Next we study representations π with $n=\operatorname{cond}\pi\geq 1$. Let v_0 be its normalized new vector in the Kirillov model $\mathbf{K}_\pi(\psi)$. We know that $v_0(t)$ depends only on the order of t, $\operatorname{Supp}v_0\subset\mathcal{P}^{-n(\psi)}$ and $v_0(\varpi^{-n(\psi)})=1$. Consider the function

$$v=\left(\sum_{\substack{u\in\mathcal{O}/\mathcal{P}\\u\notin\mathcal{P}}}\frac{1}{q-1}\pi\left(\begin{pmatrix}1&u\varpi^{-1}\\0&1\end{pmatrix}\right)v_0\right)-v_0.$$

One checks from (2.3) that

$$v(t)=\begin{cases}-\frac{q}{q-1}v_0(t) & \text{if } t\in\mathcal{U}\varpi^{-n(\psi)},\\ 0 & \text{otherwise,}\end{cases}$$

so that $v = a\chi_0 U^{-n(\psi)}$, where $a = -q/(q-1)$. On the other hand, for $u \in \mathcal{U}$ and $c \in \mathcal{P}^{n+1}$, we have

$$\begin{pmatrix} 1 & -u\varpi^{-1} \\ 0 & 1 \end{pmatrix}\begin{pmatrix} 1 & 0 \\ c & 1 \end{pmatrix}\begin{pmatrix} 1 & u\varpi^{-1} \\ 0 & 1 \end{pmatrix} = \begin{pmatrix} 1 - cu\varpi^{-1} & -cu^2\varpi^{-2} \\ c & 1 + cu\varpi^{-1} \end{pmatrix} \in G'_n,$$

hence $\pi\left(\begin{pmatrix} 1 & 0 \\ \mathcal{P}^{n+1} & 1 \end{pmatrix}\right) v = v$. We have shown that $\chi_0 U^{-n(\psi)} \in X_{n+1}$, and, by Theorem 5, $\chi_0 U^{-n(\psi)} = v_0 - b\pi\left(\begin{pmatrix} \varpi^{-1} & 0 \\ 0 & 1 \end{pmatrix}\right) v_0$ for some constant b. From the above description of v_0, we may express it as $v_0 = \sum_{i \geq -n(\psi)} c_i \chi_0 U^i$ with $c_{-n(\psi)} = 1$. Then $\pi\left(\begin{pmatrix} \varpi^{-1} & 0 \\ 0 & 1 \end{pmatrix}\right) v_0 = \sum c_i \chi_0 U^{i+1}$ and the above relation implies $c_i = b^{i+n(\psi)}$ for $i \geq -n(\psi) + 1$. If $b \neq 0$, then $\Phi(1_2, U, W_{v_0})$ has a simple pole. From Theorem 4 we conclude that v_0 is $\chi_0 U^{-n(\psi)}$, i.e., $b = 0$, if $L(\pi, U) = 1$, in which case $\Phi(1_2, U, W_{v_0}) = (q^{\frac{1}{2}} U)^{-n(\psi)}$. Conversely, suppose $v_0 = \chi_0 U^{-n(\psi)}$, we want to show that $L(\pi, U) = 1$. Suppose otherwise, then $L(\pi, U)$ has a simple pole. In view of the discussion before Theorem 4, there is a function $v = \sum_{i > -\infty} c_i \chi_0 U^i$ in $\mathbf{K}_\pi(\psi)$ such that $\Phi(1_2, U, W_v)$ has denominator $L(\pi, U)^{-1}$, thus $c_i \neq 0$ for infinitely many i. Replacing v by $\pi\left(\begin{pmatrix} \varpi^{-m} & 0 \\ 0 & 1 \end{pmatrix}\right) v$ for a large integer m if necessary, we may assume that $\operatorname{Supp} v \subset \mathcal{P}^{-n(\psi)}$, i.e., $v \in V_0$. By the smoothness of π, we have $v \in X_m$ for some m. On the other hand, by Theorem 5, v is a linear combination of $\pi\left(\begin{pmatrix} \varpi^{-i} & 0 \\ 0 & 1 \end{pmatrix}\right) v_0 = \chi_0 U^{-n(\psi)+i}$ with $0 \leq i \leq m - n$, which implies that $c_i = 0$ for $i > m - n - n(\psi)$, a contradiction. Hence $L(\pi, U) = 1$. We have shown that $L(\pi, U) = 1$ if and only if $v_0 = \chi_0 U^{-n(\psi)}$, and in which case $\Phi(1_2, U, W_{v_0}) = (q^{\frac{1}{2}} U)^{-n(\psi)} L(\pi, U)$. In the remaining case where $L(\pi, U)$ has a simple pole, we have $v_0 = \sum_{i=0}^{\infty} b^i \chi_0 U^{-n(\psi)+i}$ and $\Phi(id, U, W_{v_0}) = (q^{\frac{1}{2}} U)^{-n(\psi)} L(\pi, U)$ with $L(\pi, U) = (1 - bq^{\frac{1}{2}} U)^{-1}$. We summarize the above discussion in

Theorem 6. *Let π be an infinite-dimensional admissible irreducible representation of $GL_2(F)$ with central character ω. Let ψ be a nontrivial additive character of F with order $n(\psi)$. Denote by v_0 the normalized new vector of π in its Kirillov model $\mathbf{K}_\pi(\psi)$. Then $L(\pi, U)$ has 2 poles if and only if $\pi = \pi(\mu, \nu)$ is a principal series induced from two unramified quasi-characters μ, ν of $F^\times$, which happens if and only if π is unramified. In this case $v_0 = \sum_{i=0}^{\infty} c_i \chi_0 U^{-n(\psi)+i}$ with $c_0 = 1$, $\lambda c_i = c_{i+1} + q^{-1}\omega(\varpi) c_{i-1}$ for $i \geq 0$, in other words, v_0 is an eigenfunction of the Hecke operator with eigenvalue λ. Further, $L(\pi, U)^{-1} = (1 - \mu(\varpi)U)(1 - \nu(\varpi)U) = 1 - \lambda q^{\frac{1}{2}} U + \omega(\varpi) U^2$. Next, $L(\pi, U)$ has 1 pole if and only if $\pi = \pi(\mu, \nu)$ is a principal series with exactly one of μ, ν unramified or $\pi = \sigma(\mu, \nu)$ with $\mu\nu^{-1} = |\ |^{\pm 1}$*

and μ, ν unramified. The new vector v_0 in this case has the form $\sum_{i=0}^{\infty} b^i \chi_0 U^{-n(\psi)+i}$ with $L(\pi, U)^{-1} = 1 - bq^{\frac{1}{2}}U$. Finally, $L(\pi, U) = 1$ if and only if the new vector $v_0 = \chi_0 U^{-n(\psi)}$. In all cases we have

$$\begin{aligned}\Phi(1_2, U, W_{v_0}) &= \int_{F^\times} W_{v_0}\left(\begin{pmatrix} a & 0 \\ 0 & 1 \end{pmatrix}\right) |a|^{-\frac{1}{2}} U^{\operatorname{ord} a} d^\times a \\ &= \int_{F^\times} v_0(a) |a|^{-\frac{1}{2}} U^{\operatorname{ord} a} d^\times a = (q^{\frac{1}{2}}U)^{-n(\psi)} L(\pi, U).\end{aligned}$$

---

§3 Representations of $GL_2(F)$ for F an archimedean local field

Denote by G the group $GL_2(\mathbf{R})$ or $GL_2(\mathbf{C})$, viewed as the real points of an algebraic group defined over $\mathbf{R}$. A representation (π, H) of G on a separable Hilbert space H is a homomorphism $\pi : G \to GL(H)$, the group of all invertible bounded operators of H, such that the map from $G \times H \to H$ given by $(g, v) \mapsto \pi(g)v$ is continuous. Let K be the standard maximal compact subgroup of G. The inner product

$$\langle v, v' \rangle = \int_K (\pi(k)v, \pi(k)v') dk$$

obtained from averaging over K the original inner product $(\ ,\ )$ on H induces the same topology on H. Hence we may assume that π restricted to K is unitary. Recall that all irreducible unitary representations of K are finite-dimensional. To each class γ of irreducible unitary representation of K, let H_γ be the γ-isotypic subspace of H, that is, H_γ is the Hilbert space direct sum of the subspaces of H on which the action of $\pi(K)$ is isomorphic to γ. The representation (π, H) is said to be *admissible* if H_γ is finite-dimensional for all γ in the set $\widehat{K}$ of classes of irreducible unitary representations of K.

Given an admissible (π, H), write H_0 for the algebraic sum of the H_γ, $\gamma \in \widehat{K}$. In other words, H_0 consists of the vectors v in H such that the $\pi(K)$-invariant subspace generated by v is finite-dimensional. Write $\mathcal{G}$ for the Lie algebra of G. By a theorem of Harish-Chandra, an admissible representation (π, H) of G yields an action π of $\mathcal{G}$ on H_0 as follows:

for $X \in \mathcal{G}$ and $v \in H_0$, $\pi(X)v = \frac{d}{dt}(\pi(\exp tX)v)\ |_{t=0}$ (which again lies in H_0). This action has the following two properties.

(i) If $k \in K$ and $X \in \mathcal{G}$, then $\pi(k)\pi(X)v = \pi(Ad(k)X)\pi(k)v$ for v in H_0;

(ii) For $X, Y \in \mathcal{G}$, $\pi[X, Y] = \pi(X)\pi(Y) - \pi(Y)\pi(X) = [\pi(X), \pi(Y)]$ on H_0.

This leads to the following definition of a $(\mathcal{G}, K)$-module. A $(\mathcal{G}, K)$-module V is a complex vector space V such that

(3.1) V is a $\mathcal{G}$-module, i.e., there is a linear map from $\mathcal{G} \otimes V$ to V given by $X \otimes v \mapsto X \cdot v$ satisfying $[X, Y] \cdot v = X \cdot Yv - Y \cdot Xv$ for all $X, Y \in \mathcal{G}$, $v \in V$.

(3.2) V is a K–module, i.e., there is a map from $K \times V$ to V, linear in V, such that $1 \cdot v = v$ and $k_1 \cdot (k_2 v) = (k_1 k_2) \cdot v$ for all $k_1, k_2 \in K$, $v \in V$.

(3.3) The K–invariant subspace generated by any $v \in V$ is finite–dimensional, and if W is a finite–dimensional K–invariant subspace of V, then it decomposes as a direct sum of irreducible K–submodules and the map $K \times W \to W$ is continuous (hence real analytic).

(3.4) If $X \in \mathcal{K}$, the Lie algebra of K, and $v \in V$, then

$$X \cdot v = \frac{d}{dt} \exp tX \cdot v \,|_{t=0} \,.$$

(3.5) For $k \in K$, $X \in \mathcal{G}$, $v \in V$, $k \cdot X \cdot v = \big(Ad(k)X\big) \cdot (k \cdot v)$.

A $(\mathcal{G}, K)$ module V is *admissible* if the γ–isotypic subspace of V is finite–dimensional for all $\gamma \in \widehat{K}$. It is called *irreducible* if the only $\mathcal{G}$– and K–invariant subspaces are $\{0\}$ and V. The relation between $(\mathcal{G}, K)$–modules and representations of G is clarified by Harish-Chandra and Casselman.

Theorem 7. *(Harish–Chandra) (i) Let (π, H) be an admissible representation of G (so that H_0 is an admissible $(\mathcal{G}, K)$–module). Then (π, H) is irreducible if and only if H_0 is an irreducible $(\mathcal{G}, K)$–module.*

(ii) Let (π_i, H_i), $i = 1, 2$, be two unitary admissible representations of G. Then π_1 and π_2 are unitarily equivalent if and only if the $(\mathcal{G}, K)$–modules $H_{1,0}$ and $H_{2,0}$ are isomorphic.

(iii) Let V be an admissible $(\mathcal{G}, K)$–module. A necessary and sufficient condition for $V = H_0$ from an admissible unitary representation (π, H) of G is that V be unitary.

Hence, the determination of all (unitary) equivalence classes of admissible irreducible unitary representations of G is the same as finding the isomorphism classes of admissible irreducible unitary $(\mathcal{G}, K)$ modules. The following result of Casselman says that the same is true if we drop "unitarity".

Theorem 8. *(Casselman) Let V be an admissible irreducible $(\mathcal{G}, K)$–module. Then V is isomorphic to H_0 from some admissible irreducible representation (π, H) of G.*

The discussion above is also valid for G a real reductive Lie group. For more details, the reader is referred to the article by Wallach in [19].

Thus instead of studying representations of G, we shall study $(\mathcal{G}, K)$–modules. Other equivalent ways include studying representations of the universal enveloping algebra $\mathcal{U}(\mathcal{G})$ of $\mathcal{G}$, and the representations of the Hecke algebra $\mathcal{H}$ of G. Sometimes, by abuse of language, we shall call them representations of G.

We summarize below main results for representations of $G = GL_2(\mathbf{R})$ and $GL_2(\mathbf{C})$. The reader is referred to [10] for more details. By a representation of

G we shall mean an admissible irreducible representation as in the previous section. We shall find that the main statements hold without much changes.

Recall that a quasi-character μ of $\mathbf{R}^\times$ is given by $\mu(t) = |t|_{\mathbf{R}}^r(\operatorname{sgn} t)^m$, where $r \in \mathbf{C}$ and $m = 0$ or 1, while a quasi-character ω of $\mathbf{C}^\times$ can be written in the form $\omega(z) = |z|_{\mathbf{C}}^r z^m \bar{z}^n$, where $r \in \mathbf{C}$ and m, n are non-negative integers with product $mn = 0$. (Recall that $|z|_{\mathbf{C}} = z\bar{z}$.) A pair of quasi-characters μ, ν of $\mathbf{R}^\times$ induces an irreducible representation $\pi = \pi(\mu, \nu)$ of $GL_2(\mathbf{R})$ if $\mu\nu^{-1}(t) \neq |t|_{\mathbf{R}}^p \operatorname{sgn} t$, where $p \in \mathbf{Z} - \{0\}$. In the case that $\mu\nu^{-1}(t) = |t|_{\mathbf{R}}^p \operatorname{sgn} t$ for some nonzero integer p, the induced representation is not irreducible, but it contains a unique infinite-dimensional irreducible constituent, denoted by $\sigma(\mu, \nu)$. Using the method of Weil representations, one also gets, for each quasi-character ω of $\mathbf{C}^\times$, an infinite-dimensional irreducible representation $\pi(\omega)$ of $GL_2(\mathbf{R})$. It can be shown that $\pi(\omega)'s$ include all $\sigma(\mu, \nu)'s$, and further, any infinite-dimensional representation of $GL_2(\mathbf{R})$ is either a $\pi(\mu, \nu)$ or a $\sigma(\mu, \nu)$.

Let $\psi(x) = e^{2\pi i u x}$ be a nontrivial additive character of $\mathbf{R}$. It follows from the local theory for representations of GL_1 as developed in Tate's thesis [18] that for each quasi-character $\mu(t) = |t|_{\mathbf{R}}^r(\operatorname{sgn} t)^m$ of $\mathbf{R}^\times$, there are attached L-factor

$$L(\mu, s) = \Gamma_{\mathbf{R}}(s + r + m) \quad \text{where } \Gamma_{\mathbf{R}}(s) = \pi^{-\frac{s}{2}}\Gamma(\frac{s}{2}),$$

and ε-factor

$$\varepsilon(\mu, \psi, s) = (i \operatorname{sgn} u)^m |u|_{\mathbf{R}}^{s+r-\frac{1}{2}};$$

and for each quasi-character $\omega(z) = |z|_{\mathbf{C}}^r z^m \bar{z}^n$ of $\mathbf{C}^\times$ as above, the attached L-factor and ε-factor are

$$L(\omega, s) = 2\Gamma_{\mathbf{C}}(s + r + m + n), \quad \text{where } \Gamma_{\mathbf{C}}(s) = (2\pi)^{-s}\Gamma(s),$$

$$\varepsilon(\omega, \varphi, s) = i^{m+n}\omega(w)|w|_{\mathbf{C}}^{s-\frac{1}{2}}, \quad \text{where } \varphi(z) = e^{4\pi i Re(wz)} = e^{2\pi i Tr(wz)}.$$

Define the L- and ε-factors attached to a representation π of $GL_2(\mathbf{R})$ as follows :

$$\begin{aligned} \text{For } \pi = \pi(\mu, \nu), \quad & L(\pi, s) = L(\mu, s)L(\nu, s) \\ & \varepsilon(\pi, \psi, s) = \varepsilon(\mu, \psi, s)\varepsilon(\nu, \psi, s); \\ \text{For } \pi = \pi(\omega), \quad & L(\pi, s) = L(\omega, s) \\ & \varepsilon(\pi, \psi, s) = (i \operatorname{sgn} u)\varepsilon(\omega, \psi \circ \operatorname{Tr}_{\mathbf{C}/\mathbf{R}}, s), \end{aligned}$$

where $\psi(x) = e^{2\pi i u x}$, $u \in \mathbf{R}^\times$. In the case that $\pi(\omega) = \pi(\mu, \nu)$, the two definitions agree, as a consequence of the complement formula satisfied by the Γ-function.

Jacquet and Langlands showed that the class of an infinite-dimensional representation π of $GL_2(\mathbf{R})$ is determined by the γ-function

$$\gamma_\pi(\chi, \psi, s) = \frac{L(\tilde{\pi} \otimes \chi^{-1}, 1 - s)}{L(\pi \otimes \chi, s)}\varepsilon(\pi \otimes \chi, \psi, s), \quad \chi \text{ a quasi-character of } \mathbf{R}^\times.$$

Further, the Whittaker model and local functional equation exist as before. More precisely,

Theorem 9. *(Local theory for representations of $GL_2(\mathbf{R})$ [10]) Let π be an infinite-dimensional admissible irreducible representation of $GL_2(\mathbf{R})$. Let ψ be a nontrivial additive character of $\mathbf{R}$, and ω be the central character of π. Then*

(1) *There exists exactly one space $\mathbf{W}_\pi(\psi)$ of Whittaker functions on $GL_2(\mathbf{R})$ satisfying*

 (a) *For $W \in \mathbf{W}_\pi(\psi)$, $W\left(\begin{pmatrix} 1 & x \\ 0 & 1 \end{pmatrix} g\right) = \psi(x)W(g)$, $x \in \mathbf{R}$, $g \in GL_2(\mathbf{R})$.*

 (b) *For $W \in \mathbf{W}_\pi(\psi)$, there is a positive number N so that*

$$W\left(\begin{pmatrix} t & 0 \\ 0 & 1 \end{pmatrix}\right) = O\left(|t|_{\mathbf{R}}^N\right) \quad \text{as } |t|_{\mathbf{R}} \to \infty,$$

 (c) *Each $W \in \mathbf{W}_\pi(\psi)$ is a continuous function, and the action of $(\mathcal{G}, K)$ on $\mathbf{W}_\pi(\psi)$ with K acting by right translations and $X \in \mathcal{G}$ acting by $X \cdot W(g) = \frac{d}{dt}W(g \exp tX)\mid_{t=0}$ is isomorphic to the $(\mathcal{G}, K)$-module obtained from π.*

(2) *For $W \in \mathbf{W}_\pi(\psi)$, define*

$$\Phi(g, s, W) = \int_{\mathbf{R}^\times} W\left(\begin{pmatrix} a & 0 \\ 0 & 1 \end{pmatrix} g\right) |a|^{s-\frac{1}{2}} d^\times a,$$

$$\widetilde{\Phi}(g, s, W) = \int_{\mathbf{R}^\times} W\left(\begin{pmatrix} a & 0 \\ 0 & 1 \end{pmatrix} g\right) \omega^{-1}(a)|a|^{s-\frac{1}{2}} d^\times a.$$

Then $\Phi, \widetilde{\Phi}$ are absolutely convergent for $\mathrm{Re}\, s$ large enough. Further, the quotients $\Phi(g, s, W)/L(\pi, s)$ and $\widetilde{\Phi}(g, s, W)/L(\widetilde{\pi}, s)$ can be analytically continued to the whole s-plane as meromorphic functions, and satisfy the functional equation

$$\frac{\widetilde{\Phi}(wg, 1-s, W)}{L(\widetilde{\pi}, 1-s)} = \varepsilon(\pi, \psi, s)\frac{\Phi(g, s, W)}{L(\pi, s)} \quad \text{for } g \in GL_2(\mathbf{R}),\ W \in \mathbf{W}_\pi(\psi).$$

Moreover, with W fixed, $\Phi(g, s, W)$ remains bounded as g varies in a compact set and s varies in the region obtained by removing discs centered at the poles of $L(s, \pi)$ from a vertical strip of finite width. Finally, there is a function W in $\mathbf{W}_\pi(\psi)$ such that $\Phi(1_2, s, W) = L(\pi, s)$ times an exponential function of s.

The representation theory for $GL_2(\mathbf{C})$ is similar to $GL_2(\mathbf{R})$, but simpler. Every infinite-dimensional representation π of $GL_2(\mathbf{C})$ is a $\pi(\mu, \nu)$ induced from two quasi-characters μ, ν of $\mathbf{C}^\times$. The L- and ε-factors attached to $\pi = \pi(\mu, \nu)$ are

$$L(\pi, s) = L(\mu, s)L(\nu, s),$$
$$\varepsilon(\pi, \psi, s) = \varepsilon(\mu, \psi, s)\varepsilon(\nu, \psi, s).$$

The same statements as in Theorem 9 with $\mathbf{R}$ replaced by $\mathbf{C}$ hold.

§4 Automorphic representations of GL_2

Let F be a global field. Write $GL_2(A_f)$ for the group of finite adèlic points of $GL_2(A_F)$. It is the restricted product of $\{GL_2(F_v)\}$ with respect to $\{GL_2(\mathcal{O}_v)\}$ as v runs through all finite places of F. Denote by $GL_2(A_\infty)$ the product $\prod_{\sigma \text{ arch.}} GL_2(F_\sigma)$, which is trivial if F is a function field. Clearly, the Lie algebra $\mathcal{G}_\infty$ of $GL_2(A_\infty)$ is $\prod_\sigma \mathcal{G}_\sigma$, the product of the Lie algebra $\mathcal{G}_\sigma$ of $GL_2(F_\sigma)$, and put K_∞ to be $\prod_\sigma K_\sigma$, the product of the standard maximal compact subgroup of $GL_2(F_\sigma)$. A representation of $GL_2(A_f)$ is *smooth* if every vector in the representation space is stabilized by an open compact subgroup of $GL_2(A_f)$. A representation V of $GL_2(A_F)$ is both a $(\mathcal{G}_\infty, K_\infty)$–module and a $GL_2(A_f)$–module such that

(4.1) The action of $GL_2(A_f)$ commutes with the action of $\mathcal{G}_\infty$ and K_∞;

(4.2) For each class of continuous irreducible representation γ of $K = K_\infty K_f$, where $K_f = \prod_{v \text{finite}} GL_2(\mathcal{O}_v)$, the γ–isotypic subspace V_γ of V is finite–dimensional.

An admissible irreducible $(\mathcal{G}_\infty, K_\infty)$–module decomposes as a tensor product over the archimedean places σ of F of admissible irreducible $(\mathcal{G}_\sigma, K_\sigma)$–modules. But a smooth representation V_f of $GL_2(A_f)$ requires a further explanation.

It follows from the definition of smoothness that every vector in V_f is stabilized by $GL_2(\mathcal{O}_v)$ for almost all v. This observation leads to the following construction. Suppose that, at each finite place v of F, a representation π_v of $GL_2(F_v)$ on space V_v is given such that the space of $GL_2(\mathcal{O}_v)$ invariant vectors is nontrivial for almost all v. Denote by Σ_0 the set of exceptional finite places, namely, those finite places for which there are no nontrivial $GL_2(\mathcal{O}_v)$ invariant vectors. For each finite place $v \notin \Sigma_0$, fix a nonzero vector v_v invariant by $GL_2(\mathcal{O}_v)$. For each finite set S of finite places of F containing Σ_0, put $V_S = \bigotimes_{v \in S} V_v$.

If S and S' are two such sets with $S \subset S'$, imbed V_S into $V_{S'}$ by identifying $x \in V_S$ with $x \bigotimes_{v \in S' - S} v_v$. The projective limit $\varinjlim_S V_S$ as S runs through all finite subsets of finite places of F containing Σ_0 is called the restricted tensor product of $\{V_v\}$ with respect to $\{v_v\}$, denoted by $\bigotimes_{\{v_v\}}{}' V_v$. It is generated by vectors $\bigotimes_v x_v$, $x_v \in V_v$ and $x_v = v_v$ for almost all v. The group $GL_2(A_f)$ acts on $\bigotimes_{\{v_v\}}{}' V_v$ in the obvious way via π_v acting on V_v. Although the space $\bigotimes_{\{v_v\}}{}' V_v$ depends on the choice of v_v, the resulting $GL_2(A_f)$–modules are isomorphic if we change the base vectors v_v. Call this representation the restricted tensor product $\bigotimes_v' \pi_v$ of $\{\pi_v\}$. It can be shown that if each π_v is admissible irreducible, then $\bigotimes_v' \pi_v$ is smooth and irreducible, in fact, it is also admissible, that is, for each open compact subgroup H of K_f, the space of vectors stabilized by H is finite–dimensional. Therefore, given admissible irreducible $(\mathcal{G}_\sigma, K_\sigma)$–modules π_σ for σ archimedean and admissible irreducible $GL_2(F_v)$–modules π_v for v finite, the restricted tensor product $\pi = \bigotimes_{\sigma} \pi_\sigma \otimes \bigotimes_{v \text{ finite}}{}' \pi_v = \bigotimes_{\text{all } v}{}' \pi_v$ is an admissible irreducible representation of $GL_2(A_F)$. Conversely, it can be shown that every admissible irreducible representation of $GL_2(A_F)$ arises this way. See [5] for further discussion on the decomposition of a global representations into restricted tensor product of local representations.

By the regular representation of $GL_2(A_F)$ on the space $\mathcal{A}(GL_2(A_F), \omega)$ of automorphic functions on $GL_2(A_F)$ with central character ω we mean that $K_\infty GL_2(A^f)$ acts by right translations and $\mathcal{G}_\infty$ acts infinitesimally, that is, $X \in \mathcal{G}_\infty$ sends an automorphic form f to $X \cdot f(g) = \frac{d}{dt} f(g \exp tX) \mid_{t=0}$. An admissible irreducible representation π of $GL_2(A_F)$ is called an *automorphic representation* if there are two $GL_2(A_F)$–invariant subspaces V and W of $\mathcal{A}(GL_2(A_F), \omega)$, $V \subset W$, such that π is isomorphic to the induced action on the quotient space W/V. We also say that π is a *constituent* of the regular representation. When $V = \{0\}$, π is called a subrepresentation, and when $V \neq \{0\}$, π is called a subquotient. If V and W are contained in the space $\mathcal{A}_0(GL_2(A_F), \omega)$ of cusp forms, then π is called a *cuspidal representation* of $GL_2(A_F)$.

We summarize below main results on automorphic representations of GL_2, the reader is referred to the Lecture Notes by Jacquet and Langlands [10] and articles in the proceedings of the conference held in Corvallis, Oregan 1977, on automorphic forms [1].

Theorem 10. *([10, p.340]) Let $\pi = \bigotimes_v{}' \pi_v$ be an admissible irreducible automorphic representation of $GL_2(A_F)$. Suppose that it is not cuspidal. Then there are two idèle class characters μ, ν of $I_F/F^\times$ such that, at each place v, the representation π_v is a constituent of the induced representation $\rho(\mu_v, \nu_v)$.*

This is the representation–theoretic analogue of the statement that the space of modular forms with given parameters decomposes into a direct sum of the space of Eisenstein series and the space of cusp forms.

Jacquet and Langlands further showed that the space $\mathcal{A}_0(GL_2(A_F), \omega)$ of cusp forms decomposes as a direct sum of irreducible cuspidal representations, each occurring with finite multiplicity. In other words, cuspidal representations occur as subrepresentations. Our main concern is the cuspidal representations. Let φ be a nonzero cusp form occurring in the space of a cuspidal representation $\pi = \bigotimes_v{}' \pi_v$. If F is a function field, denote by X the set of elements $y \in A_F$ such that φ is right invariant by $\begin{pmatrix} 1 & y \\ 0 & 1 \end{pmatrix}$. Then X is an open subgroup of A_F. By Corollary 4 in Chapter 4, there is a constant $c_1 > 0$ such that $A_F = F + aX$ for any idèle $a \in I_F$ with $|a| > c_1$. This implies that, for any $z = ay$ where $y \in X$, $\varphi\left(\begin{pmatrix} 1 & z \\ 0 & 1 \end{pmatrix}\begin{pmatrix} a & x \\ 0 & 1 \end{pmatrix}\right) = \varphi\left(\begin{pmatrix} a & x \\ 0 & 1 \end{pmatrix}\begin{pmatrix} 1 & y \\ 0 & 1 \end{pmatrix}\right) = \varphi\begin{pmatrix} a & x \\ 0 & 1 \end{pmatrix}$. Further, the formula also holds with $z \in F$ and hence for all $z \in A_F$. Thus

$$\varphi\left(\begin{pmatrix} a & x \\ 0 & 1 \end{pmatrix}\right) = \frac{1}{\text{meas}\,(A_F/F)} \int_{A_F/F} \varphi\left(\begin{pmatrix} 1 & z \\ 0 & 1 \end{pmatrix}\begin{pmatrix} a & x \\ 0 & 1 \end{pmatrix}\right) dz$$
$$= 0 \qquad \text{for } |a| > c_1, \quad x \in A_F.$$

This shows that φ is compactly supported on $GL_2(F)\mathcal{Z}(A_F)\backslash GL_2(A_F)$. Moreover, as a function in $a \in I_F$, $\varphi\left(\begin{pmatrix} a & 0 \\ 0 & 1 \end{pmatrix}\right)$ is compactly supported in $I_F/F^\times$. This is because $\varphi\left(\begin{pmatrix} a & 0 \\ 0 & 1 \end{pmatrix}\right) = \varphi\left(w\begin{pmatrix} a & 0 \\ 0 & 1 \end{pmatrix}\right) = \omega(a)^{-1}\varphi\left(\begin{pmatrix} a^{-1} & 0 \\ 0 & 1 \end{pmatrix} w\right)$ and the right translation of φ by w is also a cusp form, thus there is a constant $c_2 > 0$ such that $\varphi\left(\begin{pmatrix} a^{-1} & 0 \\ 0 & 1 \end{pmatrix} w\right) = 0$ if $|a^{-1}| > c_2$, i.e., $|a| < c_2^{-1}$. If F is a number field, the corresponding statement is

(*Growth condition*) for any real number M_1 there is a real number M_2 such that

$$\left|\varphi\left(\begin{pmatrix} a & 0 \\ 0 & 1 \end{pmatrix}\right)\right| \le M_2|a|^{M_1} \quad \text{for all } a \in I_F.$$

As a consequence of the growth condition on φ, each π_v is infinite–dimensional.

Let $\pi = \bigotimes_v{}' \pi_v$ be an automorphic irreducible representation of $GL_2(A_F)$ such that all components π_v are infinite–dimensional. Let $\psi = \prod_v \psi_v$ be a nontrivial additive character of A_F trivial on F. As discussed in §§2 and 3, each π_v has a Whittaker model $\mathbf{W}_{\pi_v}(\psi_v)$. Let $\mathbf{W}_\pi(\psi)$ be the restricted tensor product of $\{\mathbf{W}_{\pi_v}(\psi_v)\}$ with respect to the normalized new vectors $\{W_v\}$ chosen at the places v where π_v is unramified. Then each $W \in \mathbf{W}_\pi(\psi)$ is a continuous function on $GL_2(A_F)$ satisfying

$$W\left(\begin{pmatrix} 1 & x \\ 0 & 1 \end{pmatrix} g\right) = \psi(x)W(g) \quad \text{for all } x \in A_F, \quad g \in GL_2(A_F).$$

Further, the action of $GL_2(A_F)$ on $\mathbf{W}_\pi(\psi)$ is isomorphic to π. One can show that such a space $\mathbf{W}_\pi(\psi)$ is unique and, moreover, it does not exist if one component π_v is finite–dimensional.

For π as above, we defined in §§2 and 3 local L and ε–factors, Now we put them together to define global L– and ε–factors. Let

$$L(\pi, s) = \prod_v L(\pi_v, s), \qquad L(\tilde{\pi}, s) = \prod_v L(\tilde{\pi}_v, s),$$

where at a nonarchimedean place v, $L(\pi_v, s)$ and $L(\tilde{\pi}_v, s)$ stand for $L(\pi_v, (Nv)^{-s})$ and $L(\tilde{\pi}_v, (Nv)^{-s})$), respectively, and

$$\varepsilon(\pi, \psi, s) = \prod_v \varepsilon(\pi_v, \psi_v, s)$$

with $\varepsilon(\pi_v, \psi_v, s)$ representing $\varepsilon(\pi_v, \psi_v, (Nv)^{-s})$ for v finite. As the central character ω of π is an idèle class character, this fact together with how a local ε–factor varies with respect to the change of additive character implies that $\varepsilon(\pi, \psi, s)$ is independent of the choice of ψ as long as it is trivial on F (and nontrivial on A_F), hence we shall write it as $\varepsilon(\pi, s)$. The analytic properties of L are summarized in the following theorem, proved by Jacquet and Langlands in [10], p. 350.

Theorem 11. *(Global functional equation) Let $\pi = \bigotimes_v{}' \pi_v$ be an admissible irreducible automorphic representation of $GL_2(A_F)$. Then the $L(\pi, s)$, $L(\widetilde{\pi}, s)$ defined above as infinite products converge absolutely in a right half plane, and they can be analytically continued to the whole s–plane as meromorphic functions. They are entire if π is a cuspidal representation. If F is a number field, they have only finitely many poles and are bounded at infinity in each vertical strip of finite width. If F is a function field with q elements in its field of constants, then they are rational functions in q^{-s}. Finally, they satisfy the functional equation*

$$L(\pi, s) = \varepsilon(\pi, s) L(\widetilde{\pi}, 1 - s).$$

We first examine the case where π is not cuspidal. By Theorem 10, there are two quasi–characters μ, ν of $I_F/F^\times$ such that π_v is a constituent of $\rho(\mu_v, \nu_v)$. Since π_v is unramified for almost all v, we have $\pi_v = \pi(\mu_v, \nu_v)$ for almost all v. Assume that $\pi_v = \pi(\mu_v, \nu_v)$ for all v. Then

$$L(\pi, s) = L(\mu, s)L(\nu, s), \qquad L(\widetilde{\pi}, s) = L(\mu^{-1}, s)L(\nu^{-1}, s),$$

the stated analytic properties follow from the analytic properties of the L–function attached to idèle class characters. For F a function field, this was proved in Chapter 5, and for F a number field, this can be found in Tate thesis [18]. If $\pi_v = \sigma(\mu_v, \nu_v)$ for finitely many places v, then $L(\pi, s)$ differs from $L(\mu, s)L(\nu, s)$ by finitely many factors and $L(\pi, s)L(\mu, s)^{-1}L(\nu, s)^{-1}$ is holomorphic, which, in the case that F is a number field, is bounded in each vertical strip of finite width, and in the case that F is a function field, is a polynomial in q^{-s}. Moreover, it follows from the definition of local ε–factor that

$$\frac{L(\pi, s)}{L(\mu, s)L(\nu, s)} = \frac{\varepsilon(\pi, s)}{\varepsilon(\mu, s)\varepsilon(\nu, s)} \frac{L(\widetilde{\pi}, 1 - s)}{L(\mu^{-1}, 1 - s)L(\nu^{-1}, 1 - s)}.$$

Hence Theorem 11 also holds for this case.

Next suppose that π is cuspidal, occurring in $\mathcal{A}_0(GL_2(A_F), \omega)$. Let φ be a cusp form in the space of π. We saw from §1 that its Fourier transform

$$W(g) = \int_{F \backslash A_F} \varphi\left(\begin{pmatrix} 1 & x \\ 0 & 1 \end{pmatrix} g\right) \psi(-x) dx, \qquad g \in GL_2(A_F),$$

is a Whittaker function in the space $\mathbf{W}_\pi(\psi)$, and further,

$$\varphi(g) = \sum_{\alpha\in F^\times} W\left(\begin{pmatrix}\alpha & 0\\ 0 & 1\end{pmatrix} g\right) \qquad \text{for all } g\in GL_2(A_F).$$

Hence we work on the space $\mathbf{W}_\pi(\psi)$ of Whittaker functions. At each finite place v, let W_v be the normalized new vector in $\mathbf{W}_{\pi_v}(\psi_v)$ so that

$$\int_{F_v^\times} W_v\left(\begin{pmatrix}a_v & 0\\ 0 & 1\end{pmatrix}\right)|a_v|_v^{s-\frac{1}{2}}d^\times a_v = (Nv)^{-n(\psi_v)(\frac{1}{2}-s)}L(\pi_v,(Nv)^{-s}),$$

where $n(\psi_v)$ is the order of ψ_v, and at each archimedean place v, let W_v be a function in $\mathbf{W}_{\pi_v}(\psi_v)$ such that

$$\int_{F_v^\times} W_v\left(\begin{pmatrix}a_v & 0\\ 0 & 1\end{pmatrix}\right)|a_v|_v^{s-\frac{1}{2}}d^\times a_v = (\text{exponential function in } s)L(\pi_v,s).$$

Put $W = \bigotimes_v W_v$, which lies in $\mathbf{W}_\pi(\psi)$. Let $\varphi(g) = \sum_{\alpha\in F^\times} W\left(\begin{pmatrix}\alpha & 0\\ 0 & 1\end{pmatrix} g\right)$, $g\in GL_2(A_F)$, which is a cusp form. Denote by $d^\times a$ the Haar measure on I_F arising from $\prod_v d^\times a_v$. Consider the integral

$$\begin{aligned}\Phi(1_2,s,W) &= \int_{F^\times\backslash I_F}\varphi\left(\begin{pmatrix}a & 0\\ 0 & 1\end{pmatrix}\right)|a|^{s-\frac{1}{2}}d^\times a\\ &= \int_{F^\times\backslash I_F}\sum_{\alpha\in F^\times} W\left(\begin{pmatrix}\alpha & 0\\ 0 & 1\end{pmatrix}\begin{pmatrix}a & 0\\ 0 & 1\end{pmatrix}\right)|a|^{s-\frac{1}{2}}d^\times a\\ &= \int_{I_F} W\left(\begin{pmatrix}a & 0\\ 0 & 1\end{pmatrix}\right)|a|^{s-\frac{1}{2}}d^\times a\\ &= \prod_v\int_{F_v^\times} W_v\left(\begin{pmatrix}a_v & 0\\ 0 & 1\end{pmatrix}\right)|a_v|_v^{s-\frac{1}{2}}d^\times a_v\\ &= \prod_v \Phi_v(1_2,s,W_v)\\ &= (\text{exponential function in } s)L(\pi,s).\end{aligned}$$

Here the last equality follows from the fact that $n(\psi_v)=0$ for almost all v. The growth condition satisfied by φ not only shows that $L(\pi,s)$ as defined by the infinite product converges absolutely for $\operatorname{Re}s \gg 0$, but also implies that the integral of φ over $F^\times\backslash I_F$ actually defines a holomorphic function on the s-plane, which is bounded in each vertical strip of finite width if F is a number field, and is a finite Laurent series in q^{-s} if F is a function field. In fact, when F is a function field, we get $L(\pi,s) = (q^{-s})^M$ times a polynomial in q^{-s} with nonzero constant term. By taking into account the behavior that $L(\pi,s)\to 1$ as $\operatorname{Re}s\to\infty$, we conclude that $M = 0$, that is, $L(\pi, s)$ is a polynomial in q^{-s} with nonzero constant term. We remark that the absolute convergence of $L(\pi, s)$ for $\mathrm{Re}\, s \gg 0$ can also be seen from the fact that $\pi_v = \pi(\mu_v, \nu_v)$ with unramified quasi-characters μ_v, ν_v of $F_v^\times$ satisfying

$$|\omega_v(\varpi_v)|^{1/2}(Nv)^{-\frac{1}{2}} \le |\mu_v(\varpi_v)|, \qquad |\nu_v(\varpi_v)| \le |\omega_v(\varpi_v)|^{1/2}(Nv)^{1/2}$$

for almost all v.

Similarly, with $w = \begin{pmatrix} 0 & 1 \\ -1 & 0 \end{pmatrix}$, the integral

$$\begin{aligned}
\widetilde{\Phi}(w, s, W) &= \int_{F^\times \backslash I_F} \varphi\left(\begin{pmatrix} a & 0 \\ 0 & 1 \end{pmatrix} w\right) \omega(a)^{-1} |a|^{s-\frac{1}{2}} d^\times a \\
&= \prod_v \int_{F_v^\times} W_v\left(\begin{pmatrix} a_v & 0 \\ 0 & 1 \end{pmatrix} w_v\right) \omega_v(a_v)^{-1} |a_v|_v^{s-\frac{1}{2}} d^\times a_v \\
&= \prod_v \widetilde{\Phi}_v(w_v, s, W_v) \\
&= (\text{exponential function in } s) L(s, \widetilde{\pi})
\end{aligned}$$

gives a holomorphic continuation of $L(\widetilde{\pi}, s)$, $\widetilde{\pi}$ being the contragredient of π. To get the functional equation, we combine the local theory studied in §§2 and 3 and the automorphicity of φ. More precisely, it follows from

$$\varphi\left(\begin{pmatrix} a & 0 \\ 0 & 1 \end{pmatrix}\right) = \varphi\left(w \begin{pmatrix} a & 0 \\ 0 & 1 \end{pmatrix}\right) = \varphi\left(\begin{pmatrix} 1 & 0 \\ 0 & a \end{pmatrix} w\right) = \omega(a) \varphi\left(\begin{pmatrix} a^{-1} & 0 \\ 0 & 1 \end{pmatrix} w\right)$$

that $\Phi(1_2, s, W) = \widetilde{\Phi}(w, 1-s, W)$. On the other hand, by Theorems 4 and 9, we have

$$\frac{\widetilde{\Phi}_v(w_v, 1-s, W_v)}{L(\widetilde{\pi}_v, 1-s)} = \varepsilon(\pi_v, \psi_v, s) \frac{\Phi_v(1_2, s, W_v)}{L(\pi_v, s)} \quad \text{for all places } v.$$

Now we obtain the desired functional equation by taking product over all places v of the above equation and cancelling $\prod_v \widetilde{\Phi}_v(w_v, 1-s, W_v) = \widetilde{\Phi}(w, 1-s, W)$ with $\prod_v \Phi_v(1_2, s, W_v) = \Phi(1_2, s, W)$. This completes the proof of Theorem 11.

Remark. When $F = \mathbf{Q}$, choose ψ to be the standard additive character of $A_{\mathbf{Q}}/\mathbf{Q}$ so that $n(\psi_v) = 0$ for all finite places v. There is a positive integer N such that $\varepsilon(\pi, s) = \mathrm{const} \cdot N^{-s}$. Call N the conductor of the representation π. In the case that π_∞ is a discrete series representation of $GL_2(\mathbf{R})$ with the action of $r(\theta) \in K_\infty$ being multiplication by $r(k\theta)$, the resulting function φ arising from $W = \bigotimes_v W_v$ with W_v normalized new vector for v finite as above is a newform of weight k level N and character ω. Moreover, the L-function attached to φ', normalized φ as defined in the preceding chapter, is the L-function attached to π under a change of variable: $L(\pi, s) = L(\varphi', s + \frac{k-1}{2})$. Conversely, given a normalized (cuspidal) newform φ' there is a unique cuspidal representation π of $GL_2(A_{\mathbf{Q}})$ such that $L(\pi, s) = L(\varphi', s + \frac{k-1}{2})$. The "push-ups" of φ' defined in Chapter 7 all lie in the space of the corresponding π.

Just like what was asked of modular forms, we can ask the following "converse question": given, at each place v of F, an admissible irreducible representation π_v of $GL_2(F_v)$ such that at almost all places v, $\pi_v = \pi(\mu_v, \nu_v)$ is unramified with $(Nv)^{-c} \leq |\mu_v(\varpi_v)|$, $|\nu_v(\varpi_v)| \leq (Nv)^c$ for some constant c independent of v, so that the restricted tensor product $\pi = \bigotimes_v' \pi_v$ is an admissible irreducible representation of $GL_2(A_F)$ and the associated $L(\pi, s)$ converges absolutely on a right half-plane. We would like to know when π is an automorphic representation. The answer for a cuspidal representation in terms of the analytic behavior of the L-function attached to π as well as its twists by all characters of $I_F/F^\times$ is given by Jacquet and Langlands [10] as follows.

Theorem 12. *(Converse theorem for GL_2) The representation π defined above is a cuspidal representation of $GL_2(A_F)$ if and only if for all characters χ of $I_F/F^\times$, the L-functions $L(\pi \otimes \chi, s)$ and $L(\tilde{\pi} \otimes \chi^{-1}, s)$ have holomorphic continuations to the whole s-plane, are bounded in each vertical strip of finite width if F is a number field, and are polynomials in q^{-s} if F is a function field with q elements in its field of constants, and satisfy the functional equation*

$$L(\pi \otimes \chi, s) = \varepsilon(\pi \otimes \chi, s) L(\tilde{\pi} \otimes \chi^{-1}, 1 - s).$$

Two remarks are in order here. The first is that only the analytic behaviors of L attached to $\pi \otimes \chi$ with conductor χ dividing the conductor of π are needed for sufficiency. Secondly, a similar criterion holds for automorphic, but noncuspidal representations. For more details, see [14].

We end this section by pointing out that Jacquet and Langlands actually proved

Theorem 13. *(Multiplicity one theorem) Let π be an irreducible cuspidal representation occurring in $\mathcal{A}_0(GL_2(A_F), \omega)$. Then it occurs there with multiplicity one.*

In fact, a stronger result holds :

Theorem 14. *(Strong multiplicity one theorem) Let $\pi_i = \bigotimes_v' \pi_{v,i}$, $i = 1, 2$, be two irreducible cuspidal representations occurring in $\mathcal{A}_0(GL_2(A_F), \omega)$. Suppose that $\pi_{v,1}$ and $\pi_{v,2}$ are isomorphic for almost all places v of F. Then $\pi_{v,1}$ and $\pi_{v,2}$ are isomorphic for all v. Consequently, π_1 and π_2 are isomorphic.*

This theorem is also valid for representations of GL_N (due to Shalika [17]) provided that $\pi_{v,1}$ and $\pi_{v,2}$ are isomorphic at all archimedean places if $n \geq 3$. The reader is referred to the article by Piatetski-Shapiro [16]. C. Moreno, using analytical method, has further strengthened the above theorem by requiring only the two global representations to agree locally at a suitable finite set of nonarchimedean places [15].

---

§5 Representations of quaternion groups

Let F be a field with characteristic $\neq 2$. Given two nonzero elements a, b of F, denote by $F\{a,b\}$ the degree 4 algebra over F generated by two elements i, j satisfying $i^2 = a$, $j^2 = b$ and $ij = -ji$. It is a simple algebra with center F.

Example 1. $F = \mathbf{R}$, $a = b = -1$. Then $\mathbf{R}\{-1,-1\}$ is the familiar Hamiltonian quaternion algebra.

Example 2. $a = 1$, $b = -1$. Let $i = \begin{pmatrix} 1 & 0 \\ 0 & -1 \end{pmatrix}$, $j = \begin{pmatrix} 0 & 1 \\ -1 & 0 \end{pmatrix}$. We have $i^2 = \begin{pmatrix} 1 & 0 \\ 0 & 1 \end{pmatrix}$, $j^2 = \begin{pmatrix} -1 & 0 \\ 0 & -1 \end{pmatrix}$, $ij = \begin{pmatrix} 0 & 1 \\ 1 & 0 \end{pmatrix} = -ji$. Thus $F\{1,-1\}$ is isomorphic to the matrix algebra $M_2(F)$.

In general it can be shown that if $F\{a,b\}$ contains a nontrivial zero divisor, then it is isomorphic to the matrix algebra $M_2(F)$; otherwise, every nonzero element has a multiplicative inverse and the algebra $F\{a,b\}$ is called a quaternion algebra. In the case that F is a local field and $F \neq \mathbf{C}$, there is only one quaternion algebra up to isomorphism.

Next consider the case where F is a global field and $a, b \in F^\times$. Then a, b are also in the completion $F_v^\times$ for all places v of F. Hence we have $F\{a,b\}$ and $F_v\{a,b\}$. Clearly, if $F\{a,b\}$ is isomorphic to $M_2(F)$, then each $F_v\{a,b\}$ is isomorphic to $M_2(F_v)$. On the other hand, if $F\{a,b\}$ is a quaternion algebra, then $F_v\{a,b\}$ may or may not be a quaternion algebra. In fact, it is described by the Hilbert symbol $(a,b)_v$ defined as

$$(a,b)_v = \begin{cases} 1 & \text{if } ax^2 + by^2 = z^2 \text{ has a nontrivial solution in } F_v, \\ -1 & \text{otherwise.} \end{cases}$$

Exercise 4. Show that $ax^2 + by^2 = z^2$ has a nontrivial solution in F_v if and only if $a \in N_{F_v(\sqrt{b})/F_v}(F_v(\sqrt{b}))$, which is equivalent to $b \in N_{F_v(\sqrt{a})/F_v}(F_v(\sqrt{a}))$.

It can be shown that $(a,b)_v = -1$ if and only if $F_v\{a,b\}$ is a quaternion algebra, in which case we say that the global quaternion algebra $F\{a,b\}$ is ramified at v. Since $(a,b)_v \neq 1$ only at finitely many places v, and the product formula

$$\prod_{\text{all places } v} (a,b)_v = 1$$

holds, we conclude that $F\{a,b\}$ is ramified at even number of places. Conversely, given a global field F and a finite set S of places of F with even cardinality and such that S does not contain any complex places of F, then, up to isomorphism, there is exactly one quaternion algebra over F which is ramified at the places in S and unramified outside S. For example, the Hamiltonian algebra $\mathbf{Q}\{-1,-1\}$ is ramified at ∞ and 2, and unramified elsewhere.

Exercise 5. Granting the above assertions, prove

(i) if a is a square in F, then $F\{a,b\}$ is isomorphic to $M_2(F)$;

(ii) if $F\{a,b\}$ is a quaternion algebra, then there is a quadratic field extension L of F such that $F\{a,b\}\otimes_F L = L\{a,b\}$ is isomorphic to $M_2(L)$;

(iii) if H is a quaternion algebra over a local field F, then H contains all quadratic extensions of F (up to isomorphism).

On a quaternion algebra $H = F\{a,b\}$ over F with basis $1,i,j,ij$ (where $i^2 = a$, $j^2 = b$), there is a canonical involution $^-$ mapping $x = \alpha + \beta i + \gamma j + \delta ji$, where $\alpha,\beta,\gamma,\delta \in F$, to $\bar{x} = \alpha - \beta i - \gamma j - \delta ji$. It is an F–linear anti–automorphism on H. Using $^-$, we define the reduced norm of x to be $Nrd\,x = x\bar{x} = \bar{x}x = \alpha^2 - \beta^2 a - \gamma^2 b + \delta^2 ab \in F$ and the reduced trace of x to be $Trd\,x = x + \bar{x} = 2\alpha$. Clearly, Nrd is a multiplicative homomorphism and Trd an additive homomorphism. Choosing a quadratic extension, say, $L = F(\sqrt{a})$, we may imbed $H = F\{a,b\}$ into $M_2(L)$ by sending i to $\begin{pmatrix} \sqrt{a} & 0 \\ 0 & -\sqrt{a} \end{pmatrix}$, j to $\begin{pmatrix} 0 & b \\ 1 & 0 \end{pmatrix}$ so that $x = \alpha + \beta i + j(\gamma + \delta i)$ is sent to $\begin{pmatrix} \alpha+\beta\sqrt{a} & b(\gamma-\delta\sqrt{a}) \\ \gamma+\delta\sqrt{a} & \alpha-\beta\sqrt{a} \end{pmatrix}$. Observe that $Trd\,x$ and $Nrd\,x$ are the trace and determinant of the matrix representing x.

Let H be a quaternion algebra over a nonarchimedean local field F with normalized valuation $|\ |_F$. Then $|\ |_F$ extends to a valuation $|\ |_H$ on H by

$$|x|_H = |Nrd\,x|_F^{1/2} \qquad \text{for } x \in H.$$

Denote by $\mathcal{O}_H$ the collection of elements in H with valuation ≤ 1 and by $\mathcal{P}_H$ the set of those with valuation < 1. Then $\mathcal{P}_H$ is the only maximal ideal in $\mathcal{O}_H$ which is also principal. Further, $\mathcal{O}_H \supset \mathcal{O}_F$, $\mathcal{P}_H \supset \mathcal{P}_F$, and $\mathcal{O}_H/\mathcal{P}_H$ is a field extension of $\mathcal{O}_F/\mathcal{P}_F$ of degree 2. Just like F, $\{\mathcal{P}_H^n\}_{n\ge1}$ is a neighborhood system of 0 in H and $\mathcal{O}_H$ is both open and compact. The group of units in H is $\mathcal{U}_H = \mathcal{O}_H - \mathcal{P}_H$, and $\{1+\mathcal{P}_H^n\}_{n\ge1}$ is a neighborhood system of 1 in $H^\times$, which we denote by $D(F)$. For an element x in $D(F)$, $\mathrm{ord}_H\, x$ denotes the smallest integer n such that x is contained in $\mathcal{P}_H^n$. The center of $D(F)$ is $\mathcal{Z}'(F) \cong F^\times$, and $D(F)/\mathcal{Z}'(F)$ is compact. Thus a smooth representation of $D(F)$ is finite–dimensional, with kernel being an open subgroup of $D(F)$. It is automatically admissible. We are interested in admissible irreducible representations of $D(F)$.

Choose a nontrivial additive character ψ of F of order zero. Let d_Hx be the Haar measure on H self-dual with respect to the additive character $\psi \circ Trd$. Put $d_Dx =$

$|Nrd\,x|_F^{-1} d_H x$, which is a Haar measure on $D(F)$. For an admissible irreducible representation ρ of $D(F)$ and a quasi-character χ of $F^\times$, consider the formal power series

$$\sum_{n\in\mathbf{Z}} \left(\int_{\substack{x\in D(F)\\ \mathrm{ord}_H x=n}} \rho(x)\chi(Nrd\,x)\psi(Trd\,x)d_D x \right) U^n.$$

The coefficient of each U^n is invariant under conjugation by $\rho(g)$ for all $g \in D(F)$ since Nrd and Trd are class functions. As ρ is irreducible, by Schur's lemma, each coefficient of U^n is a scalar operator; we denote this formally by

$$-\gamma_\rho(\chi,\psi,U)Id = \int_D \rho(x)\chi(Nrd\,x)\psi(Trd\,x)U^{\mathrm{ord}_H x} d_D x,$$

where γ_ρ is a Laurent series in powers of U. Taking trace of both sides yields

$$\gamma_\rho(\chi,\psi,U) = -\int_D c_\rho(x)\chi(Nrd\,x)\psi(Trd\,x)U^{\mathrm{ord}_H x} d_D x, \tag{5.1}$$

where $c_\rho = tr\rho/\deg\rho$ is called the reduced trace of ρ. A direct computation shows that if ρ is one-dimensional, then it is given by $\xi \circ Nrd$ for some quasi-character ξ of $F^\times$, and $\gamma_\rho(\chi,\psi,U) = (q-1)^2 q^{-2}\Gamma(\xi\chi,\psi,q^{-1}U)\Gamma(\xi\chi,\psi,U)$; while if ρ has degree ≥ 2, then $\gamma_\rho(\chi,\psi,U)$ is a monomial in U. Therefore, $\gamma_\rho(\chi,\psi,U)$ can be viewed as a rational function on $\mathcal{A}(F^\times)$ with at most one pole.

The restriction of ρ to the center $\mathcal{Z}'(F)$ of $D(F)$ is a quasi-character ω of $F^\times$. Let c be a function on $D(F)$. The following properties characterize c being the reduced trace c_ρ of an admissible irreducible representation ρ of $D(F)$ of central character ω :

(5.2) c is a locally constant class function on $D(F)$;

(5.3) c has type ω, that is, $c(ux) = \omega(u)c(x)$ for all $u \in \mathcal{Z}'(F)$ and $x \in D(F)$;

(5.4) c satisfies the functional equation

$$\int_{D(F)/\mathcal{Z}'(F)} c(xzyz^{-1})d^{\cdot}z = c(x)c(y) \quad \text{for all } x,y \in D(F).$$

Here $d^{\cdot}z = d_D x/d^\times x$ with $d^\times x$ being the Haar measure on $F^\times$ such that $\mathcal{U}_F$ has measure 1.

We elaborate more on class functions on $D(F)$. Two elements in $D(F)$ are conjugate if and only if they have the same reduced trace and reduced norm. In other words, the conjugacy classes of $D(F)$ are parametrized by the image of the map $(Nrd, Trd) : D(F) \longrightarrow F^\times \times F$. An element $(\nu,\tau) \in F^\times \times F$ does not lie in the image if and only if it is hyperbolic, that is, the equation $x^2 - \tau x + \nu = 0$ has two distinct roots in F. Hence we may regard a class function c on $D(F)$ as a function on $F^\times \times F$ vanishing on the hyperboic elements. With this interpretation, we may regard $\gamma_\rho(\chi,\psi,U)$ as the Fourier transform of c_ρ. Further, one can show that a class function c satisfies (5.4) if and only if its Fourier transform $\gamma(\chi,\psi,U)$ defined by (5.1) with c replacing c_ρ satisfies the multiplicative formula (MF) introduced in §2. In particular, this implies that $\gamma_\rho(\chi, \psi, U)$ satisfies (MF). And hence we immediately obtain the correspondence from the equivalence classes of admissible irreducible representations ρ of $D(F)$ to the classes of infinite–dimensional admissible irreducible representations π of $GL_2(F)$ such that the corresponding representations have the same γ–factor. Since $\gamma_\rho(\chi, \psi, U)$ has at most one pole, we know that the image of this correspondence is contained in discrete series representations of $GL_2(F)$. In fact, the image is all discrete series representations of $GL_2(F)$. Indeed, for a special representation $\pi = \sigma(\mu| \ |, \mu)$, it corresponds to the degree 1 representation $\mu| \ |^{-\frac{1}{2}} \circ Nrd$ of $D(F)$, as seen from comparing the associated γ–factor. For a supercuspidal representation π, by performing the Fourier inversion of (5.1) with γ_ρ replaced by γ_π, one obtains a locally constant function c on $F^\times \times F$.

From the fact that γ_π is a monomial and satisfies (MF) one proves that c vanishes on the hyperbolic elements in $D(F)$ and hence can be viewed as a class function on $D(F)$ satisfying (5.2) and (5.3). The third condition (5.4) is met since γ_π satisfies (MF). This proves that $c = c_\rho$ and hence the following

Theorem 15. *(Characterization of γ_ρ [9]) Let F be a nonarchimedean local field, and let $\gamma(\chi, \psi, U)$ be a rational function on $\mathcal{A}(F^\times)$. Then $\gamma = \gamma_\rho$ for an admissible irreducible representation ρ of $D(F)$ with central character ω if and only if γ satisfies (MF) with the same ω and it has at most one pole. Furthermore, γ_ρ determines the reduced trace c_ρ of ρ.*

Combined with Theorem 3, the above theorem yields

Theorem 16. *(Local correspondence) Let F be a nonarchimedean local field. There is a bijection from the classes of admissible irreducible representations of $D(F)$ to discrete series representations of $GL_2(F)$ such that the corresponding representations have the same γ–factors. More precisely, the degree 1 representations of $D(F)$ correspond to special representations, and degree ≥ 2 representations of $D(F)$ correspond to supercuspidal representations of $GL_2(F)$.*

The first assertion in Theorem 16 is also valid for the case $F = \mathbf{R}$. For more detailed discussion of Theorems 15 and 16, the reader is referred to [8] and [9].

Although an admissible irreducible representation ρ of a quaternion group $D(F)$ over a local field F does not have a Whittaker model, one can still define $L(\rho, s)$ and $\varepsilon(\rho, \psi, s)$ for a nontrivial additive character ψ of F; they arise from local zeta functions attached to matrix coefficients of ρ and Schwartz functions on the quaternion algebra over F, similar to the case of GL_1. Like integrals attached to Whittaker functions, these local zeta functions also have analytic continuations and satisfy local functional equations involving L– and ε–factors. In particular, we have

$$\gamma_\rho(\chi, \psi, s) = \frac{L(\check{\rho} \otimes \chi^{-1}, 1-s)}{L(\rho \otimes \chi, s)} \varepsilon(\rho \otimes \chi, \psi, s),$$

where $\gamma_\rho(\chi, \psi, s)$ stands for $\gamma_\rho(\chi, \psi, q^{-s})$ if F is a nonarchimedean local field with q elements in its residue field.

Now let F be a global field and let H be a quaternion algebra defined over F. Denote by D the quaternion group $H^\times$ and by $\mathcal{Z}'$ the center of D. At a place v where H is ramified, $D(F_v)$ is a quaternion group, and at a place v where H is unramified, $D(F_v)$ is isomorphic to $GL_2(F_v)$. Fix such an isomorphism locally. The adèlic group $D(A_F)$ is the restricted product of $\{D(F_v)\}$ with respect to $\{D(\mathcal{O}_v)\}$, where, at a nonarchimedean place v,

$$D(\mathcal{O}_v) = \begin{cases} GL_2(\mathcal{O}_v) & \text{if } H \text{ is unramified at } v, \\ \text{the group of units in } D(F_v) & \text{if } H \text{ is ramified at } v. \end{cases}$$

Let Nrd_v denote the reduced norm at a ramified place v, and the determinant map at an unramified place v, so Nrd_v is a homomorphism from $D(F_v)$ to $F_v^\times$. The global reduced norm Nrd is the homomorphism from $D(A_F)$ to I_F mapping $x = (x_v) \in D(A_F)$ to $Nrd\, x = (Nrd_v x_v) \in I_F$.

Let ω be a quasi-character of the idèle class group $I_F/F^\times$. A continuous complex-valued function f on $D(A_F)$ is called an *automorphic form* on $D(A_F)$ of central character ω if

(5.5) $f(\gamma g z) = \omega(z) f(g)$ for all $\gamma \in D(F)$, $g \in D(A_F)$ and $z \in \mathcal{Z}'(A_F)$,

(5.6) f is right K-finite, where $K = \prod_v K_v$ is the product of standard maximal compact subgroups K_v of $D(F_v)$.

Note that no growth condition is imposed since $D(F)\backslash D(A_F)/\mathcal{Z}'(A_F)$ is compact. Denote by $\mathcal{A}(D(A_F), \omega)$ the space of automorphic forms of central character ω. Admissible representations of $D(A_F)$ are defined in a similar way. An admissible irreducible representation of $D(A_F)$ is automorphic if it is a constituent of the regular representation of $D(A_F)$ on $\mathcal{A}(D(A_F), \omega)$ for some ω.

An irreducible automorphic representation π' of $D(A_F)$ is either 1-dimensional, given by $\omega \circ Nrd$ for some quasi-character ω of $I_F/F^\times$, or infinite-dimensional. In the case that π' is infinite-dimensional, it is the restricted tensor product $\otimes'_v \pi'_v$ of local admissible irreducible representations π'_v of $D(F_v)$. If H is unramified at v, then $D(F_v)$ is isomorphic to $GL_2(F_v)$ and π'_v is infinite-dimensional; if H is ramified at v, then $D(F_v)$ is compact mod center and π'_v is finite-dimensional. To each π'_v there are attached local L- and ε-factors. Define

$$L(\pi', s) = \prod_v L(\pi'_v, s), \qquad \varepsilon(\pi, s) = \prod_v \varepsilon(\pi'_v, \psi_v, s),$$

where $\psi = \prod_v \psi_v$ is a nontrivial additive character of A_F trivial on F. Using global zeta functions attached to π' defined by Schwartz functions on $H(A_F)$, Jacquet and Langlands in [10] proved the following analytic properties of L-functions attached to π' and its contragredient $\widetilde{\pi}'$.

Theorem 17. *(Global functional equation) Let F be a global field. Let π be an infinite-dimensional admissible irreducible automorphic representation of $D(A_F)$. Then $L(\pi', s)$ and $L(\widetilde{\pi}', s)$ have holomorphic continuations to the whole s-plane, which are bounded in each vertical strip of finite width if F is a number field and polynomials in q^{-s} if F is a function field with q elements in its field of constants, and satisfy the functional equation*

$$L(\pi', s) = \varepsilon(\pi', s)L(\widetilde{\pi}', 1 - s).$$

Clearly, the L-function attached to $\pi' \otimes \chi$, the twist of π' by an idèle class character χ of $I_F/F^\times$, also has the properties asserted above. In view of Theorems 14 and 12, there is an admissible irreducible cuspidal representation $\pi = \bigotimes_v' \pi_v$ of $GL_2(A_F)$ such that $\pi_v = \pi'_v$ at the places v where H is unramified, and π_v corresponds to π'_v at the places v where H is ramified, so that $L(\pi \otimes \chi, s) = L(\pi' \otimes \chi, s)$ for all quasi-characters χ of $I_F/F^\times$. This yields an injection $\pi' \mapsto \pi$ from classes of infinite-dimensional admissible irreducible automorphic representations of $D(F)$ to classes of admissible irreducible cuspidal representations whose local components at the places where H ramifies are discrete series representations. Using trace formula, Gelbart and Jacquet [7] showed that this injection is also a surjection. We summarize this in

Theorem 18. *(Global correspondence) Let F be a global field, and let H be a quaternion algebra over F. Write $D = H^\times$. Denote by S the set of places of F where H is ramified. Then there is a bijection from the classes of admissible irreducible infinite-dimensional automorphic representations of $D(A_F)$ with central character ω to the classes of admissible irreducible cuspidal representations of $GL_2(A_F)$ with central character ω whose local components at places in S are discrete series representations such that if representations $\pi' = \bigotimes_v' \pi'_v$ of $D(A_F)$ and $\pi = \bigotimes_v' \pi_v$ of $GL_2(A_F)$ correspond, then π'_v and π_v are isomorphic at the places v outside S and π'_v and π_v correspond à la local correspondence at the places v in S. In particular, $L(\pi' \otimes \chi, s) = L(\pi \otimes \chi, s)$ and $\varepsilon(\pi' \otimes \chi, s) = \varepsilon(\pi \otimes \chi, s)$ for all quasi-characters χ of $I_F/F^\times$.*

As H varies through all quaternion algebras over F, the union of the images under global correspondences consists of all admissible irreducible cuspidal representations of $GL_2(A_F)$ of central character ω with at least two local components being discrete series representations.

References

[1] A. Borel and W. Casselman: Automorphic Forms, Representations, and L-functions, Proceedings of Symposia in Pure Math., vol. 33, Parts 1, 2, Amer. Math. Soc., Providence, 1977.

[2] A. Borel and H. Jacquet : Automorphic forms and automorphic representations. In: Automorphic Forms, Representations, and L-functions, A. Borel and W. Casselman edited, Proceedings of Symposia in Pure Math., vol. 33, Part 1, 189-202, Amer. Math. Soc., Providence, 1977.

[3] H. Carayol : Représentations cuspidales du groupe linéaire, Ann. Sci. École Norm. Sup. (4) 17 (1984), no. 2, 191-225.

[4] P. Deligne : Formes modulaires et représentations de $GL(2)$. In: Modular Functions of One Variable II, W. Kuyk and P. deligne edited, Lecture Notes in Math. 349, 55-105, Springer Verlag, Berlin-Heidelberg-New York, 1973.

[5] D. Flath : Decomposition of representations into tensor products. In: Automorphic Forms, Representations, and L-functions, A. Borel and W. Casselman edited, Proceedings of Symposia in Pure Math., vol. 33, Part 1, 179-183, Amer. Math. Soc., Providence, 1977.

[6] S. Gelbart : Automorphic Forms on Adele Groups. Annals of Math. Studies, no. 83, Princeton University Press, Princeton, N. J., 1975.

[7] S. Gelbart and H. Jacquet : Forms of $GL(2)$ from the analytic point of view. In: Automorphic Forms, Representations, and L-functions, A. Borel and W. Casselman edited, Proceedings of Symposia in Pure Math., vol. 33, Part 1, 213-251, Amer. Math. Soc., Providence, 1977.

[8] P. Gérardin and W.-C. W. Li : A functional equation for degree two local factors, Canad. Math. Soc. Bull., vol. 28(3), 355-371 (1985).

[9] P. Gérardin and W.-C. W. Li : Fourier transforms of representations of quaternions, J. reine angwandte Math. 359, 121-173 (1985).

[10] H. Jacquet and R. Langlands : Automorphic Forms on $GL(2)$, Lecture Notes in Math. 114, Springer Verlag, Berlin-Heidelberg-New York, 1970.

[11] P. Kutzko : On the supercuspidal representations of GL_N and other p-adic groups, Proc. Int. Cong. Math. 1986, 853-861, Amer. Math. Soc., Providence, R. I. (1987).

[12] P. Kutzko and D. Manderscheid : On the supercuspidal representations of GL_N, N the product of two primes, Ann. Sci. École Norm. Sup. (4) 23 (1990), no. 1, 89-121.

[13] W.-C. W. Li : Barnes identities and representations of GL_2. Part II: Nonarchimedean local dield case, J. reine angwandte Math. 345, 69-92 (1983).

[14] W.-C. W. Li : On converse theorems for $GL(2)$ and $GL(1)$, Amer. J. Math. 103, no. 5, 851-885 (1981).

[15] C. Moreno: Analytic proof of the strong multiplicity one theorem, Amer. J. Math. vol. 107, no. 1, 163-206 (1985).

[16] I. Piatetski-Shapiro : Multiplicity one theorems. In: Automorphic Forms, Representations, and L-functions, A. Borel and W. Casselman edited, Proceedings of Symposia in Pure Math., vol. 33, Part 1, 209-212, Amer. Math. Soc., Providence, 1977.

[17] J. A. Shalika : The multiplicity one theorem for $GL(n)$, Ann. of Math. (2) 100, 171-193 (1974).

[18] J. Tate : Fourier analysis in number fields and Hecke's zeta-functions. Thesis, Princeton University 1950. Published in Algebraic Number Theory,

J.W.S. Cassels and A. Fröhlich ed., Thompson, Washington D.C. (1967), republished by Academic Press, London.

[19] N. Wallach : Representations of reductive Lie groups. In: Automorphic Forms, Representations, and *L*-functions, A. Borel and W. Casselman ed., Proceedings of Symposia in Pure Math., vol. 33, Part 1, 71-86, Amer. Math. Soc., Providence, 1977.

[20] A. Weil : Dirichlet Series and Automorphic Forms. Lecture Notes in Math. 189, Springer Verlag, Berlin-Heidelberg-New York, 1971.

CHAPTER 9

Applications

§1 Expanders, property T and eigenvalues

In this chapter we illustrate how the number–theoretic results discussed in the first 8 chapters can be applied to solving problems in real life. The problem we are concerned with arises from communication networks : it is to construct efficient networks at a cost not exceeding a fixed amount.

Represent a communication network by a finite graph G. The network is efficient if the information carried at certain vertices can be quickly spread out all over the network. To formulate this mathematically, define, for a subset X of the vertices of G, the boundary ∂X of X to be the set of vertices not in X but adjacent to X. The *magnifying constant* of G is

$$c = \min_X \frac{|\partial X|}{|X|},$$

where X runs through all subsets of vertices of G not exceeding half of the size of G. Clearly, c measures the efficiency of the network. For example, if G is the complete graph on n vertices, then

$$c = \begin{cases} 1 & \text{if } n \text{ is even,} \\ \frac{n+1}{n-1} & \text{if } n \text{ is odd.} \end{cases}$$

In general, $0 \leq c \leq 1$. If G has n vertices, and every vertex has at most k neighbors, we say that G is an (n, k, c')-expander for every $0 \leq c' \leq c$.

Although the transmission of information is most efficient if the network is a complete graph, from realistic and economical viewpoint, this is not welcomed. Instead, we would like to search for expanders with fixed number of vertices n and maximal number of neighbors k, but large c. Or even better, construct an infinite family of good expanders with the same k and large c. Probabilistically speaking, if a graph is a random graph, then its magnifying constant is usually good; on the other hand, to determine a given graph's magnifying constant is a very difficult problem. Therefore it is desirable to have explicit constructions of good expanders.

For this, two systematic methods emerged in the past two decades. We give a brief survey below.

The first method uses Kazhdan's property T. A locally compact group Γ is said to have property T if the trivial representation is isolated from other unitary representations of Γ. To measure how far apart the trivial representation is from the other representations, there is a so-called Kazhdan's constant κ. The topology on the representations is similar to the "compact-open" topology, with the role of compact sets played by sets of generators of the group Γ. Thus the "distance" from the trivial representation to other representations, which is κ, depends on the choice of generators of Γ.

The connection between Kazhdan's property T and constructions of expanders was first noticed by Margulis in 1975. In [19] he used $\Gamma = SL_2(\mathbf{Z}) \ltimes \mathbf{Z}^2$ to construct an infinite family of bipartite graphs with degree $k = 5$ and expanding constant $\kappa^2/2$, where κ is Kazhdan's constant arising from the four standard generators of Γ, which were used to define his graphs. Unfortunately, he was not able to compute κ. In 1981, Gabber and Galil [12] followed the same method but took another set of four generators to construct a new family of graphs with $k = 5$, and they succeeded in computing the expanding constant explicitly. Since then, there were similar constructions with improved expanding constants.

The second method uses the smallest positive eigenvalue λ_1 of the Laplacian operator on G. More precisely, regard G as a one-dimensional simplicial complex. Define a linear map d from $C^0(G)$, the space of functions on vertices of G, to $C^1(G)$, the space of functions on edges of G, via

$$(df)(e) = f(e^+) - f(e^-)$$

for any edge e of G. Here we have chosen an orientation on each edge, e^+ (resp. e^-) denotes the ending (resp. starting) point of e. Denote the adjoint of d by d^*, in our case it is also the transpose of d. The Laplacian operator Δ is defined to be d^*d. An easy computation shows that if we choose as a basis of $C^0(G)$ the characteristic functions of the vertices of G, then Δ is represented by the matrix $D - A$, where D and A are square matrices with rows and columns parametrized by the vertices of G, D is called the *degree matrix* of G with off-diagonal entries zero and the diagonal entries equal to the number of neighbors (called the *degree*) of the vertices; A is called the *adjacency matrix* with the entry at xy equal to the number of edges from x to y.

Exercise 1. Show that A may be regarded as a linear operator on the space $\mathcal{F}(G)$ of real-valued functions on (vertices of) G which sends f to Af given by

$$(Af)(x) = \sum_y f(y),$$

where y runs through all outneighbors of x in G.

In particular, Δ is independent of the choice of the orientation put on the edges of G. Now if f is a nonzero eigenfunction of Δ with eigenvalue λ, then using the inner product $<,>$ on $C^0(G)$, we find from

$$\lambda < f, f > = < \Delta f, f > = < df, df >$$

that $\lambda \geq 0$. Moreover, the constant functions are eigenfunctions with eigenvalue 0. Hence we may list the n eigenvalues of G as

$$0 = \lambda_0 \leq \lambda_1 \leq \cdots \leq \lambda_{n-1}.$$

Here n is the size of G. Assume G connected so that 0 is an eigenvalue of multiplicity one, in other words, $\lambda_1 > 0$.

Tanner in 1984 gave an explicit lower bound of the magnifying constant c in terms of λ_1 :

Theorem 1. *([33]) Assume that the degree at each vertex does not exceed k. Then $c \geq \frac{2\lambda_1}{k+2\lambda_1}$.*

Therefore, the larger λ_1 is, the more expanding G is.

Conversely, there is a lower bound of λ_1 in terms of c proven by Alon and Milman in 1985 :

Theorem 2. *([1])* $\lambda_1 \geq \frac{c^2}{4+2c^2}$.

In view of the above discussions, we see that the method using Kazhdan's property T suffers from two defects: there is no known explicit bound(s) of the magnifying constant c in terms of the Kazhdan constant κ like Theorem 1, and it is difficult to compute κ. Thus the second method appears to be more appealing: as soon as we can bound λ_1, we get an estimate for c. In section 4 we shall survey various ways of constructing graphs with large λ_1. On the other hand, Alon and Milman found a connection between λ_1 and κ, which we now explain. Let Γ be a countable discrete group with property T, let $S = S^{-1}$ be a symmetric set of generators of Γ. Denote by κ the Kazhdan constant derived from S. Let $\overline{\Gamma}$ be a finite quotient of Γ and $\overline{S}$ the image of S under the projection $\Gamma \to \overline{\Gamma}$. The graph G to be considered is the Cayley graph $\overline{\Gamma}$ with generators $\overline{S}$. In other words, the vertices of G are elements in $\overline{\Gamma}$ and the neighbors of $x \in G$ are $x\overline{S}$.

Theorem 3. *([1])* $\lambda_1(G) \geq \kappa$.

While this can be regarded as an upper bound estimate of the mysterious κ, it can also be viewed as a way to construct an infinite family of graphs whose smallest positive eigenvalues are uniformly bounded from below. We refer to Bien's article [3] and Sunada's papers [31], [32] for more detailed discussions of materials in this and next sections.

Digression : Laplacian and property T for manifolds

Many ideas and results discussed in the previous section actually came from results on manifolds. Here we mention a couple. Let M be a compact Riemannian manifold of dimension n. Analogous to the magnifying constant of a graph, there is a *Cheeger constant* of M defined by

$$c = \inf_S \frac{\text{area } (S)}{\min\left(\text{vol } (M_1), \text{ vol } (M_2)\right)}$$

Here S runs through $(n-1)$-dimensional submanifolds of M which divide M into two parts M_1 and M_2 with at least one part having finite volume (so that S is the common boundary of M_1 and M_2), and the "area form" on S is induced from the volume form on M.

The Laplacian operator Δ on M is also defined by d^*d with d the extension by continuity to L^2 functions on M of the differential operator on differentiable functions; and λ_1 is defined as

$$\lambda_1 = \lambda_1(M) = \inf_f \frac{\int_M \|df\|^2}{\int_M |f|^2},$$

where f ranges over nonzero functions perpendicular to the constant function on M, that is, $\int_M f = 0$. In 1970 Cheeger showed the following inequality :

Theorem 4. *([6])* $\lambda_1 \geq \frac{c^2}{4}$.

The bound in Theorem 2 for graphs is fairly close to the above form.

Concerning Kazhdan's property T for manifolds, Brooks [4] proved the following result which is similar to Theorem 3 although not as precise.

Theorem 5. *([4]) Let M be a compact Riemannian manifold. Suppose that its fundamental group $\pi_1(M)$ has Kazhdan's property T. Then there is a constant $c > 0$ such that $\lambda_1(M') \geq c$ for all finite coverings M' of M.*

---

§2 Spectra of regular graphs

For simplicity, consider only *k–regular graphs*, that is, each vertex has indegree = outdegree = k. The Laplacian $\Delta = D - A$ on a k–regular graph G then has the degree matrix D equal to the scalar matrix kI so that

$$\lambda_1 = k - \text{the second largest eigenvalue of } A.$$

The eigenvalues of $A = A(G)$ are called the *spectrum* of G. When G is k–regular, each row and each column of A have entries add up to k, thus k is an eigenvalue of A with constant functions on G being eigenfunctions. This corresponds to the 0 eigenvalue of the Laplacian $\Delta = kI - A$. Observe that any eigenvalue λ of $A(G)$ satisfies $|\lambda| \leq k$. Indeed, let f be a nontrivial eigenfunction of $A = A(G)$ with eigenvalue λ. Assume that the maximum absolute value of f is achieved at the vertex x. Then $f(x) \neq 0$. Further,

$$\lambda f(x) = (Af)(x) = \sum_{x \to y} f(y).$$

Taking absolute value of both sides yields

$$|\lambda|\,|f(x)| \leq \sum_{x \to y} |f(y)| \leq k|f(x)|,$$

which implies $|\lambda| \leq k$ since $f(x) \neq 0$.

A group G is called *r–partite* if the vertices of G can be partitioned into r disjoint sets $V_1, \cdots, V_r$ such that the outedges from vertices in V_i end in V_{i+1} for $i \in \mathbf{Z}/r\mathbf{Z}$. Suppose that G is r–partite and let λ be an eigenvalue of G with f being a nontrivial eigenfunction. Let ζ be any rth root of unity. Define a new function $\widetilde{f}$ on vertices of G by $\widetilde{f}(x) = \zeta^i f(x)$ for vertices x in V_i. Then

$$\begin{aligned}(A\widetilde{f})(x) &= \sum_{x \to y} \widetilde{f}(y) = \zeta^{i+1} \sum_{x \to y} f(y) = \zeta^{i+1}(Af)(x) = \zeta^{i+1}\lambda f(x)\\ &= \zeta\lambda\widetilde{f}(x) \quad \text{for } x \in V_i.\end{aligned}$$

As this holds for all $i \in \mathbf{Z}/r\mathbf{Z}$, we have shown that $\widetilde{f}$ is an eigenfunction of A with eigenvalue $\zeta\lambda$.

In particular, if G is a k–regular r–partite graph, then ζk is an eigenvalue of $A(G)$ for all rth roots of unity ζ. Call ζk's *trivial eigenvalues* of G, and the remaining eigenvalues nontrivial. For a k–regular graph G, let

$$\lambda(G) = \text{the maximal nontrivial eigenvalue of } A(G) \text{ in absolute value.}$$

Then $\lambda(G) \geq k - \lambda_1$, and it suffices to construct k–regular graphs with small $\lambda(G)$.

$\lambda(G)$ not only is intimately related to the magnifying constant of G as we saw in the last section, but also provides other important information about G. For instance, it gives an upper bound of the diameter of G, which is the distance of two farthest apart vertices. If G represents a communication network, then its diameter can be thought of as transmission delay. The precise statement is the following theorem proved by Chung.

Theorem 6. *([8]) Let G be a k–regular undirected graph with n vertices. Then the diameter of $G \leq \frac{\log n-1}{\log \frac{k}{\lambda(G)}}$.*

Hence the smaller $\lambda(G)$ is, the smaller the diameter is. This gives another reason to construct regular graphs with small λ.

How small can $\lambda(G)$ be ? If G is k–regular with $A = A(G)$ diagonalizable by a unitary matrix, then the trace of AA^t is nk and the eigenvalues of AA^t are the absolute value of the eigenvalues of A squared. Thus if G is also r–partite, then $n \geq rk$ and $rk^2 + (n-r)\lambda(G)^2 \geq nk$, which yields the following trivial lower bound of $\lambda(G)$:

$$\lambda(G) \geq \left(\frac{n-rk}{n-r}\right)^{1/2} \sqrt{k}.$$

A nontrivial lower bound for undirected graphs is given by Alon and Boppana.

Theorem 7. *([18]) For k–regular undirected graphs G, we have*

$$\liminf \lambda(G) \geq 2\sqrt{k-1}$$

as $|G| \to \infty$.

The same lower bound also holds for k–regular directed graphs G with $A = A(G)$ diagonalizable by unitary matrices. This is because the bipartite undirected graph with adjacency matrix $\begin{pmatrix} 0 & A \\ A^t & 0 \end{pmatrix}$ has eigenvalues $\pm|\lambda|$, where λ runs through eigenvalues of A. In view of the above theorem, we say that a k–regular graph G has small eigenvalues if $\lambda(G) \leq 2\sqrt{k-1}$. Following Lubotzky–Phillips–Sarnak [18], we call a graph G *Ramanujan graph* if

(2.1) G is k–regular;

(2.2) $\lambda(G) \leq 2\sqrt{k-1}$;

(2.3) $A(G)$ is diagonalizable by a unitary matrix.

Since a directed k–regular r–partite Ramanujan graph will give rise to r bipartite undirected Ramanujan graphs, therefore we include also directed graphs in the definition above.

Remark. The third condition is automatically satisfied if G is an undirected graph, for $A(G)$ is symmetric and thus diagonalizable by orthogonal matrices. On the other hand, Theorem 7 no longer holds if we drop the condition that $A(G)$ is diagonalizable by a unitary matrix and assume merely that $A(G)$ is diagonalizable. In fact, the following result is proved by Feng and Li.

Theorem 8. *([11]) Fix positive integers r and k. Then there exists an infinite family of k–regular r–partite directed graphs whose adjacency matrices are diagonalizable and whose nontrivial eigenvalues all lie on the unit circle. In particular,* $\liminf_{|G|\to\infty} \lambda(G) = 1$ *as G runs through all k–regular r–partite graphs with diagonalizable adjacency matrices.*

Ramanujan graphs have wide applications in communication networks, extremal graph theory, and computational complexity. Despite of the fact that a randomly picked up k-regular graph G has a high probability to have $\lambda(G)$, if not already $\le 2\sqrt{k-1}$, not much bigger than $2\sqrt{k-1}$, it is difficult to verify if a given graph is indeed Ramanujan. Hence it is desirable to have explicit constructions of Ramanujan graphs. Up to date, there are three known systematic methods, all number-theoretic, to construct such graphs, which we discuss in the next three sections.

---

§3 Ramanujan graphs based on quaternion groups

The Ramanujan graphs constructed in this section have vertices which are double cosets of adèlic points of certain quaternion groups and they are $(q+1)$-regular graphs with q being a prime or a prime power. The first explicit construction was given by Margulis [20] and, independently, by Lubotzky, Phillips and Sarnak [18] using Hamiltonian quaternion group, although the method can be traced back to Eichler and Brandt.

Fix a prime p. Let H be a quaternion algebra defined over $\mathbf{Q}$ which is ramified at ∞ and unramified at p, and let D be the multiplicative group $H^\times$. Let $\prod_q K_q$ be an open congruence subgroup of the standard maximal compact subgroup of $D(A^f_{\mathbf{Q}})$ such that $K_p = D(\mathbf{Z}_p)$ and $Nrd(\prod_q K_q) = \prod_q \mathcal{U}_q$. It follows from the strong approximation theorem for the subgroup SD of D consisting of elements of reduced norm 1 that

$$D(A_{\mathbf{Q}}) = D(\mathbf{Q}) \cdot D(\mathbf{R}) D(\mathbf{Q}_p) \prod_q K_q$$

so that the global double coset space

$$X = D(\mathbf{Q}) \backslash D(A_{\mathbf{Q}}) / D(\mathbf{R}) \prod_q K_q \mathcal{Z}'(A_{\mathbf{Q}})$$

can also be expressed as local double coset space

$$X = \tilde{\Gamma} \backslash D(\mathbf{Q}_p) / D(\mathbf{Z}_p) \mathcal{Z}'(\mathbf{Q}_p) = \tilde{\Gamma} \backslash GL_2(\mathbf{Q}_p) / GL_2(\mathbf{Z}_p) \mathcal{Z}(\mathbf{Q}_p),$$

where $\tilde{\Gamma} = D(\mathbf{Q}) \cap \prod_{q \ne p} K_q$ is a congruence subgroup of $D(\mathbf{Z}[\frac{1}{p}]) = D(\mathbf{Q}) \cap \prod_{q \ne p} D(Z_q)$, which is a discrete subgroup of $D(\mathbf{Q}_p) = GL_2(\mathbf{Q}_p)$. As H is ramified at ∞, the set X is finite. To define a graph on X, we first study $GL_2(\mathbf{Q}_p)/GL_2(\mathbf{Z}_p)\mathcal{Z}(\mathbf{Q}_p)$.

Each matrix $\begin{pmatrix} a & b \\ c & d \end{pmatrix} \in GL_2(\mathbf{Q}_p)$ represents a rank two $\mathbf{Z}_p$-module, with basis $e_1 = \begin{pmatrix} a \\ c \end{pmatrix}$, $e_2 = \begin{pmatrix} b \\ d \end{pmatrix}$, that is, a lattice $L = \mathbf{Z}_p e_1 \oplus \mathbf{Z}_p e_2$ in $\mathbf{Q}_p^2$. Two matrices g_1, g_2 in $GL_2(\mathbf{Q}_p)$ represent the same lattice if and only if $g_1^{-1} g_2 \in GL_2(\mathbf{Z}_p)$. Thus we may regard $GL_2(\mathbf{Q}_p)/GL_2(\mathbf{Z}_p)$ as the set of lattices in $\mathbf{Q}_p^2$. Two lattices L_1, L_2 are equivalent if there is a nonzero element $x \in \mathbf{Q}_p$ such that $L_1 = xL_2$.

Thus $GL_2(\mathbf{Q}_p)/GL_2(\mathbf{Z}_p)\mathcal{Z}(\mathbf{Q}_p)$ parametrizes equivalence classes of lattices in $\mathbf{Q}_p^2$. View each equivalence class as a vertex; two vertices are adjacent if they can be represented by two lattices L_1, L_2 such that L_2 is a sublattice of L_1 with index p. If so, then pL_1 is a sublattice of L_2 with index p, hence we obtain an undirected graph. It can be shown that the neighbors of the class of $g = \begin{pmatrix} a & b \\ c & d \end{pmatrix}$ are the classes of $g = \begin{pmatrix} p & u \\ 0 & 1 \end{pmatrix}$, $0 \leq u \leq p-1$ and $g = \begin{pmatrix} 1 & 0 \\ 0 & p \end{pmatrix}$. Thus $GL_2(\mathbf{Q}_p)/GL_2(\mathbf{Z}_p)\mathcal{Z}(\mathbf{Q}_p)$ is an infinite $(p+1)$–regular tree, which is the universal cover of any $(p+1)$–regular undirected graphs. For more details, see [28]. The group $\widetilde{\Gamma}$ acts on $GL_2(\mathbf{Q}_p)/GL_2(\mathbf{Z}_p)\mathcal{Z}(\mathbf{Q}_p)$ by left translations, and the quotient graph is the graph structure we put on X. Then X is a $(p+1)$–regular graph counting multiplicities since $\widetilde{\Gamma}$ may contain torsion elements.

Given a function f on X, we may view it as an automorphic form F on $D(A_{\mathbf{Q}})$ right invariant under $D(\mathbf{R})\prod_q K_q\mathcal{Z}'(A_{\mathbf{Q}})$. The action of the adjacency matrix $A(X)$ on f is nothing but the Hecke operator T_p on F, and hence the eigenvalues of $A(X)$ are the eigenvalues of T_p on forms on $D(\mathbf{Q})\backslash D(A_{\mathbf{Q}})/D(\mathbf{R})\prod_q K_q\mathcal{Z}'(A_{\mathbf{Q}})$. Note that such automorphic forms all have trivial central character, and the space of forms on $D(\mathbf{Q})\backslash D(A_{\mathbf{Q}})/D(\mathbf{R})\prod_q K_q\mathcal{Z}'(A_{\mathbf{Q}})$ decomposes as a direct sum of subspaces, each subspace generates an automorphic irreducible representation π' of $D(A_{\mathbf{Q}})$. If $\pi' = \omega \circ \mathrm{Nrd}$ is 1–dimensional, then ω is trivial on $\mathrm{Nrd}\left(D(\mathbf{R})\prod_q K_q\mathcal{Z}'(A_{\mathbf{Q}})\right)$, which contains $\mathbf{R}_{>0}\cdot\prod_q \mathcal{U}_q$. As $I_{\mathbf{Q}} = \mathbf{Q}^\times\cdot\mathbf{R}_{>0}\cdot\prod_q\mathcal{U}_q$, ω is the trivial character and the corresponding space of automorphic forms is 1–dimensional, consisting of the constant functions on $D(A_{\mathbf{Q}})$. They are eigenfunctions of T_p with eigenvalue $p+1$. The remaining representations π' occurring in the decomposition are infinite–dimensional. Let $\pi' = \bigotimes_v' \pi_v'$ be such a representation. Then π_∞' is the trivial representation of $D(\mathbf{R})$, which, according to the local correspondence, corresponds to a discrete series representation π_∞ of $GL_2(\mathbf{R})$ of highest weight 2. Thus the global representation $\pi = \bigotimes_v' \pi_v$ of $GL_2(A_{\mathbf{Q}})$ corresponding to π' by Theorem 18 in §5, chapter 8, arises from a cuspidal newform f of weight 2. The eigenvalues λ of T_p on the forms on $D(\mathbf{Q})\backslash D(A_{\mathbf{Q}})/D(\mathbf{R})\prod_q K_q\mathcal{Z}'(A_{\mathbf{Q}})$ lying in the space of π' are the same as the eigenvalue of T_p on f, which, in view of the Ramanujan–Petersson conjecture proved by Deligne (Theorem 6 in §3, Chapter 7), satisfies

$$|\lambda| \leq 2\sqrt{p} = 2\sqrt{k-1}.$$

(In fact, the Ramanujan-Petersson conjecture for cusp forms of weight 2 was proved by Eichler and Shimura, Deligne extended it to forms of weight greater than 2.) We have shown that $A(X)$ has one eigenvalue equal to $k = p+1$, and the remaining eigenvalues have absolute value $\leq 2\sqrt{k-1}$. This proves

Theorem 8. *The graph X is a $(p+1)$-regular Ramanujan graph.*

The $(p+1)$–regular Ramanujan graphs studied in [22] by Mestre and Oesterlé were constructed by choosing $H = H_\ell$ ramified only at ∞ and at a prime $\ell \neq p$, with the double coset space always modulo the product of real points and the standard maximal compact subgroups at nonarchimedean places on the right; and by letting ℓ tend to infinity, they obtained infinitely many such graphs. On the other hand, Margulis [20] and, independently, Lubotzky, Phillips and Sarnak [18] (see also [7]) took H to be the Hamiltonian quaternion, they obtained an infinite family of $(p+1)$–regular Ramanujan graphs by taking congruence subgroups of Γ. In general, one may both vary quaternion algebras and take congruence subgroups to construct infinitely many Ramanujan graphs. Taking congruence subgroups of high level has another advantage of avoiding multiple edges, that is, the resulting quotient graphs are truely $(q+1)$-regular.

The same argument works when the base field $\mathbf{Q}$ is replaced by a function field of one variable over a finite field. In that case the resulting graphs are Ramanujan because Drinfeld [9] has proved the Ramanujan conjecture for GL_2 over function fields. This is done in Morgenstern's paper [23]. Note that the graphs so constructed have valency $k = q+1$ with q a power of a prime. In [27] Pizer constructed $(p+1)$–regular Ramanujan graphs allowing multiple edges by using the action of the classical Hecke operator at p on spaces of certain theta series of weight 2.

---

§4 Ramanujan graphs based on finite abelian graphs

The Ramanujan graphs constructed in this section have vertices which are elements of abelian groups. We start with general constructions of k–regular graphs based on abelian groups. Let G be a finite abelian group and let S be a k–element subset of G. Using S we define two k-regular graphs, called sum graph $X_s(G,S)$ and difference graph $X_d(G,S)$ on G. For $x \in G$, the out–neighbors of x in $X_s(G,S)$ (resp. $X_d(G,S)$) are those $y \in G$ such that $x+y \in S$ (resp. $y-x \in S$), i.e., $-x+S$ (resp. $x+S$). It follows from definition that the sum graph is undirected and the difference graph is usually directed; it is undirected if and only if S is symmetric, that is, $S=-S$. The sum graphs and difference graphs constructed this way have the following nice properties. For a character ψ of G, put $e(\psi, S) = \sum_{s\in S} \psi(s)$.

Proposition 1. *(i) Each character ψ of G is an eigenfunction of the adjacency matrix of $X_d(G,S)$ with eigenvalue $e(\psi,S)$.*

(ii) If $e(\psi,S)=0$, then ψ and ψ^{-1} are both eigenfunctions with eigenvalue 0 of the adjacency matrix $A(X_s)$ of $X_s(G,S)$; if $e(\psi,S)\neq 0$, then $|e(\psi,S)|\psi \pm e(\psi,S)\psi^{-1}$ are two eigenfunctions of $A(X_s)$ with eigenvalues $\pm|e(\psi,S)|$, respectively.

Proposition 2. *(i) The adjacency matrices of $X_d(G,S)$ and $X_s(G,S)$ are diagonalizable by unitary matrices.*

(ii) The absolute values of the eigenvalues of the adjacency matrices of $X_d(G,S)$ and $X_s(G,S)$ are the same, they are $\left|\sum_{s\in S}\psi(s)\right|$, where ψ runs through all characters of G.

Exercise 2. Prove Propositions 1 and 2.

Thus if we can find a suitable finite abelian group G and a suitable k-element subset S of G such that

$$\left|\sum_{s\in S}\psi(s)\right| \le 2\sqrt{k-1}$$

for all nontrivial characters ψ of G, then $X_s(G,S)$ and $X_d(G,S)$ are Ramanujan graphs. This reduces a combinatorial problem to the problem of character sum estimates. We appeal to the estimates obtained in §3, Chapter 6.

We recall some notations. Let F be a finite field with q elements, F_n be the degree n field extension of F, and N_n be the kernel of the norm map $N_{F_n/F}$. Assume $n \ge 2$. Let t be an element in F_n such that $F_n = F(t)$. Define the set

$$S_n = \left\{\frac{t^q+a}{t+a} : a \in F \cup \{\infty\}\right\}.$$

Then $|S_n| = q+1$. It then follows from Theorem 6 in §3, Chapter 6, that $X_s(N_n,S_n)$, and $X_d(N_n,S_n)$ are Ramanujan graphs for $n = 3,4$. In fact, when $n = 3$, the bound is $\sqrt{q}$, so we may enlarge S by randomly joining in up to $\sqrt{q}$ more elements in N_3 and still get Ramanujan graphs. As a consequence of Theorem 7′ in Chapter 6, $X_s(F_2,S_2)$ and $X_d(F_2,S_2)$ are Ramanujan graphs. Further, $X_s(N_2\times F_2,S_2)$ and $X_d(N_2\times F_2,S_2)$ are also Ramanujan graphs by Theorem 9 in Chapter 6. Theorem 11 there implies that $X_s(F_2^\times,Y)$ and $X_d(F_2^\times,Y)$ are Ramanujan graphs for $Y = \{t+a : a\in F\}$ and so are $X_s(F^\times\times F^\times,Y')$ and $X_d(F^\times\times F^\times,Y')$ for $Y' = \{(a-c,b-c) : c\neq a,b,\ c\in F\}$. Here a,b are unequal elements in F.

Finally we remark that the character sum estimates obtained in Chapter 6, §3 are derived from idèle class group characters of function fields over F with genus 0. Likewise, given an elliptic curve E with $k\ge q+1$ points over F and at least one F_2– but not F–rational point (the latter condition is always satisfied provided $q\ge 5$), one can construct k–regular Ramanujan graphs defined on $F_2\times E(F)$. We know from the Riemann hypothesis for curves that $q+1\le k\le q+1+2\sqrt{q}$. In the case that q is a prime, for each integer $q+1\le k\le q+1+2\sqrt{q}$ there exists an elliptic curve with k points over F. Hence for each prime $p\ge 5$ and each $p+1\le k\le p+1+2\sqrt{p}$, there exists a k–regular Ramanujan graph with p^2k vertices constructed using elliptic curves. Judging from the data for small k, one is tempted to make the following

Conjecture. *For every $k \geq 6$, there is a k–regular Ramanujan graph constructed using elliptic curves.*

In view of our discussion above, the conjecutre can be restated in terms of gaps between two consecutive primes. Denote by p_n the nth prime. Then the conjecture above is equivalent to

Conjecture′. $p_{n+1} - p_n \leq 2\sqrt{p_n}$ *for n large.*

The reader is referred to the articles [8] and [15] for more details on the Ramanujan graphs constructed in this section.

---

§5 Ramanujan graphs based on finite nonabelian groups

In this section we shall construct Ramanujan graphs based on finite nonabelian groups, which enjoy certain properties already seen in the previous two methods. Again, we start with a general approach.

Let G be a finite group and let K be a subgroup of G. Denote by $L(G)$ the space of complex–valued functions on G and by $L(G/K)$ the subspace of functions invariant under right translations by K. For a K–double coset $S = KsK$, define the operator A_S on $L(G)$ by sending a function f on G to

$$(A_S f)(x) = \sum_{y \in S} f(xy), \qquad x \in G.$$

Let λ be a nonzero eigenvalue of A_K, and let f be a nonzero eigenfunction with eigenvalue λ. Define h_f on G to be the average of f over K :

$$h_f(x) = \sum_{k \in K} f(xk), \qquad x \in G.$$

Then $h_f \in L(G/K)$ and $f = (f - \frac{1}{|K|}h_f) + \frac{1}{|K|}h_f$ with $f - \frac{1}{|K|}h_f$ lying in the 0–eigenspace L_0 of A_K.

We have shown that

$$L(G) = L(G/K) \oplus L_0,$$

where each subspace is invariant under the left translation by G. Further, A_S acts on L_0 as 0 operator, and $L(G/K)$ is A_S–invariant.

We assume that the operators A_S are diagonalizable and mutually commutative. As $A_S f$ is nothing but the convolution of f with the characteristic function of S^{-1}, our assumption implies that the convolution algebra $L(K\backslash G/K)$ of the bi-K–invariant functions on G is commutative. Note that the assumption is satisfied if each double coset S is symmetric, that is, $S = S^{-1}$. Then $L(G/K)$ is a direct sum of common eigenspaces of A_S, and each common eigenspace is invariant under left translations by G, that is, $g \in G$ acts on f by $(g \cdot f)(x) = f(g^{-1}x)$ for $x \in G$. Hence $L(G/K)$ decomposes as a direct sum of irreducible representations (V, π) of G such that on each space V the operators A_S act by multiplication by scalars λ_S. We want to find a way to compute the λ_S. Denote by e the identity of G.

Proposition 3. *Let (V, π) be an irreducible representation of G occurring in the space $L(G/K)$. Then*

(i) *If h is a bi-K-invariant function in V with $h(e) = 0$, then $h = 0$.*

(ii) *There exists a unique bi-K-invariant function h in V with $h(e) = 1$.*

(iii) *The function h in (ii) satisfies*

$$\frac{1}{|K|}\sum_{k\in K} h(xky) = h(x)h(y) \qquad \text{for all } x, y \in G$$

and for any K-double coset $S = KsK$,

$$\lambda_S = \frac{|K|^2}{|\operatorname{Stab} s|} h(s),$$

where Stab s *consists of $k \in K$ such that $Ksk = Ks$.*

Proof. Let f be a nonzero function in V with $f(g) = 1$ for some $g \in G$. Replacing f by $\pi(g^{-1})f$ if necessary, we may assume $f(e) = 1$. Define

$$h(x) = \frac{1}{|K|}\sum_{k\in K} f(kx) = \frac{1}{|K|}\sum_{k\in K} (\pi(k)f)(x) \qquad \text{for } x \in G.$$

Then h lies in V, is bi-K-invariant, and satisfies $h(e) = f(e) = 1$. This proves the existence part of (ii). The uniqueness part of (ii) will follow from (i).

Let h be a bi-K-invariant function in V. Then $A_S h = \lambda_S h$ for all K-double cosets $S = KsK$. Fix $s \in G$. We have, for all $x \in G$,

$$\frac{1}{|K|}\sum_{k\in K} h(xks) = \frac{|\operatorname{Stab} s|}{|K|^2}\sum_{y\in S} h(xy) = \frac{|\operatorname{Stab} s|}{|K|^2}(A_S h)(x) = \frac{|\operatorname{Stab} s|}{|K|^2}\lambda_S h(x).$$

Setting $x = e$, we obtain

$$h(s) = \frac{|\operatorname{Stab} s|}{|K|^2}\lambda_S h(e).$$

Hence $h(s) = 0$ for all $s \in G$ if $h(e) = 0$. This proves (i). Suppose $h(e) = 1$. The above equation yields

$$h(s) = \frac{|\operatorname{Stab} s|}{|K|^2}\lambda_S$$

and hence

$$\frac{1}{|K|}\sum_{k\in K} h(xks) = h(x)h(s) \qquad \text{for all } x, s \in G$$

and λ_S is as asserted. □

Remark. If we take a bi-K-invariant function h in L_0, then the same proof shows that $h = 0$. Hence L_0 contains no nontrivial bi-K-invariant function. This proves that representations occurring in L_0 have no K-invariant vectors.

The proposition above implies that in V, the $\pi(K)$-invariant subspace is 1-dimensional, generated by h in (ii); and further, as remarked above, the representations occurring in $L(G/K)$ are characterized by having a nonzero K-invariant vector. Fix $x \in G$ and consider the operator $\sum_{k\in K} \pi(kx^{-1}) =: \rho(x)$ on V. Let $f \in V$. Then $\rho(x)f$ is a bi-K-invariant function in V whose value at e is equal to

$$(\rho(x)f)(e) = \sum_{k\in K} f(xke) = |K|f(x).$$

Hence $\rho(x)f = |K|f(x)h$. Choose a basis $f_1 = h, f_2, \cdots, f_r$ of V. Then the matrix representation of $\rho(x)$ with respect to this basis is

$$\begin{pmatrix} |K|h(x) & |K|f_2(x) & |K|f_3(x) & \cdots & |K|f_r(x) \\ 0 & 0 & 0 & & 0 \\ \vdots & \vdots & \vdots & & \vdots \\ 0 & 0 & 0 & \cdots & 0 \end{pmatrix}$$

so that

$$\frac{1}{|K|} tr\, \rho(x) = \frac{1}{|K|}\sum_{k\in K} tr\, \pi(kx^{-1}) = h(x).$$

We summarize the above discussion in

Theorem 9. *The group G acts on the space $L(G/K)$ of right K-invariant functions on G by left translations. The space $L(G/K)$ decomposes into a direct sum of irreducible subspaces (V, π) such that on each subspace V the operators A_S act by scalar multiplications by $\lambda_{S,\pi}$ for all K-double cosets S. Further, on each irreducible subspace V, the space of left K-invariant functions is one-dimensional, generated by*

$$h_\pi(x) := \frac{1}{|K|}\sum_{k\in K} tr\, \pi(kx^{-1}). \tag{5.1}$$

Such a function is bi-K-invariant, satisfies $h_\pi(e) = 1$ and

$$\frac{1}{|K|}\sum_{k\in K} h_\pi(xky) = h_\pi(x)h_\pi(y) \qquad \text{for all } x, y \in G.$$

Moreover, the eigenvalues $\lambda_{S,\pi}$ *are given by*

$$\lambda_{S,\pi} = \frac{|K|^2}{|\,\mathrm{Stab}\, s|} h_\pi(s) \qquad \text{for any } s \in S,$$

where Stab s *consists of* $k \in K$ *such that* $Ksk = Ks$.

Now we turn to the construction of graphs. Let T be a finite union of K–double cosets. Define a graph $X = X(G/K, T/K)$ on cosets G/K such that the outneighbors of xK are xg_iK, $i = 1, \cdots, k$, where $T = \bigcup_{i=1}^{k} g_iK$ is a disjoint union of K-cosets. This is a k–regular directed graph, and it is undirected if $T = T^{-1}$. When G is abelian and K is trivial, this is the difference graph defined in the previous section. When $G = PGL_2(\mathbf{Q}_p)$, $K = PGL_2(\mathbf{Z}_p)$ and $T = K \begin{pmatrix} p & 0 \\ 0 & 1 \end{pmatrix} K$, the resulting graph is the $(p+1)$–regular infinite tree $\mathcal{T}$ associated to $PGL_2(\mathbf{Q}_p)$ in §2.

The adjacency matrix $\overline{A}_T$ of $X = X(G/K, T/K)$ is $\frac{1}{|K|} \sum_{S \subset T} A_S$. By the theorem above, the eigenvalues of $\overline{A}_T$ are given by

$$\sum_{S \subset T} \frac{|K|}{|\,\mathrm{Stab}\, s|} h_\pi(s), \qquad \text{for any element } s \text{ in } S,$$

as π runs through the irreducible representations of G occurring in $L(G/K)$.

As remarked before, an irreducible representation π of G occurs in $L(G/K)$ if and only if it has a nonzero K–invariant vector. Thus to construct Ramanujan graphs of the above type, it suffices to find suitable G, K and T such that $L(K\backslash G/K)$ is a commutative algebra, A_S is diagonalizable for each K–double coset S, and for all nontrivial irreducible representations π of G containing a nonzero K–invariant vector, the function h_π defined by (5.1) satisfies

$$\left| \sum_{S \subset T} \frac{|K|}{|\,\mathrm{Stab}\, s|} h_\pi(s) \right| \leq 2\sqrt{k-1},$$

where

$$k = \sum_{S \subset T} \frac{|K|}{|\,\mathrm{Stab}\, s|}, \qquad \text{and } s \text{ is any element in } S.$$

A family of $(q+1)$–regular Ramanujan graphs was constructed and studied by Terras and her students [2, 5 and references therein]. They took a finite field F of q elements with q odd, $G = GL_2(F)$. Choose a nonsquare element δ in F, imbed $E^\times = F(\sqrt{\delta})^\times$ into G by representing multiplication by $x \in E^\times$ via its matrix with respect to the basis $\{1, \sqrt{\delta}\}$. Denote the imbedded image by

$$K = \left\{ \begin{pmatrix} a & b\delta \\ b & a \end{pmatrix} \in G : a, b \in F \right\}.$$

The coset space G/K can be represented by

$$H = \left\{ \begin{pmatrix} y & x \\ 0 & 1 \end{pmatrix} : y \in F^{\times} \text{ and } x \in F \right\}$$

so that it is analogous to the classical Poincaré upper half-plane. The group G has q double cosets, of which $K \begin{pmatrix} 1 & 0 \\ 0 & 1 \end{pmatrix} K = K$ and $K \begin{pmatrix} -1 & 0 \\ 0 & 1 \end{pmatrix} K = \begin{pmatrix} -1 & 0 \\ 0 & 1 \end{pmatrix} K$, and each of the remaining $q-2$ double cosets KtK is the union of $(q+1)$ K–cosets. More precisely, each double coset KtK is associated to an ellipse $x^2 = ay + \delta(y-1)^2$, where $a \in F$, $a \neq 0, 4\delta$, such that $KtK = \bigcup \begin{pmatrix} y & x \\ 0 & 1 \end{pmatrix} K$ with (x, y) running through all the F–points of the ellipse. We denote this double coset KtK by S_a. Choose $T = S_a$, $a \neq 0, 4\delta$, so that we get a $(q+1)$–regular graph $X = X(G/K,\ T/K)$. Because all K–double cosets S are symmetric, i.e., $S = S^{-1}$, the graph $X = X(G/K,\ T/K)$ is undirected and the algebra $L(K\backslash G/K)$ is commutative. Further, $\operatorname{Stab} t$ is the subgroup of diagonal matrices in K, hence $|\operatorname{Stab} t| = q - 1$ and $|K|/|\operatorname{Stab} t| = q + 1 = k$.

One checks from the known table of representations of G in [26] that there are two types of irreducible nontrivial representations of G containing a nonzero K–invariant vector. The first type arises from the $q+1$–dimensional representation of G induced from the 1–dimensional representation of the Borel subgroup given by

$$\begin{pmatrix} a & b \\ 0 & d \end{pmatrix} \longmapsto \chi(a)\chi(d)$$

for a character χ of $F^{\times}$. If χ has order greater than 2, then the above induced representation is irreducible, denote it by π_χ; while if χ has order 2, denote by π_χ the q-dimensional irreducible subrepresentation of the above induced representation. The realization of π_χ in $L(G/K)$ is the subspace generated by the left translations of the function f on G/K defined by

$$f\left(\begin{pmatrix} y & x \\ 0 & 1 \end{pmatrix} K \right) = \chi(y).$$

(The other constituent of the induced representation is 1–dimensional, which can be realized in L_0.)

The eigenvalue $\lambda_{T,\chi}$ of $\overline{A}_T$ on the space of π_χ can be easily seen to be

$$\lambda_{T,\chi} = \sum_{\substack{y \in F \\ ay+\delta(y-1)^2 = x^2}} \chi(y).$$

It was shown by R. Evans and by H. Stark that

$$|\lambda_{T,\chi}| \leq 2\sqrt{q} = 2\sqrt{k-1} \quad \text{for } \chi \text{ nontrivial},$$

using Weil's estimate [34], which follows from the Riemann hypothesis for projective curves over F. Soto–Andrade [30] computed h_χ using (5.1) with $\pi = \pi_\chi$ and arrived at the same expression for $\lambda_{T,\chi}$. Here we note that since $T = T^{-1}$, we have

$$\lambda_{T,\pi} = \frac{|K|}{q-1} h_\pi(t^{-1}) = \frac{1}{q-1} \sum_{k \in K} tr\, \pi(kt)$$

for all representations π occurring in $L(G/K)$.

The second type of irreducible representations of G in question, denoted by π_ω, is the representation of G associated to a multiplicative character ω of the quadratic extension $F(\sqrt{\delta})$ of F. Here $\omega \neq \omega^q$, that is, ω is not trivial on the kernel of the norm map from $F(\sqrt{\delta})$ to F. The eigenvalue $\lambda_{T,\omega}$ on the space of π_ω was computed by Soto–Andrade [30] to be

$$\lambda_{T,\omega} = \sum_{\substack{z = x + y\sqrt{\delta} \in F(\sqrt{\delta}) \\ x^2 - \delta y^2 = 1}} \varepsilon\left(\frac{a}{\delta} - 2 + 2x\right)\omega(z),$$

where ε is the function on F equal to 1 on squares in $F^\times$, -1 on nonsquares in $F^\times$, and 0 at 0. Using étale cohomology and algebraic geometry, N. Katz [14] showed that

$$|\lambda_{T,\omega}| \leq 2\sqrt{q} = 2\sqrt{k-1}.$$

In fact, the above inequality can also be proved directly using Weil's estimate. We derive the estimates for $\lambda_{T,\chi}$ and $\lambda_{T,\omega}$ in the proposition below via the techniques developed in Chapter 6.

Theorem 10. *Let δ be a nonsquare in the finite field F of q elements, q odd, and $a \in F$, $a \neq 0, 4\delta$. Let χ be a nontrivial character of $F^\times$ and let ω be a multiplicative character of the quadratic extension $F(\sqrt{\delta})$ of F such that $\omega \neq \omega^q$. Let, as above,*

$$\lambda_{T,\chi} = \sum_{\substack{y \in F \\ ay + \delta(y-1)^2 = x^2}} \chi(y) \quad \text{and} \quad \lambda_{T,\omega} = \sum_{\substack{z = x + y\sqrt{\delta} \in F(\sqrt{\delta}) \\ x^2 - \delta y^2 = 1}} \varepsilon\left(\frac{a}{\delta} - 2 + 2x\right)\omega(z),$$

where ε is the function on F equal to 1 on squares in $F^\times$, -1 on nonsquares in $F^\times$ and 0 at 0. Then

$$|\lambda_{T,\chi}| \leq 2\sqrt{q} \quad \text{and} \quad |\lambda_{T,\omega}| \leq 2\sqrt{q}.$$

Proof. Let $f(y) = \delta(y-1)^2 + ay$. Assume first that f is irreducible over F. Denote by w the place of degree 2 determined by the roots of $f(y)$. By theorem 4 in Chapter 6, §2, there is an idèle class character η of $F(t)$ such that at a finite place $v \neq w$ with uniformizing element $\pi_v = P_v(t) = \prod_{j=1}^{\deg v} (t - \beta_{j,v})$,

$$\eta_v(\pi_v) = \varepsilon\left(\prod_{j=1}^{\deg v} f(\beta_{j,v})\right), \quad \eta_v(\mathcal{U}_v) = 1,$$

and at $v = w$ or ∞, η_v is trivial on $1+\mathcal{P}_v$. Further, the conductor of η is $w+\infty$. On the other hand, there is an idèle class character ξ of $F(t)$ with conductor $0+\infty$ such that ξ_0 on $\mathcal{U}_0/1+\mathcal{P}_0$ is given by $\xi_0(a) = \chi^{-1}(a)$ for $a \in F^\times$, and ξ_∞ on $\mathcal{U}_\infty/1+\mathcal{P}_\infty$ is given by $\xi_\infty(a) = \chi(a)$ for $a \in F^\times$ and $\xi_\infty(t^{-1}) = \xi_\infty(\pi_\infty) = 1$. Then $\xi\eta$ is an idèle class character of the rational function field $F(t)$ with conductor $w+0+\infty$ or $w+0$ which has degree 4 or 3. At a finite place v of degree 1 with $\pi_v = t-\beta$ where ξ and η are unramified, we have

$$\xi_v(\pi_v) = \xi_\infty(t-\beta)^{-1}\xi_0(t-\beta)^{-1} = \chi(-\beta) \quad \text{and} \quad \eta_v(\pi_v) = \varepsilon(f(\beta)).$$

In the case that the conductor of $\eta\xi$ is $w+0+\infty$, Corollary 2 in Chapter 6, §1 gives

$$\left|\sum_{\substack{v\neq 0,\infty \\ \deg v=1}} \eta_v\xi_v(\pi_v)\right| = \left|\sum_{y\in F^\times} \chi(-y)\varepsilon(f(y))\right| = \left|\sum_{y\in F} \chi(y)\varepsilon(f(y))\right| \leq 2\sqrt{q}.$$

In the case that $\eta\xi$ has conductor $w+0$, the same corollary gives

$$\left|\eta_\infty\xi_\infty(\pi_\infty) + \sum_{y\in F^\times} \chi(-y)\varepsilon(f(y))\right| \leq \sqrt{q},$$

which also gives rise to the same bound

$$\left|\sum_{y\in F} \chi(y)\varepsilon(f(y))\right| \leq \sqrt{q}+1 \leq 2\sqrt{q}.$$

Since χ is nontrivial, so $\sum_{y\in F} \chi(y) = 0$ and hence

$$\sum_{y\in F} \chi(y)\varepsilon(f(y)) = \sum_{y\in F} \chi(y)\varepsilon(f(y)) + \sum_{y\in F} \chi(y) = 2\lambda_{T,\chi}.$$

This together with the above inequality yields $|\lambda_{T,\chi}| \leq \sqrt{q}$.

If f is reducible over F, then it has two distinct roots because $a \neq 0, 4\delta$. Denote by v_1, v_2 the two degree one places of F which are roots of f. As δ is a nonzero, we know $v_1, v_2 \neq 0$. The character η defined the same way as above has conductor equal to $v_1+v_2+\infty$ so that $\xi\eta$ has conductor $0+v_1+v_2+\infty$ or $0+v_1+v_2$. The same Corollary 2 gives the estimate

$$\left|\sum_{y\in F} \chi(y)\varepsilon(f(y))\right| \leq 2\sqrt{q}$$

in both situations. But we find

$$\sum_{y\in F}\chi(y)\varepsilon(f(y)) = \sum_{y\in F}\chi(y)\varepsilon(f(y)) + \sum_{y\in F}\chi(y) = 2\lambda_{T,\chi} - \chi(v_1) - \chi(v_2),$$

which implies $|\lambda_{T,\chi}| \le \sqrt{q}+1 \le 2\sqrt{q}$.

Finally, we estimate $\lambda_{T,\omega}$. Denote by w the place of degree 2 of $F(t)$ with a uniformizer $\pi_w = t^2 - \delta$. Elements in the residue field at w, which is $F(\sqrt{\delta})$, of norm 1 can be written as $z = \frac{-\sqrt{\delta}-b}{\sqrt{\delta}-b} = \frac{b^2+\delta+2b\sqrt{\delta}}{b^2-\delta}$, where $b \in F \cup \{\infty\}$. ($z = 1$ when $b = \infty$.) So we may express $\lambda_{T,\omega}$ as

$$\lambda_{T,\omega} = \sum_{b\in F\cup\{\infty\}} \varepsilon\Big(\frac{a}{\delta} - 2 + 2\frac{b^2+\delta}{b^2-\delta}\Big)\omega\Big(\frac{-\sqrt{\delta}-b}{\sqrt{\delta}-b}\Big) = \sum_{b\in\mathbf{P}^1(F)} \varepsilon\big(f(b)\big)\omega\Big(\frac{-\sqrt{\delta}-b}{\sqrt{\delta}-b}\Big),$$

where $f(t) = \frac{f_1(t)}{f_2(t)}$ with $f_1(t) = \frac{a}{\delta}t^2 - a + 4\delta$ and $f_2(t) = t^2 - \delta$. As shown in the proof of Theorem 6 in Chapter 6, §3, there is an idèle class character ξ of $F(t)$, unramified outside w, such that ξ_w is trivial on $F^\times(1+\mathcal{P}_w)$ and on $\mathcal{U}_w/F^\times(1+\mathcal{P}_w)$ it is given by $\xi_w(\alpha+\beta t) = \omega\Big(\frac{\alpha+\beta\sqrt{\delta}}{\alpha-\beta\sqrt{\delta}}\Big)$, where $\alpha, \beta \in F$ and $\alpha+\beta t \in \mathcal{U}_w$, and ξ_∞ is trivial. Since ω is nontrivial on elements in $F(\sqrt{\delta})$ with norm 1, ξ has conductor w. At any place v of $F(t)$ of degree 1 with $\pi_v = t - b$ or $v = \infty$ with $\pi_\infty = \frac{1}{t}$, we have $\xi_v(\pi_v) = \omega\Big(\frac{-\sqrt{\delta}-b}{\sqrt{\delta}-b}\Big)$, where $b = \infty$ when $v = \infty$. Recall that the support of f is the union of places of $F(t)$ where f has either zero or pole. With $f(t) = f_1(t)/f_2(t)$ as given above, $\operatorname{supp} f = \{w_1, w\}$ or $\{v_1, v_2, w\}$, where w_1 is the place of degree 2 of $F(t)$ determined by the roots of $f_1(t)$ if $f_1(t)$ is irreducible, and v_1, v_2 are the two places of degree 1 of $F(t)$ which are the two distinct roots of $f_1(t)$ when it is reducible. Note that $f_1(t)$ does not have repeated roots since $a \ne 4\delta$. The character η_v defined by the same formula as above at finite places v of $F(t)$ outside $\operatorname{supp} f$, and by $\eta_\infty(\pi_\infty) = \varepsilon\big(f(\infty)\big) = \varepsilon(a\delta^{-1})$, $\eta_\infty(\mathcal{U}_\infty) = 1$, extends to an idèle class character η of $F(t)$ with $\operatorname{cond}\eta = \sum_{v\in\operatorname{supp} f} v$. Therefore $\eta\xi$ has conductor of degree 2 or 4, and by Corollary 2 in Chapter 6, §1, we have

$$\Bigg|\sum_{\substack{\deg v=1\\ \eta_v\xi_v \text{ unram.}}} \eta_v(\pi_v)\xi_v(\pi_v)\Bigg| = \Bigg|\sum_{b\in\mathbf{P}^1(F)} \varepsilon\big(f(b)\big)\omega\Big(\frac{-\sqrt{\delta}-b}{\sqrt{\delta}-b}\Big)\Bigg| = |\lambda_{T,\omega}| \le 2\sqrt{q}.$$

This completes the proof of Theorem 10. □

Theorem 11. *The graphs* $X = X(G/K,\ S_a/K)$, $a \in F$, $a \ne 0, 4\delta$, *are* $(q+1)$*-regular Ramanujan graphs.*

This is the third explicit construction of Ramanujan graphs, which uses both representations of GL_2 over a finite field and character sum estimates resulting from the Riemann hypothesis for curves over finite fields. The reader is referred to [2], [5], [10], [16] and [17] for more detailed discussion.

---

§6 Two proofs of the Alon-Boppana Theorem

In this section we give two proofs of Theorem 7, which states that the liminf of the largest nontrivial eigenvalue in absolute value, $\lambda(G)$, of k-regular graphs G is at least $2\sqrt{k-1}$ when the size of the graph $|G|$ tends to infinity. This will be seen as consequences of two different theorems. The results proved in this section can be generalized to hypergraphs, which was done by Feng and Li [11]. The first way is from a result of Nilli, which relates the diameter of G to $\lambda_2(G)$, the second largest eigenvalue of G. Note that $\lambda(G) \geq \lambda_2(G)$.

Theorem 12. *([24]) Suppose that G is a k–regular graph. If the diameter of G is* $\geq 2l+2 \geq 4$, *then*

$$\lambda_2(G) > 2\sqrt{q} - \frac{2\sqrt{q}-1}{l}.$$

Write D for the diameter of G. Then

$$|G| \leq 1 + k + k \cdot q + \cdots + k \cdot q^{D-1} < 1 + k + \cdots + k^D,$$

which implies

$$D \geq \frac{\log |V(G)|}{\log k} - O(1).$$

Hence the diameter of G tends to infinity as $|G|$ approaches infinity. It is then obvious that $\liminf \lambda_2(G)$ and hence $\liminf \lambda(G)$ is at least $2\sqrt{q}$ when $|G|$ tends to infinity.

Proof. We may assume that G is connected. Regard A as a linear operator on the space $\mathcal{F}(G)$ of real-valued functions on (vertices of) G which sends f to Af given by

$$(Af)(x) = \sum_y f(y),$$

where y runs through all vertices in G adjacent to x. Let Δ be the Laplace operator $kI - A$ on $\mathcal{F}(G)$. There is a natural inner product $<,>$ on $\mathcal{F}(G)$:

$$< f_1, f_2 >= \sum_{x \in V(G)} f_1(x) f_2(x).$$

$\mathcal{F}(G)$ has an orthonormal basis which are eigenfunctions of Δ. Clearly, the constant function $f_0 \equiv 1$ is an eigenfunction with eigenvalues 0, and other eigenvalues of Δ

are positive with eigenfunctions perpendicular to the constant function f_0. The second smallest eigenvalue α of Δ is $k - \lambda_2(G)$, which is also equal to

$$\alpha = \min_{\substack{f\neq 0,\ f\in\mathcal{F}(G)\\ <f,f_0>=0}} <\Delta f, f> / <f,f> .$$

We shall obtain an upper bound of α from a special choice of f.

By assumption, the diameter of G is $\geq 2l+2 \geq 4$, hence there are two vertices u and v in G with dist $(u,v) \geq 2l+2$. For $i \geq 0$ define

$$U_i = \{\text{vertices } x \text{ in } G: \quad \text{dist } (x,u) = i\}, \quad \text{and}$$
$$V_i = \{\text{vertices } x \text{ in } G: \quad \text{dist } (x,v) = i\}.$$

Then $U_0, \cdots, U_l, V_0, \cdots, V_l$ are disjoint, and no vertex in $U = \bigcup_{i=0}^{l} U_i$ is adjacent to any vertex in $V = \bigcup_{i=0}^{l} V_i$. Define a function f on G by

$$f(x) = \begin{cases} a & \text{if} \quad x \in U_0 \cup U_1, \\ a\,q^{-(i-1)/2} & \text{if} \quad x \in U_i, \quad 2 \leq i \leq l, \\ -b & \text{if} \quad x \in V_0 \cup V_1, \\ -b\,q^{-(i-1)/2} & \text{if} \quad x \in V_i, \quad 2 \leq i \leq l, \\ 0 & \text{otherwise.} \end{cases}$$

Here a, b are any choice of positive numbers such that f is perpendicular to f_0.

First we estimate $<f,f> = \sum_{x\in U} f(x)^2 + \sum_{x\in V} f(x)^2$. Clearly, $|U_0| = 1$, $|U_1| = k$. For each vertex $x \in U_i$ with $i \geq 1$, among its k neighbors, at least 1 vertex lies in U_{i-1} and at most $q = k-1$ neighbors of x lie in U_{i+1}. This shows that $|U_{i+1}| \leq q|U_i|$ for $i = 1, \cdots, l-1$. Hence

$$\begin{aligned} A_1 &= \sum_{x\in U} f(x)^2 \\ &= a^2\Big(1 + |U_1| + \sum_{i=2}^{l} |U_i|\,q^{-(i-1)}\Big) \geq a^2\Big(1 + l\,\frac{|U_l|}{q^{l-1}}\Big). \end{aligned} \tag{6.1}$$

Similarly, $B_1 = \sum_{x\in V} f(x)^2 \geq b^2\Big(1 + l\,\frac{|V_l|}{q^{l-1}}\Big)$.

Next we estimate $<\Delta f, f>$. Denote by $E(G)$ the edge set of G. Observe that

$$\begin{aligned} \sum_{\{x,y\}\in E(G)} \big(f(x)-f(y)\big)^2 &= \sum_{\{x,y\}\in E(G)} f(x)^2 - 2f(x)f(y) + f(y)^2 \\ &= k\sum_{x\in G} f(x)^2 - \sum_{x\in G} f(x) \sum_{\substack{y\in G\\ \{y,x\}\in E(G)}} f(y) \\ &= k <f,f> - <Af,f> = <\Delta f, f> . \end{aligned}$$

Write $<\Delta f, f> = A_2 + B_2$, where

$$A_2 = \sum_{\substack{\{x,y\}\in E(G)\\ \text{at least one of } x,y\in U}} (f(x)-f(y))^2$$

and B_2 is defined analogously with V replacing U. In view of our definition of f and the fact that each $x \in U_i$ has at most q neighbors in U_{i+1}, we have

$$\begin{aligned}
A_2 &\le \sum_{i=1}^{l-1} |U_i| q \left(q^{-\frac{i-1}{2}} - q^{-\frac{i}{2}}\right)^2 a^2 + |U_l| q \cdot q^{-(l-1)} a^2 \\
&= (\sqrt{q}-1)^2 \left(|U_1| + |U_2| q^{-1} + \cdots + |U_{l-1}| q^{-(l-2)} + |U_l| q^{-(l-1)}\right) a^2 \\
&\quad + a^2 (2\sqrt{q}-1) |U_l| q^{-(l-1)} \\
&\le (\sqrt{q}-1)^2 (A_1 - a^2) + (2\sqrt{q}-1) \cdot \frac{A_1 - a^2}{l} \qquad \text{by (6.1)} \\
&< \left(1 + q - 2\sqrt{q} + \frac{2\sqrt{q}-1}{l}\right) A_1.
\end{aligned}$$

Similar, we get

$$B_2 < \left(1 + q - 2\sqrt{q} + \frac{2\sqrt{q}-1}{l}\right) B_1.$$

Therefore

$$k - \lambda_2(G) = \alpha \le \frac{A_2 + B_2}{A_1 + B_1} < 1 + q - 2\sqrt{q} + \frac{2\sqrt{q}-1}{l},$$

which yields

$$\lambda_2(G) > 2\sqrt{q} - \frac{2\sqrt{q}-1}{l},$$

as desired. □

The second way to obtain the Alon-Boppana theorem is to see it as an immediate consequence of the following theorem asserting the distribution of large eigenvalues of a regular graph proved by Serre [29], which is also implicitly contained in [18].

Theorem 13. *Given $\varepsilon > 0$, there exists a positive constant c, depending only on ε and k, such that for every k–regular graph G, the number of eigenvalues λ of G with $\lambda \ge (2-\varepsilon)\sqrt{k-1}$ is at least $c|G|$.*

Before proving this theorem, we first introduce a measure μ_G attached to a k-regular graph G and discuss its main properties which will be used later in the proof. Write q for $k-1$. The measure μ_G is supported on $[-M, M]$, $M = k/\sqrt{q}$, given by

$$\mu_G = \frac{1}{|G|} \sum_{\lambda} \delta_{\lambda/\sqrt{q}},$$

where λ runs through all eigenvalues of G. A measure on the real line is determined by its value on a family of polynomials $\{X_n(x)\}$ with deg $X_n = n$. For convenience, we shall choose $X_n(x)$ as follows : $X_0(x) = 1$, $X_1(x) = x$, and $X_i(x)$ for $i \geq 2$ are defined by the recursive formula $X_{i+1}(x) = xX_{i-1}(x) - X_i(x)$. These polynomials are well-known, in fact, $X_m(2x)$ are the Chebychev polynomials of the second kind. Listed below are some properties of $X_m(x)$ which will be used.

(6.2) As a formal power series in t, we have

$$\sum_{m=0}^{\infty} X_m(x)t^m = \frac{1}{1 - xt + t^2}.$$

(6.3) $X_m(x) = \prod_{j=1}^{m} (x - 2\cos \frac{j\pi}{m+1})$.

Hence each X_m has m distinct real roots located in the interval $(-2, 2)$ and the largest root of $X_m(x)$ approaches 2 as m tends to infinity.

(6.4) $\{X_m(x)\}$ is orthonormal with respect to the Sato-Tate measure $\rho(x)$ on $[-2, 2]$ defined as

$$\rho(x) = \frac{1}{\pi}\sqrt{1 - \frac{x^2}{4}}\, dx.$$

That is,

$$\int_{-2}^{2} X_i(x)\, X_j(x)\, \rho(x) = \delta_{ij}.$$

Let $a = \frac{1}{\sqrt{q}}$. We define a new family of polynomials : $Y_0(x) = 1$, and $Y_i(x) = X_i(x) + aX_{i-1}(x)$ for $i \geq 1$. In view of the recursive formula for $X_m(x)$, we get

(6.5) $\{Y_m(x)\}_{m \geq 0}$ satisfies the formal power series in t :

$$\sum_{m=0}^{\infty} Y_m(x)\, t^m = \frac{1 + at}{1 - xt + t^2}.$$

Further, one can prove inductively

(6.6) $Y_i(x)\, Y_j(x) = \big(Y_{i+j}(x) + Y_{i+j-2}(x) + \cdots\big) + a\big(Y_{i+j-1}(x) + Y_{i+j-3}(x) + \cdots\big)$.

It follows immediately from (6.3) and the definition of $Y_m(x)$ that

(6.7) deg $Y_m(x) = m$, and $Y_m(x)$ has m distinct real roots, with the largest real root α_m located between $2\cos\frac{\pi}{m}$ and $2\cos\frac{\pi}{m+1}$.

Thus $\alpha_m \in (-2, 2)$ and α_m increases to 2 as m tends to infinity.

To express the values $\mu_G(Y_m)$, we consider two families of operators on the space $\mathcal{F}(G)$ of functions on G. For $l \geq 0$, define the operator $U_l = U_l(G)$ on $\mathcal{F}(G)$ by sending a function f to $U_l f$ given by

$$(U_l f)(x) = \sum_p f(y(p)),$$

where p runs through all (directed) paths in G starting at x of length l and with no backtracking, and $y(p)$ denotes the ending vertex of the path p. Then we find $U_0 = I$ (=identity),

$$U_1 = A(G) := A, \quad U_2 = U_1 A - kI,$$

and, for $i \geq 3$,

$$U_i = U_{i-1}A - qU_{i-2}.$$

This can be expressed in terms of the formal power series

$$\sum_{m=0}^{\infty} U_m t^m = \frac{(1-t)(1+t)}{1 - At + qt^2}.$$

Define another family of operators $\{T_m\}_{m\geq 0}$ on $\mathcal{F}(G)$ by

$$\sum_{m=0}^{\infty} T_m t^m = \frac{1+t}{1 - At + qt^2}. \tag{6.8}$$

In other words, $T_m = U_m + U_{m-1} + \cdots + U_1 + U_0$.

Comparing (6.5) and (6.8), we get

$$q^{-m/2}T_m = Y_m\left(\frac{A}{\sqrt{q}}\right) \quad \text{for} \quad m \geq 0. \tag{6.9}$$

As A is diagonalizable and the trace of an operator is invariant by conjugation, we see from the definition of μ_G and (6.9) that

$$\mu_G(Y_m) = \frac{1}{|G|}\sum_{\lambda} Y_m = \frac{1}{|G|}q^{-m/2}Tr(T_m) \quad \text{for} \quad m \geq 0. \tag{6.10}$$

This is the key relation we search for. Further, by taking a basis of $\mathcal{F}(G)$ consisting of the characteristic functions of each vertex, we get immediately that the operators U_m and hence T_m have nonnegative trace. Combined with (6.10), this yields

$$\mu_G(Y_m) \geq 0 \quad \text{for} \quad m \geq 0. \tag{6.11}$$

Now we begin the proof of Theorem 13. The strategy is to find a nonnegative linear combination of $Y_m(x)$ whose values are under control. For this purpose, we first show

Proposition 3. *Denote by α_m the largest real root of $Y_m(x)$. Let $Z_m(x) = \frac{Y_m(x)^2}{x-\alpha_m}$. Then $Z_m(x) = \sum_{j=0}^{2m-1} y_j Y_j(x)$, where, for $0 \le i < [m/2]$,*

$$y_{2m-(2i+1)} = y_{2i+1} = Y_0(\alpha_m) + Y_2(\alpha_m) + \cdots + Y_{2i}(\alpha_m) > 0,$$
$$y_{2m-(2i+2)} = y_{2i+2} = Y_1(\alpha_m) + Y_3(\alpha_m) + \cdots + Y_{2i+1}(\alpha_m) > 0,$$

and $y_0 = 0$.

Proof. Clearly $Z_m(x)$ is a polynomial of degree $2m-1$, hence is a linear combination of $Y_0(x), \cdots, Y_{2m-1}(x)$ with coefficients $y_0, \cdots, y_{2m-1}$, respectively. The question is to show that $y_i's$ are as expressed above. We have

$$\begin{aligned} Y_m(x)^2 &= (x-\alpha_m)Z_m(x) = (x-\alpha_m)\sum_{j=0}^{2m-1} y_j Y_j(x) \\ &= y_0\big(xY_0(x) - \alpha_m\big) + \sum_{j=1}^{2m-1} y_j\big(xY_j(x) - \alpha_m Y_j(x)\big) \\ &= y_0\big(Y_1(x) - a - \alpha_m\big) + \sum_{j=1}^{2m-1} y_j\big(Y_{j+1}(x) + Y_{j-1}(x) - \alpha_m Y_j(x)\big) \end{aligned}$$

since $xY_j(x) = Y_{j+1}(x) + Y_{j-1}(x)$ for $j \ge 1$, while the left hand side is equal to

$$\begin{aligned} Y_m^2(x) = &\,(Y_{2m}(x) + Y_{2m-2}(x) + \cdots + Y_2(x) + Y_0(x)) \\ &+ a\,(Y_{2m-1}(x) + Y_{2m-3}(x) + \cdots + Y_3(x) + Y_1(x)) \end{aligned}$$

by (6.6). Comparing coefficients of $Y_0, \cdots, Y_{2m}$, we get, for $1 \le j \le m$,

$$y_{2j-1} - \alpha_m y_{2j} + y_{2j+1} = 1, \qquad (a)_j$$
$$y_{2j-2} - \alpha_m y_{2j-1} + y_{2j} = a, \qquad (b)_j$$

and

$$y_1 - (a+\alpha_m)y_0 = 1. \qquad (c)$$

Here we have set $y_{2m} = y_{2m+1} = 0$. From $(a)_m$ we obtain $y_{2m-1} = 1 = Y_0(\alpha_m)$. This together with $(b)_m$ yields $y_{2m-2} = a + \alpha_m = Y_1(\alpha_m)$. We proceed to compute $y_{2m-(2i+1)}$ and $y_{2m-(2i+2)}$ inductively for $i = 1, \cdots, [m/2]-1$, having checked that the formulae hold for $i = 0$. Suppose the formulae hold for $i-1$, that is,

$$y_{2m-(2i-1)} = Y_0(\alpha_m) + Y_2(\alpha_m) + \cdots + Y_{2i-2}(\alpha_m)$$

and

$$y_{2m-2i} = Y_1(\alpha_m) + Y_3(\alpha_m) + \cdots + Y_{2i-1}(\alpha_m).$$

It then follows from $(a)_{m-i}$ and $(b)_{m-i}$ respectively that

$$\begin{aligned} y_{2m-2i-1} &= 1+\alpha_m y_{2m-2i} - y_{2m-2i+1} \\ &= 1+\alpha_m\left(Y_1(\alpha_m)+Y_3(\alpha_m)+\cdots+Y_{2i-1}(\alpha_m)\right) \\ &\quad -\left(Y_0(\alpha_m)+Y_2(\alpha_m)+\cdots+Y_{2i-2}(\alpha_m)\right) \\ &= Y_0(\alpha_m)+Y_2(\alpha_m)+\cdots+Y_{2i}(\alpha_m) \end{aligned}$$

and

$$\begin{aligned} y_{2m-2i-2} &= a+\alpha_m y_{2m-2i-1} - y_{2m-2i} \\ &= a+\alpha_m\left(Y_0(\alpha_m)+Y_2(\alpha_m)+\cdots+Y_{2i}(\alpha_m)\right) \\ &\quad -\left(Y_1(\alpha_m)+Y_3(\alpha_m)+\cdots+Y_{2i-1}(\alpha_m)\right) \\ &= Y_1(\alpha_m)+Y_3(\alpha_m)+\cdots+Y_{2i+1}(\alpha_m), \end{aligned}$$

as desired. To compute the remaining y_j's, we have to distinguish two cases according to m even or odd. Since the computation is similar, we shall do the case m even. We have, from $i = [m/2]-1$, that

$$y_{m+1} = Y_0(\alpha_m)+Y_2(\alpha_m)+\cdots+Y_{m-2}(\alpha_m)$$

and

$$y_m = Y_1(\alpha_m)+\cdots+Y_{m-1}(\alpha_m).$$

Equations $(a)_{m/2}$ and $(b)_{m/2}$ give rise to

$$\begin{aligned} y_{m-1} &= 1+\alpha_m y_m - y_{m+1} \\ &= 1+\alpha_m\left(Y_1(\alpha_m)+\cdots+Y_{m-1}(\alpha_m)\right) \\ &\quad -\left(Y_0(\alpha_m)+Y_2(\alpha_m)+\cdots+Y_{m-2}(\alpha_m)\right) \\ &= Y_0(\alpha_m)+Y_2(\alpha_m)+\cdots+Y_{m-2}(\alpha_m) = y_{m+1} \quad \text{since } Y_m(\alpha_m)=0 \end{aligned}$$

and

$$\begin{aligned} y_{m-2} &= a+\alpha_m y_{m-1} - y_m \\ &= a+\alpha_m\left(Y_0(\alpha_m)+Y_2(\alpha_m)+\cdots+Y_{m-2}(\alpha_m)\right) \\ &\quad -\left(Y_1(\alpha_m)+\cdots+Y_{m-1}(\alpha_m)\right) \\ &= Y_1(\alpha_m)+Y_3(\alpha_m)+\cdots+Y_{m-3}(\alpha_m) = y_{m+2}. \end{aligned}$$

Again, we prove inductively as before that $y_{2m-(2i+1)} = y_{2i+1}$ and $y_{2m-(2i+2)} = y_{2i+2}$ for $0 \le i < [m/2]$. Thus we have determined the coefficients from y_{2m-1} to y_1. Finally, we get from $(b)_1$ that

$$y_0 = a+\alpha_m y_1 - y_2 = a+\alpha_m Y_0(\alpha_m) - Y_1(\alpha_m) = 0,$$

which satisfies (c) since $y_1 = Y_0(\alpha_m) = 1$. □

Observe that $Z_m(\alpha_m)=0$, $Z_m(x)>0$ for $x>\alpha_m$ and $Z_m(x)\le 0$ for $x<\alpha_m$. Given $\varepsilon>0$, there are positive integers m, m' and positive constants z, z' such that

$\alpha_m, \alpha_{m'} > 2-\varepsilon$ and the polynomial $Z(x) = zZ_m(x) + z'Z_{m'}(x)$ has the following properties :

(6.13) $Z(x) = \sum_{i\geq 0} z_i Y_i(x)$ is a nonnegative linear combination of $Y_i(x)$,

(6.14) $Z(x) \leq -1$ for $x \leq 2-\varepsilon$,

(6.15) $Z(x) > 0$ for $x \geq 2$.

Let Q be the maximum of $Z(x)$ on the interval $[2-\varepsilon, M]$. Then $Q > 0$. Let g', g be the characteristic function of the interval $[-M, 2-\varepsilon)$ and $[2-\varepsilon, M]$, respectively. We have

$$\mu_G(Z) = \sum_i z_i\, \mu_G(Y_i) \geq 0 \quad \text{since} \quad \mu_G(Y_i) \geq 0 \quad \text{by} \quad (6.11).$$

On the other hand,

$$\mu_G(Z) = \mu_G(Z(g'+g)) = \mu_G(Zg') + \mu_G(Zg) \leq -1\cdot\mu_G(g') + Q\mu_G(g).$$

As $\mu_G(g') = 1 - \mu_G(g)$, the above two inequalities yield

$$\mu_G(g) \geq \frac{1}{Q+1}.$$

Note that $\mu_G(g)$ is the proportion of eigenvalues λ of G satisfying $\lambda \geq (2-\varepsilon)\sqrt{q}$, hence we may choose the constant c to be $\frac{1}{Q+1}$, which is independent of G. This completes the proof of Theorem 13.

---

§7 A limit distribution

A *primitive cycle* in a graph is a cycle with no backtracking and containing no proper subcycles. Denote by $c_l(G)$ the number of primitive cycles in G of length l. The following theorem of McKay says that in a family of regular graphs with increasing size, if the number of primitive cycles has moderate growth, then the sequence of measures attached to these graphs converges to a nice limit measure.

Theorem 14. *[21] Let $\{G_m\}$ be a family of connected k–regular hypergraphs with $|G_m| \to \infty$ as $m \to \infty$. Assume*

(7.1) *For each integer $l \geq 1$, $c_l(G_m)/|G_m| \to 0$ as $m \to \infty$.*

Then the sequence of measures $\{\mu_{G_m}\}$ converges weakly to the measure

$$\mu = \frac{1+\frac{1}{q}}{(1+\frac{1}{q}-\frac{x}{\sqrt{q}})(1+\frac{1}{q}+\frac{x}{\sqrt{q}})}\cdot\frac{1}{\pi}\sqrt{1-\frac{x^2}{4}}\,dx$$

supported on $[-2,2]$, where $q = k-1$ as before, and dx is the Lebesque measure on the real line.

Theorem 14 gives an overall asymptotic distribution of eigenvalues of G_m for m large. In particular, since the limit measure of any real number α is zero, we get immediately the following consequence which will be used in the next section.

Corollary 1. *Let $\{G_m\}$ be as in Theorem 14 satisfying (7.1). Then for each real number α, the multiplicity of α as an eigenvalue of G_m is $o(|G_m|)$ as $m \to \infty$.*

The proof below is adapted from Feng-Li [11], where the result is extended to hypergraphs. We first give another interpretation of the assumption (7.1)

Proposition 4. *The condition (7.1) is equivalent to*

(7.2) *For each integer $l \geq 1$, $TrU_l(G_m)/|G_m| \to 0$ as $m \to \infty$.*

Proof. As each primitive cycle of length l in a graph G contributes 1 to $TrU_l(G)$, we see immediately that $TrU_l(G) \geq c_l(G)$, the number of primitive cycles of length l in G. Hence (7.2) implies (7.1). Conversely, suppose (7.1) holds. Since a path of length l in G without backtracking and contributing to $TrU_l(G)$ is necessarily a cycle in G, so it contains at least one primitive cycle of length $j \leq l$ in G; further, such a primitive cycle is contained in at most $(l-j+1)q^{l-j}$ cycles in G of length l and without backtracking, so in G there are at most $\sum_{j=1}^{l}(l-j+1)q^{l-j}c_j(G)$ cycles of length l without backtracking. We have shown that

$$TrU_l(G) \leq \sum_{j=1}^{l}(l-j+1)k^{l-j}c_j(G).$$

It is then clear that (7.1) implies (7.2). □

We proceed to prove Theorem 14 under the assumption (7.2). Write μ_m for μ_{G_m}. As analyzed in the previous section, to show that limit μ_m exists, we have to show that $\lim_{m\to\infty} \mu_m(Y_l)$ exists for each $l \geq 0$. From (6.10) and the definition of T_l given in (6.8) we note that, for $l \geq 1$,

$$q^{l/2}\mu_m(Y_l) - q^{(l-1)/2}\mu_m(Y_{l-1}) = \frac{1}{|G_m|}TrU_l(G_m).$$

Thus the assumption (7.2) implies $\lim_{m\to\infty} q^{l/2}\mu_m(Y_l) - q^{(l-1)/2}\mu_m(Y_{l-1}) = 0$ for $l \geq 1$. Combined with $\mu_m(Y_0) = 1$, the above then implies $\lim_{m\to\infty} \mu_m(Y_l)$ exists and is equal to $q^{-l/2}$ for $l \geq 0$. In other words, $\lim_{m\to\infty} \mu_m$ exists, denote it by μ; and we have $\mu(Y_l) = q^{-l/2}$ for $l \geq 0$. To compare μ with the Sato-Tate measure ρ given in (6.4), we have to compute $\mu(X_l)$. One checks inductively from the definition of Y_l that

$$X_l = Y_l - aY_{l-1} + a^2Y_{l-2} - a^3Y_{l-3} + \cdots + (-a)^lY_0 \quad \text{for } l \geq 0.$$

Here $a = \frac{1}{\sqrt{q}}$. Hence

$$\mu(X_l) = \sum_{i=0}^{l}(-a)^i\mu(Y_{l-i}) = \sum_{i=0}^{l}(-1)^i q^{-\frac{l}{2}}$$
$$= \frac{1}{2}\left(q^{-\frac{l}{2}} + (-1)^l q^{-\frac{l}{2}}\right).$$

Consequently,

$$\begin{aligned}\sum_{l=0}^{\infty}\mu(X_l)X_l(x) &= \sum_{l=0}^{\infty}\frac{1}{2}X_l(x)q^{-\frac{l}{2}} + \frac{1}{2}\sum_{l=0}^{\infty}X_l(x)(-1)^l q^{-\frac{l}{2}} \\ &= \frac{1}{2}\frac{1}{1-xq^{-1/2}+q^{-1}} + \frac{1}{2}\frac{1}{1+xq^{-1/2}+q^{-1}} \qquad \text{by (6.2)} \\ &= \frac{1+\frac{1}{q}}{\left(1+\frac{1}{q}-\frac{x}{\sqrt{q}}\right)\left(1+\frac{1}{q}+\frac{x}{\sqrt{q}}\right)}.\end{aligned}$$

Hence the limit measure μ is as described. This proves Theorem 14.

---

§8 The growth of the dimension of cusp forms with integral eigenvalues at p

Fix a prime p. For a positive integer N prime to p, denote by $C'(\Gamma_0(N), 2)$ the space generated by cusp forms of weight 2 for $\Gamma_0(N)$ which are eigenfunctions of the Hecke operator T_p with integer eigenvalues. Using the combinatorial result from the previous section, we derive an estimate of the growth of the dimension of $C'(\Gamma_0(N), 2)$, originally proved in [11]. More precisely, we prove

Theorem 15. *(i) ([29]) Let $\{l_i\}$ be a sequence of primes different from p such that $l_i \to \infty$ as $i \to \infty$. Then for i large, we have*

$$dim\ C'(\Gamma_0(l_i), 2) = o(l_i).$$

(ii) ([11]) Let $\{M_i\}$ be a sequence of positive integers prime to p such that $M_i = l_i N_i$ with l_i a prime not dividing N_i and $l_i \to \infty$ as $i \to \infty$. Then for i large, we have

$$dim\ C'(\Gamma_0(M_i), 2) = o(M_i)$$

if the M_i's have a bounded number of prime factors; otherwise we have

$$dim\ C'(\Gamma_0(M_i), 2) = o(M_i \ln\ln M_i).$$

Note that in (ii) we may choose M_i to be square free and l_i to be the largest prime factor of M_i. Then as $M_i \to \infty$, so does l_i.

We shall apply Theorem 14 to a special family of graphs constructed from quaternion algebras, as we did in §3. Fix a prime p. For a prime $l \neq p$, denote by H_l the quaternion algebra over $\mathbf{Q}$ ramified only at ∞ and l. Write D_l for the quotient of the multiplicative group $H_l^\times$ by its center. The following strong approximation theorem holds for the adelic points in D_l :

$$D_l(A_{\mathbf{Q}}) = D_l(\mathbf{Q}) \cdot D_l(\mathbf{R})D_l(\mathbf{Q}_p)K,$$

where $K = \prod_{q\neq p,\infty} K_q$ is a compact open subgroup of the restricted product $\prod_{q\neq p,\infty} D_l(\mathbf{Q}_q)$ such that the image of K under the reduced norm map is the product of the group of units in $\mathbf{Q}_q$ over all finite places $q \neq p$ of $\mathbf{Q}$. At the place l where H_l ramifies, we take K_l to be the maximal compact subgroup of $D_l(\mathbf{Q}_l)$; at a place q other than l, p and ∞, H_l splits so that $D_l(\mathbf{Q}_q)$ is isomorphic to $PGL_2(\mathbf{Q}_q)$, we shall take

$$K_q = \left\{ \begin{pmatrix} a & b \\ c & d \end{pmatrix} \in GL_2(\mathbf{Z}_q) : ord_q c \geq n(q) \right\} / \text{ center of } GL_2(\mathbf{Z}_q)$$

such that $n(q) \geq 0$ for all $q \neq l, p, \infty$ and $n(q) = 0$ for almost all q. Let $N = \prod_{q\neq l,p,\infty} q^{n(q)}$ and denote the group K so chosen by $B_0(l, N)$. The intersection

$$D_l(\mathbf{Q}) \cap D_l(\mathbf{R}) D_l(\mathbf{Q}_p) B_0(l, N)$$

is a congruence subgroup of $D_l(\mathbf{Z}[\frac{1}{p}])$, which we denote by $\widetilde{\Gamma}_0(l, N)$. By the strong approximation theorem, the double coset space

$$X(l, N) = D_l(\mathbf{Q}) \backslash D_l(A_{\mathbf{Q}}) / D_l(\mathbf{R}) B_0(l, N) D_l(\mathbf{Z}_p)$$

can be seen locally at p as

$$\begin{aligned} X(l, N) &= \widetilde{\Gamma}_0(l, N) \backslash D_l(\mathbf{Q}_p) / D_l(\mathbf{Z}_p) \\ &= \widetilde{\Gamma}_0(l, N) \backslash PGL_2(\mathbf{Q}_p) / PGL_2(\mathbf{Z}_p). \end{aligned}$$

As explained in §3, $PGL_2(\mathbf{Q}_p)/PGL_2(\mathbf{Z}_p)$ has a natural structure as a $(p+1)$-regular infinite tree $\mathcal{T}$, and $\widetilde{\Gamma}_0(l, N)$, being a discrete subgroup of $PGL_2(\mathbf{Q}_p)$, is the fundamental group of the graph $X(l, N) = \widetilde{\Gamma}_0(l, N) \backslash \mathcal{T}$ if it is torsion free. This graph is finite since H_l ramifies at ∞, and is $(p+1)$-regular, counting possible loops and multiple edges. The adjacency matrix of $X(l, N)$ is nothing but the usual Hecke operator T_p acting on the space of automorphic forms on $D_l(A_{\mathbf{Q}})$ left invariant by $D_l(\mathbf{Q})$ and right invariant by $D_l(\mathbf{R}) B_0(l, N) D_l(\mathbf{Z}_p)$.

We would like to draw the same conclusion as Corollary 1 for the family of graphs $\{X(l, N)\}$ as l tends to infinity. For this purpose, we need to check the condition (7.1) or equivalently, the condition (7.2) in Proposition 4, §7.

Proposition 5. *([29]) For every $n \geq 1$, $0 \leq TrU_n(X(l, 1)) \leq C(n, p)$, where $C(n, p)$ is a constant depending only on p^n and not on l.*

Proof. The vertices of $X(l, 1)$ can be interpreted as equivalence classes of supersingular elliptic curves in characteristic l, and two classes are adjacent if and only if they are represented by two elliptic curves isogeneous of degree p. Thus the trace of U_n on $X(l, 1)$ essentially counts the number of classes of elliptic curves which are self-isogeneous via a cyclic group of order p^n. Such classes are known to be bounded and independent of l. □

Consequently, for $n \geq 1$, the number of cycles of length n with no backtracking in $X(l,1)$ is bounded by $C(n,p)$ and is independent of l. Now we study the number of such cycles in $X(l,N)$. Note that $X(l,N)$ is a covering graph of $X(l,1)$ with degree equal to the index $\left[\widetilde{\Gamma_0}(l,1):\widetilde{\Gamma_0}(l,N)\right]$. Each cycle of length n in $X(l,N)$ with no backtracking projects to a cycle with the same properties in $X(l,1)$. Thus

$$TrU_n(X(l,N)) \leq \left[\widetilde{\Gamma_0}(l,1):\widetilde{\Gamma_0}(l,N)\right] TrU_n(X(l,1))$$

and hence

$$0 \leq \frac{TrU_n(X(l,N))}{|X(l,N)|} \leq \frac{\left[\widetilde{\Gamma_0}(l,1):\widetilde{\Gamma_0}(l,N)\right] TrU_n(X(l,1))}{\left[\widetilde{\Gamma_0}(l,1):\widetilde{\Gamma_0}(l,N)\right] |X(l,1)|}$$
$$= \frac{TrU_n(X(l,1))}{|X(l,1)|} \leq \frac{C(n,p)}{|X(l,1)|}.$$

Observe that the numerator of the upper bound is independent of l and N, while the denominator tends to infinity as l approaches infinity. This proves

Proposition 6. *Given a sequence of prime numbers $\{l_i\}$ with $l_i \neq p$, $l_i \to \infty$ as $i \to \infty$, and a sequence of positive integers $\{N_i\}$ with N_i prime to pl_i, we have, for each $n \geq 1$,*

$$\frac{c_n(X(l_i,N_i))}{|X(l_i,N_i)|} \longrightarrow 0 \quad \text{as } i \longrightarrow \infty.$$

Hence Corollary 1 holds for the family of graphs $\{X(l_i,N_i)\}$. More precisely,

Proposition 7. *Let $\{l_i\}$ and $\{N_i\}$ be as in Theorem 15. Given any real number α, the multiplicity of α as an eigenvalue of the Hecke operator T_p on the space $\mathcal{A}(\widetilde{\Gamma_0}(l_i,N_i))$ of automorphic forms on $D_{l_i}(\mathbf{Q})\backslash D_{l_i}(A_{\mathbf{Q}})/D_{l_i}(\mathbf{R})D_{l_i}(\mathbf{Z}_p)B(l_i,N_i)$ is $o(|X(l_i,N_i)|)$ as $i \to \infty$.*

Next we analyze the space $\mathcal{A}(\widetilde{\Gamma_0}(l_i,N_i))$. It contains constant functions, which are eigenfunctions of T_p with eigenvalue $p+1$. Depending on the choice of the congruence subgroup, sometimes the space contains functions which take opposite values at adjacent vertices, these are eigenfunctions with eigenvalue $-(p+1)$. The space $\mathcal{A}^{\perp}(\widetilde{\Gamma_0}(l_i,N_i))$ of functions in $\mathcal{A}(\widetilde{\Gamma_0}(l_i,N_i))$ perpendicular to these functions is T_p-invariant and has codimension 1 or 2. Further, by the theory of automorphic forms for GL_2 and quaternion groups developed by Jacquet and Langlands and the correspondence proved by Gelbart and Jacquet as discussed in Chapter 8, the forms in $\mathcal{A}^\perp(\widetilde{\Gamma_0}(l_i, N_i))$ can be identified with the forms in the space $\mathcal{C}_{l_i}(\Gamma_0(l_iN_i), 2)$ generated by cusp forms for $\Gamma_0(l_iN_i)$ of weight 2 whose underlying automorphic representations of $GL_2(A_\mathbf{Q})$ have components at l_i being the unramified special representations of $GL_2(\mathbf{Q}_{l_i})$. Therefore, the eigenvalues of T_p on $\mathcal{A}^\perp(\widetilde{\Gamma_0}(l_i, N_i))$ are among the eigenvalues of T_p on the space $\mathcal{C}(\Gamma_0(l_iN_i), 2)$ of cusp forms for $\Gamma_0(l_iN_i)$ of weight 2. As we discussed in Chapter 7, the former Ramanujan-Petersson conjecture established by Deligne asserts that the eigenvalues λ_p of T_p on cusp forms of weight 2 satisfies

$$|\lambda_p| \le 2\sqrt{p}.$$

In particular, only finitely many integer eigenvalues for T_p on $\mathcal{A}^\perp(\widetilde{\Gamma_0}(l_i, N_i))$ are possible. This combined with Proposition 7 proves

Theorem 16. *Let $\{l_i\}$ be a sequence of primes $\neq p$ with $l_i \to \infty$ as $i \to \infty$. Let $\{N_i\}$ be a sequence of positive integers with N_i prime to pl_i. Then the dimension of the space $\mathcal{A}'(\widetilde{\Gamma_0}(l_i, N_i))$ generated by eigenforms of T_p in $\mathcal{A}^\perp(\widetilde{\Gamma_0}(l_i, N_i))$ with integer eigenvalues is $o(|X(l_i, N_i)|)$ as $i \to \infty$.*

As for the size of $X(l_i, N_i)$, we have

$|X(l_i, N_i)| = \dim \mathcal{A}(\widetilde{\Gamma_0}(l_i, N_i)) \le 2 + \dim \mathcal{A}^\perp(\widetilde{\Gamma_0}(l_i, N_i)) \le 2 + \dim \mathcal{C}(\Gamma_0(l_iN_i), 2)$.
On the other hand, $\dim \mathcal{C}(\Gamma_0(l_iN_i), 2)$ is the genus of the modular group $\Gamma_0(l_iN_i)$, whose dominating term is

$$\frac{1}{12}\, l_iN_i \prod_{\substack{q|l_iN_i \\ q \text{ prime}}} (1 + \frac{1}{q})$$

(cf [25]). Thus $|X(l_i, N_i)| = O\Big(l_iN_i \prod_{q|l_iN_i} (1 + \frac{1}{q})\Big)$ as $i \to \infty$.

We specialize Theorem 16 to various cases to prove Theorem 15. Firstly, in the case that $N_i = 1$ for all i, we get $\mathcal{C}_{l_i}(\Gamma_0(l_iN_i), 2) = \mathcal{C}(\Gamma_0(l_i), 2)$ since there are no cusp forms of weight 2 for $SL_2(\mathbf{Z})$. The first statement of Theorem 15 follows immediately from Theorem 16. Next consider the case $M = lN$ with l prime to N. Then by the theory of newforms explained in Chapter 7, §3, the orthogonal complement of $\mathcal{C}_l(\Gamma_0(M), 2)$ in $\mathcal{C}(\Gamma_0(M), 2)$ is T_p-invariant and is generated by $\mathcal{C}(\Gamma_0(N), 2)$ and its "push-up" at l, so that

$$\dim \mathcal{C}(\Gamma_0(M), 2) = \dim \mathcal{C}_l(\Gamma_0(M), 2) + 2 \dim \mathcal{C}(\Gamma_0(N), 2).$$

We prime a space of forms to denote its subspace generated by forms which are eigenfunctions of T_p with integral eigenvalues. Since the push-up operator at l commutes with T_p, we obtain, from the above formula,

$$\begin{aligned} \dim \mathcal{C}'(\Gamma_0(M), 2) &= \dim \mathcal{C}'_l(\Gamma_0(M), 2) + 2 \dim \mathcal{C}'(\Gamma_0(N), 2) \\ &= \dim \mathcal{A}'(\widetilde{\Gamma_0}(l, N)) + 2 \dim \mathcal{C}'(\Gamma_0(N), 2). \end{aligned}$$

As discussed before,

$$\dim \mathcal{C}'(\Gamma_0(N), 2) \le \dim \mathcal{C}(\Gamma_0(N), 2) << N \prod_{q|N} (1 + \frac{1}{q})$$

so that

$$2 \dim \mathcal{C}'(\Gamma_0(N), 2) << M \prod_{q|M} (1 + \frac{1}{q}) \cdot \frac{2}{l+1}.$$

Now let $M = M_i = l_i N_i$ with $l_i \to \infty$ when $i \to \infty$. By Theorem 16,

$$\dim \mathcal{A}'(\widetilde{\Gamma_0}(l_i, N_i)) = o\Big(M_i \prod_{q|M_i} (1 + \frac{1}{q})\Big)$$

and so is $2 \dim \mathcal{C}'(\Gamma_0(N_i), 2)$ by the above analysis, therefore

$$\dim \mathcal{C}'(\Gamma_0(M_i), 2) = o\Big(M_i \prod_{q|M_i} (1 + \frac{1}{q})\Big)$$

as $i \to \infty$. If $M_i's$ have bounded number of prime factors, then $\prod_{q|M_i} (1 + \frac{1}{q})$ is bounded; otherwise, $\prod_{q|M_i} (1 + \frac{1}{q})$ is $O(\ln \ln M_i)$ (cf. [13, p.90]), this proves the second statement of Theorem 15. The proof of Theorem 15 is completed.

As a by-product of our proof, combining Proposition 6 and Theorem 14, we have

Theorem 17. *Let $\{l_i\}$ be a sequence of primes $\neq p$ with $l_i \to \infty$ as $i \to \infty$. Let $\{N_i\}$ be a sequence of positive integers with N_i prime to pl_i. Then, as $i \to \infty$, the eigenvalues of the Hecke operator T_p, normalized by dividing by $2\sqrt{p}$, on the space $\mathcal{C}_{l_i}(\Gamma_0(l_i N_i), 2)$ is uniformly distributed with respect to the measure*

$$\mu = \frac{1 + \frac{1}{p}}{(1 + \frac{1}{p} - \frac{x}{\sqrt{p}})(1 + \frac{1}{p} + \frac{x}{\sqrt{p}})} \cdot \frac{1}{\pi} \sqrt{1 - \frac{x^2}{4}}\, dx$$

supported on $[-2, 2]$, where dx is the Lebesque measure on the real line. In particular, the normalized eigenvalues of the Hecke operator T_p on the space of cusp forms of prime level l and weight 2 are uniformly distributed with respect to μ as l tends to infinity.

References

[1] N. Alon and V. Milman: λ_1, isoperimetric inequalities for graphs and superconcentrators, J. Comb. Theory Ser.B 38 (1985), 73-88.

[2] J. Angel, N. Celniker, S. Poulos, A. Terras, C. Trimble, and E. Velasquez: Special functions on finite upper half planes, Contemp. Math., 138 (1992), 1-26.

[3] F. Bien: Constructions of telephone networks by group representations, Notices of Amer. Math. Soc., vol.36 (1989), 5-22.

[4] R. Brooks: The spectral geometry of a tower of coverings, J. Diff. Geom. 23 (1986), 97-107.

[5] N. Celniker, S. Poulos, A. Terras, C. Trimble, and E. Velasquez: Is there life on finite upper half planes? Contemp. Math., 143 (1993), 65-88.

[6] J. Cheeger: A lower bound for the smallest eigenvalue of the Laplacian, Problems in Analysis, Gunning ed., Princeton Univ. Press (1970), 195-199.

[7] P. Chiu: Cubic Ramanujan graphs, Combinatorica 12 (1992), 275-285.

[8] F. R. K. Chung: Diameters and eigenvalues, J. Amer. Math. Soc. 2 (1989), 187-196.

[9] V. G. Drinfeld: The proof of Petersson's conjecture for $GL(2)$ over a global field of characteristic p, Functional Anal. Appl. 22 (1988), 28-43.

[10] R. Evans: Character sums as orthogonal eigenfunctions of adjacency operators for Cayley graphs, preprint, 1993.

[11] K. Feng and W.-C. W. Li: Spectra of hypergraphs and applications, submitted.

[12] O. Gabber and Z. Galil: Explicit construction of linear sized superconcentrators, J. Comput. Sys. Sci. 22 (1981), 407-420.

[13] L. K. Hua: Introduction to Number Theory, Springer-Verlag, Berlin, Heidelberg, New York, 1982.

[14] N. Katz: Estimates for Soto-Andrade sums, J. reine angew. Math. 438 (1993), 143-161.

[15] W.-C. W. Li: Character sums and abelian Ramanujan graphs, J. Number Theory, 41 (1992), 199-217.

[16] W.-C. W. Li: Number theoretic constructions of Ramanujan graphs, Astérisque, French Math. Soc.(1995), to appear.

[17] W.-C. W. Li: A survey of Ramanujan graphs, to appear in the Proceedings of Arithmetic Geometry and Coding Theory, June 1993, Luminy.

[18] A. Lubotzky, R. Phillips and P. Sarnak: Ramanujan graphs, Combinatorica 8 (1988), 261-277.

[19] G. Margulis: Explicit construction of concentrators, Problems of Information Transmission 9 (1975), 325-332.

[20] G. Margulis: Explicit group theoretic constructions of combinatorial schemes and their application to the design of expanders and concentrators, J. Prob. of Info. Trans. (1988), 39-46.

[21] B. D. McKay: The expected eigenvalue distribution of a large regular graph, Linear Alg. and Its Appli. 40(1981), 203-216.

[22] J.-F. Mestre: La méthode des graphes. Exemples et applications, Proc. Int. Conf. on Class Numbers and Fundamental Units of Algebraic Number Fields, June 24-28, 1986, Katata, Japan, 217-242.

[23] M. Morgenstern: Existence and explicit constructions of $q+1$ regular Ra-

manujan graphs for every prime power q, J. Comb. Theory, series B, vol. 62 (1994), 44-62.

[24] A. Nilli: On the second eigenvalue of a graph, Disc. Math. 91(1991), 207-210.

[25] A. P. Ogg: Modular Forms and Dirichlet Series, Benjamin, New York, 1969.

[26] I. I. Piatetski-Shapiro: Complex Representations of $GL(2,K)$ for Finite Field K, Contemporary Math. 16, Amer. Math. Soc., Providence, 1983.

[27] A. Pizer: Ramanujan graphs and Hecke operators, Amer. Math. Soc. Bull. 23 (1990), 127-137.

[28] J.-P. Serre: Trees, Springer-Verlag, Berlin, Heidelberg, New York (1980).

[29] J.-P. Serre: Private letter to W.-C. W. Li, dated Oct. 8, 1990 and Nov. 5, 1990.

[30] J. Soto-Andrade: Geometrical Gel'fand models, tensor quotients and Weil representations, Proc. Symp. Pure Math. 47, Amer. Math. Soc., Providence (1987), 305-316.

[31] T. Sunada: L^2-functions in geometry and some applications, Proc. Taniguchi Symp. 1985, Lecture Notes in Math. 1201, Springer-Verlag, Berlin, Heidelberg, New York (1986), 266-284.

[32] T. Sunada: Fundamental groups and Laplacians, Lecture Notes in Math. 1339, Springer-Verlag, Berlin, Heidelberg, New York (1988), 248-277.

[33] R. Tanner: Explicit concentrators from generalized N-gons, SIAM J. Alg. Dis. Math. 5 (1984), 287-293.

[34] A. Weil: On some exponential sums, Proc. Nat. Aca. Sci., 34 (1948), 204-207.

[35] A. Lubotzky: Discrete Groups, Expanding Graphs and Invariant Measures, Progress in Math. 125, Birkhaüser Boston, Inc. 1994.

Index